TECHNICAL SHOP MATHEMATICS

JOHN G. ANDERSON

Professor of Industrial Technology
Macomb County Community College
Warren, Michigan

INDUSTRIAL PRESS INC.

200 Madison Avenue, New York, New York 10016

i

Library of Congress Cataloging in Publication Data

Anderson, John G   1912-
  Technical shop mathematics.

  1.   Shop mathematics.   I.   Title.
TJ1165.A56   510$'$.2$'$46   74-16115
ISBN 0-8311-1085-6

TECHNICAL SHOP MATHEMATICS

Contents

Preface

Early educators put great emphasis on the three "R's"; Reading, 'Riting, and 'Rithmetic. Much classroom time was spent in gaining proficiency in these basic skills as well as many agonizing hours over homework. However, time is a limited commodity and as many more subjects were added to a school's curriculum to meet the needs of our increasingly complex society, less time could be devoted to the three R's and 'Rithmetic was increasingly slighted.

Many men and women entering into business and industry have a serious handicap, namely, poor mathematical skills and comprehension. As they realize their weakness in mathematics, both older and younger people are enrolling in the expanding vocational education programs of their local high schools and community colleges to improve their skills, not only in mathematics but in other business and trade-related subjects as well. People seeking a practical, industrial-oriented mathematics education need a practical, industrial-oriented text for their studies. This textbook is designated to meet this need.

As an instructor in apprentice training, the author became concerned over the inadequate mathematical abilities of both younger and older students. Some of these students had received or retained little knowledge of basic arithmetic while older students had forgotten much of what they had studied in years past.

This book was written to up-grade these students in mathematical skills and comprehension and to sharpen their computational skills. The material is basic, but carefully selected, as a result of surveys made among training directors of both industry and major industrial trade unions. The surveys pointed out the major critical needs in mathematical skills required in the training of journeymen and technicians in scores of trades and technical occupations.

The text is designed both as a review and an introduction to mathematics. The subject matter progresses from the basic arithmetical operations through measurement systems, basic algebra for shop formula-solving skills, practical geometry with shop examples and applications, and trigonometry emphasizing its use in the shops and the trades.

Each new phase in each chapter is discussed in detail for ready comprehension with Study Examples to detail the proper steps to use in the solutions of problems. Then the Study Problems are given. These are based on typical shop applications and are given to provide the student with the necessary practice to acquire and have confidence in that particular mathematical skill.

Special attention is given, where possible, to the use of both the common English system of measurement and the S.I. (metric) system which students will be using with greater frequency in the future. To aid in the conversion of basic units from one system to the other, comprehensive tables are included, along with many Study Problems involving conversions.

The material presented in this book may, with judicious selection, be used for a one-year course or, in its entirety, as a two-year text.

The author would like to take this opportunity to acknowledge the help and encouragement given him by many people over many years, which made this work possible. First, to my parents and early elementary school teachers who perceived my mechanical aptitudes and directed my early education into mathematics and science. Then, to my teachers in college and in industry who taught me basic skills and their proper application in the trades and industry. Also I would like to acknowledge the help and encouragement of my present colleagues at Macomb County Community College.

I must also express my deep appreciation to Karl Hans Moltrecht of Industrial Press Inc., for suggesting, in the first place, that I write the book and then for his most valuable and continuing counsel and help in the technicalities of producing an accurate, readable, and useful text.

Also helpful in the planning and production of this book was Mr. Holbrook L. Horton of Industrial Press Inc., who permitted me to use much valuable data, charts, and tables from their reference *Machinery's Handbook* and from Mr. Horton's fine book, *Mathematics at Work*.

Lastly, but not least, I must pay tribute to my wife, Sallie, for her patience, help, and encouragement. It is to her and to our daughters, Lynne and Alice, that I dedicate this book.

CHAPTER 1

The Number System

Mathematics is a universal language. It is used all over the world and by its use people can communicate with each other even though they speak foreign tongues. Mathematics is extremely useful in the shop and in the trades; in science and in engineering it is essential.

Today's industry requires trained men with a good working knowledge of mathematics. In fact, their skill in mathematics is often as important as their other mechanical or electrical skills.

Perhaps you are a recent high school graduate who has had a difficult time with math; or maybe you are an older person who has been away from school a number of years. In either case you will be amazed when you begin to read and study this text at how much you remember about mathematics. This book will help you build up your knowledge of math by reviewing the fundamentals. Then it will show you how to use mathematics as a "tool" in your work.

This volume has been written for students and shop people who want to learn how to use this basic tool. The text, which is meant to be read slowly and thoughtfully, has many practical problems for you to practice on. You may have forgotten a lot of math, but if you will put the time and honest effort into carefully reading the text, reviewing the example problems, and then working out the study problems, you will learn mathematics. Although this effort may cut into your leisure time, you should spend at least one hour in study for every hour in class. Learning to use mathematics is work, but your efforts will be well repaid.

Symbols in Mathematics

Working with numbers requires a knowledge of the special "shorthand" of mathematics: symbols. There are many symbols used to represent actions and operations; even the numbers themselves are symbols. Do you remember the roman numerals? They are seen on the cornerstones of public buildings. Most people today cannot read these numbers because they did not learn roman numerals in school or use them in their work. The roman numeral system is not adapted to doing complex mathematics, and some historians even say that the Roman civilization failed for lack of mathematical science.

Our number system came to us many centuries ago. We call these numbers "arabic" numerals but they really first came from ancient India where the Hindus originated them. The Arabs, as merchants and traders of the Middle or "Dark" Ages, adopted the Hindu number system to help them in commerce and business. Western Europe was still struggling with the Latin language and the roman numerals in their schools and universities. Science was waiting for a "break-through" in communications, particularly in mathematics, and arabic numerals saved the day.

Arabic numerals are simple to form and simple to read. Systems of calculations using arabic numerals were developed that are still used today in the shop, in the office, and in the scientific laboratories. Mass production of consumer goods and programs to send space vehicles to the moon depend on mathematical calculations using a simplified number system and universal mathematical symbols. A comparison of roman and arabic numerals is shown in Fig. 1-1.

Arabic	Roman	Arabic	Roman	Arabic	Roman
1	I	12	XII	50	L
2	II	13	XIII	60	LX
3	III	14	XIV	70	LXX
4	IV	15	XV	80	LXXX
5	V	16	XVI	90	XC
6	VI	17	XVII	100	C
7	VII	18	XVIII	500	D
8	VIII	19	XIX	600	DC
9	IX	20	XX	1000	M
10	X	30	XXX	1900	MCM
11	XI	40	XL	1,000,000	$\overline{M}$

Fig. 1-1. Arabic and roman numerals.

Digits

The individual arabic numerals are called "digits" and are: 0, 1, 2, 3, 4, 5, 6, 7, 8 and 9. Larger numbers are made from these basic digits and conform to the "tens" or "decimal" method of counting. When mankind was young, like little children, they used their ten fingers to express quantities. As a result we have the decimal system, a system of "tens."

Simple counting involves adding and subtracting. Multiplication is a form of addition, and division is a form of subtraction. These are mathematical "operations," and the symbols used for these operations are:

Addition:　　　　+　　　　Subtraction:　　—

Multiplication:　× or ·　　Division:　　　÷

Grouping-Symbols

In mathematics symbols are needed to show grouping of terms and expressions. This is similar to the punctuation we use in writing. The mathematical symbols used for grouping are:

Parentheses:　( )　　　Brackets:　　[ ]　　　Braces:　　　{ }

A fourth grouping—symbol is the vinculum: ——— a line placed above terms. This is the line at the top of the long division sign: ⌐ and the square root radical: $\sqrt{}$. It is also the bar in a fraction: $\frac{3}{4}$.

Grouping-symbols are important in the solution of mathematical problems because they show what the specific relationships are between numbers and groups of numbers. Grouping-symbols may occur within one another such as:

$$6[2(3 + 2) - 3(4 - 2)] + 12$$

These groupings and the proper method of solution will be discussed later in the book.

Operational Symbols

In the shorthand of mathematics the following symbols are used to express a thought relating to an operation. These are:

Equals or "is equal to": $=$

Does not equal: $\neq$

Is approximately equal: $\approx$ or: $\cong$

Plus or minus: $\pm$ (used in tolerances in dimensioning)

Is greater than: $>$

Is less than: $<$

Is similar to: $\sim$ (used mainly in geometry)

The "Times" Table

Early in your schooling, you were required to learn the products of all the number combinations from 1 to 12. The following chart is a 20 times 20 table as used in many schools in Europe. It is necessary that you learn the products of 2×2 through 9×9. It is suggested that you review your ability to quickly and accurately multiply single digit numerals. This basic skill will greatly help you in your solution of the math problems to come.

Study Example.

1. In the "Times" table (Fig. 1-2), multiply 13 by 18.
   a. Locate 13 along the top row.
   b. Locate 18 down the left side column.
   c. Read answer, 234, at intersection of the 13 column and the 18 row.

Study Problems.

1. Use the "Times" table and look up 20 random pairs of numbers to multiply.

2. If you experience difficulty in remembering the products, make a set of "flash-cards"; use 3 by 5 cards with the multiplication on one side such as 8×7. Put the answer, 56, on the reverse side of the card. Shuffle the cards and try to answer each as you look at the problem side. If you know the answer check it by looking on the back of the card. If you do not remember the answer, look at the back of the card and memorize it. Practice, practice, practice.

Table 1-1. Multiplication Table

×	1	2	3	4	5	6	7	8	9	10	11	12	13	14	15	16	17	18	19	20
1	1	2	3	4	5	6	7	8	9	10	11	12	13	14	15	16	17	18	19	20
2	2	4	6	8	10	12	14	16	18	20	22	24	26	28	30	32	34	36	38	40
3	3	6	9	12	15	18	21	24	27	30	33	36	39	42	45	48	51	54	57	60
4	4	8	12	16	20	24	28	32	36	40	44	48	52	56	60	64	68	72	76	80
5	5	10	15	20	25	30	35	40	45	50	55	60	65	70	75	80	85	90	95	100
6	6	12	18	24	30	36	42	48	54	60	66	72	78	84	90	96	102	108	114	120
7	7	14	21	28	35	42	49	56	63	70	77	84	91	98	105	112	119	126	133	140
8	8	16	24	32	40	48	56	64	72	80	88	96	104	112	120	128	136	144	152	160
9	9	18	27	36	45	54	63	72	81	90	99	108	117	126	135	144	153	162	171	180
10	10	20	30	40	50	60	70	80	90	100	110	120	130	140	150	160	170	180	190	200
11	11	22	33	44	55	66	77	88	99	110	121	132	143	154	165	176	187	198	209	220
12	12	24	36	48	60	72	84	96	108	120	132	144	156	168	180	192	204	216	228	240
13	13	26	39	52	65	78	91	104	117	130	143	156	169	182	195	208	221	234	247	260
14	14	28	42	56	70	84	98	112	126	140	154	168	182	196	210	224	238	252	266	280
15	15	30	45	60	75	90	105	120	135	150	165	180	195	210	225	240	255	270	285	300
16	16	32	48	64	80	96	112	128	144	160	176	192	208	224	240	256	272	288	304	320
17	17	34	51	68	85	102	119	136	153	170	187	204	221	238	255	272	289	306	323	340
18	18	36	54	72	90	108	126	144	162	180	198	216	234	252	270	288	306	324	342	360
19	19	38	57	76	95	114	133	152	171	190	209	228	247	266	285	304	323	342	361	380
20	20	40	60	80	100	120	140	160	180	200	220	240	260	280	300	320	340	360	380	400

Definitions

Mathematical terms are described in a strict and rigid manner. Proper use of these descriptions is necessary in the stating of mathematical problems and in explaining the steps taken in their solutions. A number, term, or symbol must be used *exactly* as defined. The most common definitions follow:

Integer: A whole number such as: 1, 5, 45, 385. These are also referred to as: *integral numbers*.

Even Numbers: Any integer that is divisible by 2, such as: 6, 8, 46, 4506.

Odd Numbers: Any integer *not* exactly divisible by 2, such as: 5, 9, 33, 501.

Factor: A whole number that exactly divides another whole number. 4 is a factor of 12.

Prime Numbers: A whole number that has no factor other than itself or 1, such as: 5, 11, 17.

Prime Factors: The factors of a number that have been reduced to prime numbers. The prime factors of 12 are: $2 \times 2 \times 3$; of 36 are: $2 \times 2 \times 3 \times 3$.

Composite Number: A whole number that has factors other than itself or 1. In $10 = 5 \times 2$, 5 and 2 are factors of 10, and 10 is a composite number.

Common Factor or Common Divisor: A number that will equally divide each of two or more numbers. For example, 3 is a common factor of 9, 12, and 18.

Multiple: A number that is exactly divisible by a given number. For example, 18 is a multiple of 6; it is also a multiple of 2, 3, and 9.

Sum: The result of adding two or more numbers. Thus: $10 + 12 = 22$, and 22 is the sum.

Difference: The result of subtracting one number from another. Thus: $28 - 26 = 2$, and 2 is the difference.

Product: The result of multiplying two or more numbers. The product of $3 \times 4 \times 5$ is 60.

Quotient: The result of dividing one number by another. Thus, $27 \div 3 = 9$, or $36 \div 4 = 9$, and 9 is the quotient.

Divisor: The divisor is the number that divides. In $9\,\overline{)27}$ and $27 \div 9$, 9 is the divisor.

Dividend: The dividend is the number being divided. In $\frac{27}{9}$ and $27 \div 9$, 27 is the dividend.

Study Problems.

1. Add the product of 3 and 8 to the difference of $12 - 9$.
2. Subtract the quotient of $48 \div 12$ from their sum.
3. What number called a "common factor" will divide evenly into 10, 15, and 25?
4. Solve: $12 + 35 - 9 + 6$
5. Name the prime numbers from 1 to 31.
6. In the numbers from 1 to 12, what is the sum of the prime numbers?
7. Find the product of 13 and 7.
8. Multiply the quotient of $36 \div 4$ by the product of 4 and 5.
9. Subtract the sum of 3 and 2 from the quotient of $144 \div 6$.
10. Name four numbers not divisible by 2. What are these numbers called?

Operations in Arithmetic

Addition and subtraction are operations in the additive process. Multiplication and division are closely related to addition and subtraction and are part of the additive system.

When numbers are added, the sequence of addition may be taken in any order. For example, $3 + 2 + 7 + 8 = 20$; also, $2 + 3 + 8 + 7 = 20$, and $8 + 3 + 7 + 2 = 20$. Likewise, a series of all negative numbers may be added in any sequence. The nominal values of the negative numbers are added and the answer is prefixed by a negative:

$$
\begin{array}{ccc}
-\ 3 & \text{or} & -\ 8 \\
-\ 2 & & -\ 3 \\
-\ 8 & & -\ 7 \\
\underline{-\ 7} & & \underline{-\ 2} \\
-20 & & -20
\end{array}
$$

A mixed series of addition and subtraction or, more correctly, the addition of positive and negative numbers, can be done correctly by two methods: Step by step, one term to the next. For example:

$$90 - 15 + 10 - 5 =$$

Step 1. $90 - 15 = 75$

Step 2. $75 + 10 = 85$

Step 3. $85 - 5 = 80$

The alternative method is to add all positive (plus) terms:

$$90 + 10 = 100$$

then add all negative (minus) terms:

$$(-15) + (-5) = -20$$

then add the two totals:

$$100 + (-20) = 80$$

Plus (+) or minus (−) signs are used either as the sign of the numbers or they can refer to the operation to be performed, addition or subtraction. Care must be taken to properly identify which use is meant in a problem. In the alternative method, above, note that the negative numbers are placed within parentheses to avoid confusion.

Adding with Unlike Signs

When adding two numbers with unlike signs, calculate the numerical difference between the two numbers. Prefix this difference with the sign of the larger of the two numbers.

Study Examples.

Add:	25	13	−19	− 4	−348	504
	−10	−26	20	13	118	−648
	+15	−13	+ 1	+ 9	−230	−144

Study Problems.

1. Add:

+12	+2	+15	+36	−20	+74
−10	−8	−10	−40	+35	−24

2. Add:

+25	−36	+33	+45	−33	+36
+34	−58	−22	+34	−22	+14
−22	+27	+25	−42	+13	−28
− 7	+82	+ 6	−32	+45	− 8

3. Do indicated operations:

−25	42	−29	29	−17	17
13	−17	17	−17	29	−29

Subtracting with Like and Unlike Signs

Subtraction is sometimes confusing when negative numbers are concerned. When a negative number is subtracted from a *larger* negative number, the answer is the numerical difference between the two numbers with a *negative sign* prefixed to the answer (difference).

$$\begin{array}{r} (-25) \\ -(-15) \\ \hline -10 \end{array}$$

When a larger negative number is subtracted from a *smaller* negative number, the answer is the numerical difference between the two numbers, with a *positive* sign:

$$\begin{array}{r} (-15) \\ -(-25) \\ \hline +10 \end{array}$$

When a negative number is subtracted from a positive number, the answer is the *positive sum* of the two numerals:

$$\begin{array}{r} 23 \\ -(-10) \\ \hline +33 \end{array} \qquad \begin{array}{r} 12 \\ -(-15) \\ \hline +27 \end{array}$$

A handy rule with problems involving the subtraction of a negative number is: "Change *both* negative signs and add."

$$\begin{array}{cc} +45 & +45 \\ \underline{-(-13)} \quad \text{(change both signs)} & \underline{+(+13)} \\ & +58 \end{array}$$

Study Problems.

Perform operations as shown:

1.	45	2.	222	3.	(−58)	4. (−73)	5. (−35)
	−21		−(− 64)		−(−68)	57	(−24)

Series Multiplication and Division

A series of multiplications may be performed in any sequence. For example:

$$7 \times 2 \times 3 \times 5 = 210$$

also

$$2 \times 5 \times 7 \times 3 = 210$$

However, a series of divisions must be done in the sequence given: For example:

$$90 \div 15 \div 3 = 2$$

by steps:

$$90 \div 15 = 6$$
$$6 \div 3 = 2$$

If the sequence is not followed, an error will be made; for example:

$$15 \div 3 = 5$$
$$90 \div 5 = 18 \text{ (which is not correct)}.$$

Study Problems.

Perform indicated operations:

1. $9 \times 6 \times 2 =$

2. $4 \times 2 \times 6 \times 8 \times 4 =$

3. $45 \div 3 \div 5 =$

4. $98 \div 7 \div 7 =$

5. $11 \times 24 \times 4 =$

6. $128 \div 4 \div 4 \div 2 =$

7. $54 \div 3 \div 6 =$

8. $8 \times 6 \times 3 \times 7 =$

9. $6 \times 5 \times 10 \times 2 =$

10. $75 \div 5 \div 5 =$

Order of Mixed Operations

In a mixture of additions, subtractions, multiplications and divisions, such as:

$$3 + 2 \times 4 - 2 + 6 \div 3 - 2 =$$

the operations must be performed in a specific sequence, which is,

1. Multiplication

2. Division

3. Addition

4. Subtraction.

A handy memory "code" to use to remember this sequence is M.D.A.S., from the initials of the operations in proper order. This is easily remembered by the phrase: "*My Dear Aunt Sarah.*"

It is advisable to use "grouping" symbols to a problem to aid in visualizing the separate operations. The preceding problem would be grouped:

$$3 + (2 \times 4) - 2 + (6 \div 3) - 2 =$$

The first step in the solution would be multiplication: (2×4);

$$3 + 8 - 2 + (6 \div 3) - 2 =$$

The next step is the division of 6 by 3;

$$3 + 8 - 2 + 2 - 2 =$$

Addition of the plus terms follows;

$$3 + 8 + 2 = 13$$

Addition of the minus terms follows:

$$(-2) + (-2) = (-4)$$

The final step is the addition of the plus and minus sums:

$$13 + (-4) = 9$$

Study Problems.

1. $5 + 8 + 3 + 9 + 2 =$

2. $19 + 13 - 6 - 7 + 2 =$

3. $24 \div 3 + 4 \times 5 - 6 =$

4. $10 + (-8) + 4 \times 5 =$

5. $18 - 34 \times 2 - 34 =$

6. $25 \div 5 + 3 \times 6 =$

7. $24 \div 8 + 6 \times 4 - 10 \div 2 =$

8. $5 \times 7 + 6 - 4 \times 7 =$

9. $32 \div 4 + 8 \times 3 - 5 =$

10. $6 - 3 \times 8 + 6 \times 5 - 14 \div 2 + 33 \div 11 =$

Multiplication and Division of Signed Numbers

The multiplication and division of signed numbers involve two short rules: "The product or quotient of numbers with *like* signs results in a *positive* number," and "The product or the quotient of numbers with *unlike* signs results in a *negative* number." For example:

$$+9 \times (+12) = +108 \qquad +90 \div (-6) = -15$$

$$+9 \times (-12) = -108 \qquad +90 \div (+15) = +6$$

$$-9 \times (+12) = -108 \qquad +10 \times (-3) = -30$$

$$-108 \div (+12) = -9 \qquad +10 \times (+3) = +30$$

$$+108 \div (+9) = +12 \qquad -10 \times (+3) = -30$$

$$+15 \times (+6) = +90 \qquad -30 \div (+3) = -10$$

$$+15 \times (-6) = -90 \qquad +30 \div (+10) = +3$$

$$-15 \times (-6) = +90$$

In the above examples, both the positive and negative signs are shown. In practice, only negative signs would be shown.

Study Problems.

1. $14 \times (-3) =$

2. $12 \times 4 =$

3. $(-9) \times 6 =$

4. $(-5) \times 20 =$

5. $(-8) \times (-36) =$

6. $22 \times 6 =$

7. $24 \times 2 \div 8 =$

8. $(-32) \div (-4) + 6 =$

9. $75 \div 5 \times 2 =$

10. $14 \times (-3) + 4 - 24 \div 6 + 5 - (-7) =$

11. $(-6) \times (-8) \times (-3) =$

12. $57 \div 3 - 5 \times (-5) =$

Short and Long Division

Two methods of division are used in mathematics. When the divisor has one figure (other than zeros), the short division method is commonly used as it is quick and fairly accurate.

Study Example.

1. Divide 636 by 6. Write as:

$$6 \lfloor \underline{636}$$

Step 1. Determine in the mind whether the divisor (6) will divide into the first number of the dividend (636). In the above problem it will. The result of division is (1) which is placed under the (6).

$$6 \lfloor \underline{636}$$
$$1$$

Step 2. The divisor (6) will not divide into (3) and a zero is placed under the (3).

$$6\ \underline{)\,636}$$
$$10$$

Step 3. The (3) is now taken with the third number (6) to become (36). The divisor (6) divides evenly into (36) and the quotient (6) is placed under the (6). The answer is 106.

$$6\ \underline{)\,636}$$
$$106$$

When the divisor contains two or more numbers, the *long* division method is favored.

2. Divide 6048 by 56.

Step 1. Set up the problem by placing a division sign over the dividend (6048) with the divisor (56) to the left.

$$56\ \overline{)\,6048}$$

Step 2. Start from the left of the dividend and find the smallest number that the divisor will divide into. In this case, the number is 60 (60 ÷ 56 = 1 +). Place the quotient (1) above the division sign directly over the 0 of 60.

$$1$$
$$56\ \overline{)\,6048}$$

Step 3. Multiply the divisor (56) by the quotient (1) and place the answer under the smallest number (60) in the dividend found in Step 2.

$$1$$
$$56\ \overline{)\,6048}$$
$$56$$

Step 4. Subtract the product (56) from the number above it (60).

$$1$$
$$56\ \overline{)\,6048}$$
$$\underline{56}$$
$$4$$

Step 5. Bring down the next number in the dividend to form a partial remainder (44).

$$1$$
$$56\ \overline{)\,6048}$$
$$\underline{56}$$
$$44$$

Step 6. Divide the divisor (56) into the partial remainder (44) and place the quotient above the division sign. In this case, 56 cannot be divided into 44 and a zero is placed above the sign after the (1).

$$10$$
$$56\ \overline{)\,6048}$$
$$\underline{56}$$
$$44$$

Step 7. Bring down the next number in the dividend to form a new partial remainder (448).

$$10$$
$$56\ \overline{)\,6048}$$
$$\underline{56}$$
$$448$$

Step 8. Divide the divisor into the partial remainder (448) and place the quotient above the division sign.

$$108$$
$$56\ \overline{)\,6048}$$
$$\underline{56}$$
$$448$$

Step 9. Multiply the divisor (56) by the quotient (8) and place the product below the partial remainder.

$$108$$
$$56\ \overline{)\,6048}$$
$$\underline{56}$$
$$448$$
$$448$$

Step 10. Subtract the product (448) from the previous partial remainder (448). Repeat Steps 7 through 10 until all of the numbers in the dividend have been divided.

$$\begin{array}{r} 108 \\ 56\overline{)6048} \\ 56 \\ \hline 448 \\ 448 \\ \hline 0 \end{array}$$

In many problems the answer may have a "remainder"; that is, the divisor will not divide into the dividend evenly. The remainder is handled as shown below:

Study Example.

1. Divide 4789 by 25

$$\begin{array}{r} 191 \\ 25\overline{)4789} \\ 25 \\ \hline 228 \\ 225 \\ \hline 39 \\ 25 \\ \hline 14 \end{array}$$

The answer is: $191 \frac{14}{25}$

Study Problems.

Do the indicated operations.

1. $13\overline{)390}$

2. $17\overline{)9134}$

3. $128\overline{)35000}$

4. $33\overline{)1000}$

5. $24\overline{)50412}$

6. $19\overline{)1357}$

Review Problems.

1. $(-6) \times (-7) =$

2. $55 + 13 - 7 =$

3. $(-36) \div 13 =$

4. $4560 \div 4 =$

5. $36 \times 5 \div 9 =$

6. $63 + (-9) =$

7. $30\overline{)6630}$

8. $14\overline{)29806}$

9. $235\overline{)28905}$

10. $40\overline{)13000}$

11. $67\overline{)51188}$

12. $83\overline{)43243}$

13. Ten parts are to have the following size holes drilled:

No. of parts	No. 1/2" dia. holes	No. 3/4" dia. holes	No. 1" dia. holes
3	6	12	1
4	8	15	6
2	12	2	10
1	2	3	8

What is the total number of holes that must be drilled for each size?

14. A total of 48 1/2-in. dia., 96 3/8-in. dia., and 120 1/4-in. dia. holes were drilled in eight identical parts.

   a. How many holes does each part contain?

   b. How many holes for each size given does each individual part have?

15. A carpenter uses 12 16-penny nails in each 2 × 8 he places in framing a floor. The 2 × 8's are spaced on 16-in. centers and are 14 ft long. If the floor is 26 ft 5 in. by 32 ft, how many 2 × 8's, 14 ft long, will be needed to span the short dimension and how many nails will be used?

16. A stock crib inventory shows the following quantities of cold rolled rounds on hand:

Diameter	Length, Ft		
	6	8	10
	Number on Hand		
$\frac{1}{2}$ in.	8	5	6
$\frac{5}{8}$ in.	3	6	7
$\frac{3}{4}$ in.	7	2	0
$\frac{7}{8}$ in.	5	4	2
1 in.	12	7	8

   a. How many bars of each diameter are there in stock?

   b. How many 6 ft lengths?

   c. How many 8 ft lengths?

   d. How many 10 ft lengths?

17 The assembly of a special pump requires varying sizes and lengths of hex head machine screws. These are shown in the following table.

   a. What is the total number of screws required per pump?

   b. How many of each size is required per pump?

   c. How many of each length is required per pump?

Screw Size	Length, In.			
	$\frac{1}{2}$	$\frac{3}{4}$	1	$1\frac{1}{2}$
	Number Required			
$\frac{1}{4}$ - 20	6	4	8	2
$\frac{1}{2}$ - 13	4	6	3	10
$\frac{5}{8}$ - 11	0	2	4	2
$\frac{3}{4}$ - 10	0	0	6	8

18. A foundry mixes five different materials to form cores and molds. Eight different blends are listed below with their specified quantities of materials in pounds:

Blend No.	Materials				
	A	B	C	D	E
	Quantity, Lb.				
1	2	4	1	0	0
2	1	6	2	1	3
3	0	4	3	5	2
4	2	0	4	3	4
5	6	2	0	0	1
6	5	1	6	2	5
7	4	3	0	1	2
8	3	5	0	0	1

a. How many pounds does each blend weigh?

b. How much of each material is used to make up one batch of each of the 8 blends?

CHAPTER 2

Fractions and Mixed Numbers

Fractions are used extensively in industry and in construction, particularly in specifying dimensions. A common fraction expresses a part of a whole, such as one-half of an inch ($\frac{1}{2}$ inch). A common fraction is a number which indicates a division of one number by another. In the common fraction $\frac{5}{8}$, the "5" is the dividend and the 8 is the divisor. Examples of common fractions are:

$$\frac{2}{3}, \frac{11}{16}, \frac{5}{6}, \frac{21}{7} \text{ and } \frac{1}{8}$$

In a common fraction the numbers above and below the bar have descriptive names.

Numerator: The number above the bar is called the numerator. It is the *dividend* mentioned above. The numerator designates the number of parts of the whole which the fraction is a part of. For example, in $\frac{2}{3}$, the numerator "2" designates that this fraction specifies 2 parts of 3, or two pieces of pie when the pie is cut into 3 parts.

Denominator: The denominator is the number below the bar. It is the *divisor* mentioned above. It designates the number of equal parts the whole number is divided into; i.e., the parts of a whole.

Proper and Improper Fractions

Fractions are classifed into two broad groups: *proper fractions*, in which the numerator is the smaller number of the two, thus making the value of the fraction less than one; and, *improper fractions*, in which the numerator is greater than its denominator, thus making the value greater than one. Examples of proper fractions are:

$$\frac{6}{7} \quad \frac{3}{5} \quad \frac{233}{654} \quad \frac{875}{1000}$$

18

Examples of improper fractions are:

$$\frac{53}{4} \quad \frac{4}{3} \quad \frac{284}{25} \quad \frac{635}{505}$$

As improper fractions are greater than one, they are generally changed to mixed numbers by dividing the numerator by the denominator.

Study Examples.

1. Change $\frac{53}{8}$ to a mixed number.

   Write as a long division:

   $8\overline{)53}$

   Divide 53 by 8:

   $$\begin{array}{r} 6 \\ 8\overline{)53} \\ \underline{48} \\ 5 \end{array}$$

   Write answer as whole number and the remainder as a fraction: $6\frac{5}{8}$

2. Change $\frac{284}{25}$ to a mixed number:

   Write as a long division problem and solve:

   $$\begin{array}{r} 11 \\ 25\overline{)284} \\ \underline{25} \\ 34 \\ \underline{25} \\ 9 \end{array}$$

   Write answer as whole number and a fraction: $11\frac{9}{25}$

NOTE: Fractions are sometimes written using a "slash" (/) rather than a horizontal bar ——. This practice, which is used when fractions appear with text in a printed paragraph, should only be used where there is no possibility of confusion. The use of the horizontal bar is recommended.

Reduction of Fractions to Lowest Terms

When working with fractions it is usually more convenient to reduce proper fractions and the fractional part of a mixed number to *lowest terms.* A fraction in its lowest terms is one whose numerator and denominator cannot be divided evenly by the same whole number.

The process of reducing fractions to their lowest terms involves finding a whole number that can evenly divide into both the numerator and the denominator. Such a number is called a *common factor.* Factors are numbers that when multiplied together result in a *product.* There are two types of numbers: those that have factors *other* than themselves and one (5 × 1 = 5) and those that are "prime" (having no whole number factors other than themselves and one.

Prime Numbers

Prime numbers are numbers that have no factors other than themselves and one. The prime numbers from 1 to 32 are:

$$2, \ 3, \ 5, \ 7, \ 11, \ 13, \ 17, \ 19, \ 23, \ 29 \text{ and } 31.$$

The other numbers between 3 and 31 have factors. For example:

$$4 = 2 \times 2, \ 6 = 2 \times 3, \ 8 = 2 \times 2 \times 2, \ 9 = 3 \times 3.$$

When fractions are involved, any fraction that contains a prime number in either the numerator or the denominator cannot be reduced.

Reduction of Fractions Using Prime Factors

Fractions may be reduced to lowest terms if the numerator and the denominator contain common prime factors. For example:

1. $\dfrac{9}{12} = \dfrac{3 \times 3}{4 \times 3}$, but $\dfrac{3}{3} = 1$

   therefore: $\dfrac{9}{12} = \dfrac{3 \times 3}{4 \times 3} = \dfrac{3}{4} \times 1 = \dfrac{3}{4}$

2. $\frac{17}{24}$, the numerator is a prime number and the fraction cannot be reduced.

3. $\dfrac{25}{60} = \dfrac{5 \times 5}{5 \times 12}$, but $\dfrac{5}{5} = 1$

therefore: $\dfrac{25}{60} = \dfrac{5 \times 5}{5 \times 12} = \dfrac{5}{12} \times 1 = \dfrac{5}{12}$

If a mixed number such as $15\frac{25}{30}$ is to be reduced, only the fractional part of the mixed number, $\frac{25}{30}$, is reduced. For example:

$$\dfrac{25}{30} = \dfrac{5 \times 5}{6 \times 5} = \dfrac{5}{6} \times 1 = \dfrac{5}{6}$$

therefore: $\qquad 15\frac{25}{30} = 15\frac{5}{6}$

Equivalent Fractions

If both the numerator and the denominator of a fraction are multiplied by the same number, the value of the fraction is not changed.

For example: $\dfrac{3 \times 3}{4 \times 3} = \dfrac{9}{12}$

This is the same as multiplying by a fraction $\frac{3}{3}$, which is equal to 1. Fractions remain unchanged in value when multiplied by a fraction equal to one, such as: $\frac{3}{3}$ or $\frac{6}{6}$. This is true because when both numerator and denominator are multiplied by the same number the operation is actually multiplying by one; for example:

$$\dfrac{3}{4} \times \dfrac{5}{5} = \dfrac{3 \times 5}{4 \times 5} = \dfrac{15}{20}$$

as $\frac{5}{5}$ equals 1, the value of $\frac{15}{20}$ is the same as $\frac{3}{4}$. This operation is very valuable in the addition and subtraction of fractions.

Conversion of Mixed Numbers to Improper Fractions

Often it is necessary to convert a mixed number to a fraction. To do this, the integer (the whole number) is multiplied by the denominator of the fraction and the numerator is added to the product. The result is written over the fraction bar with the original denominator beneath. For example:

Convert: $6\frac{7}{8}$ to an improper fraction.

Step 1. Multiply (6) by the denominator (8): $6 \times 8 = 48$

Step 2. Add the numerator (7) to (48): $7 + 48 = 55$

Step 3. Write (55) over (8): $\frac{55}{8}$

thus: $6\frac{7}{8} = \frac{55}{8}$

While this answer is an improper fraction, it can be used in many operations with fractions, as described on later pages.

Study Problems.

1. Write a fraction whose numerator is 8 and whose denominator is 13.

2. Write a fraction whose denominator is 36 and whose numerator is 18. Reduce to lowest terms.

3. Write a fraction whose numerator is 45 and whose denominator is 6. Convert this improper fraction to a mixed number with its fraction in lowest terms.

4. Reduce the following improper fractions to mixed numbers with the fraction in lowest terms:

$$\frac{12}{5} \qquad \frac{33}{5} \qquad \frac{148}{12} \qquad \frac{1605}{40} \qquad \frac{97}{8} \qquad \frac{58}{3} \qquad \frac{32}{8}$$

5. Convert the following mixed numbers to improper fractions;

$$6\frac{1}{2} \qquad 2\frac{3}{4} \qquad 9\frac{7}{8} \qquad 12\frac{9}{12} \qquad 25\frac{4}{5} \qquad 19\frac{11}{13}$$

6. Raise the following fractions to equivalent fractions:

$$\frac{2}{3} \text{ to } \frac{4}{12} \qquad \frac{5}{6} \text{ to } \frac{10}{12} \qquad \frac{3}{4} \text{ to } \frac{18}{24} \qquad \frac{11}{16} \text{ to } \frac{44}{64}$$

7. Reduce the following fractions to lowest terms:

$$\frac{7}{35} = \qquad \frac{104}{120} = \qquad \frac{105}{115} = \qquad \frac{68}{85} = \qquad \frac{133}{152} =$$

8. The fraction $\frac{5}{8}$ is how many 64ths?

9. The fraction $\frac{4}{5}$ is how many 85ths?

Addition and Subtraction of Fractions

Fractions can be added and subtracted. However, the methods used to add and subtract fractions are quite different from those used to add and subtract whole numbers and must be performed in a special and precise manner.

Fractions with the Same Denominator

Fractions with the same numeral in their denominators may be added by merely adding the separate numerators and placing the sum over the common denominator numeral.

Study Example.

$$\frac{3}{8} + \frac{5}{8} + \frac{7}{8} + \frac{1}{8} = \frac{3 + 5 + 7 + 1}{8} = \frac{16}{8} = 2$$

In some special cases, all the denominators may be factors of the largest numeral in the denominator. Addition is possible after each fraction is converted to an "equivalent" fraction with a denominator equal to the largest denominator.

Study Example.

$$\frac{1}{8} + \frac{1}{4} + \frac{3}{8} + \frac{3}{4} - \frac{1}{2} =$$

Step 1. Change all fractions to 8ths (largest denominator):

$$\frac{1}{8} + \frac{2}{8} + \frac{3}{8} + \frac{6}{8} - \frac{4}{8} =$$

Step 2. Add after placing numerators over the common denominator:

$$\frac{1 + 2 + 3 + 6 - 4}{8} = \frac{8}{8} = 1$$

Study Problems.

1. $\frac{3}{2} + \frac{5}{2} =$ 4. $\frac{3}{4} - \frac{5}{8} + \frac{1}{2} =$ 7. $\frac{1}{2} + \frac{1}{32} + \frac{3}{4} =$

2. $\frac{3}{4} + \frac{5}{8} =$ 5. $\frac{3}{4} + \frac{1}{2} - \frac{3}{8} =$ 8. $\frac{3}{16} + \frac{5}{32} - \frac{5}{8} =$

3. $\frac{9}{16} - \frac{1}{4} =$ 6. $\frac{7}{9} + \frac{5}{6} - \frac{1}{18} =$ 9. $\frac{3}{32} + \frac{3}{8} - \frac{3}{64} =$

Fractions with Different Denominators

When denominators are not the same or are not factors of a larger numeral denominator, a "common" denominator must be determined and used in the addition and subtraction of fractions.

A common denominator of a group of fractions is a single denominator to which all of the fractions can be raised. In a group of fractions, the product of all of the denominators when multiplied together is always a common denominator; however, this common denominator is frequently a very large number that is unwieldy to work with. Often a smaller common denominator can be found.

The smallest possible common denominator that can be used is one into which all the original denominators can be evenly divided. This is called the "Least Common Denominator," which is abbreviated to "LCD." The arithmetic of adding and subtracting fractions can be greatly reduced by using the LCD.

The method of calculating the LCD is based on the determination of the prime factors of each of the denominators of the separate fractions. Prime factors are most easily determined by doing a series of short divisions starting with the lowest prime number that will evenly divide into the denominator and continuing as shown in the following example:

Study Example.

Find the prime factors of 36.

Step 1.　Divide 36 by 2;　　　　　　　　　　　　　　$2\lfloor 36$

Step 2.　Divide 18 by 2;　　　　　　　　　　　　　　$2\lfloor 18$

Step 3.　Divide 9 by 3;　　　　　　　　　　　　　　$3\lfloor 9$
　　　　　　　　　　　　　　　　　　　　　　　　　　3

The prime factors of 36 are: 2, 2, 3, and 3.

Study Problems.

Determine the prime factors of the following numbers:

1. 74	3. 66	5. 99	7. 168	9. 64
2. 84	4. 55	6. 91	8. 52	10. 97

The LCD is the product of all the unique primes with each unique factor being used as many times as it occurs in the denominator using that factor the greatest number of times. A suggested procedure to use is as follows:

Study Example.

Find the LCD of the following group of fractions:

$$\frac{1}{20} \qquad \frac{7}{18} \qquad \frac{4}{15} \qquad \frac{1}{36}$$

Step 1. Under each denominator place its prime factors:

$$\frac{1}{20} \qquad \frac{7}{18} \qquad \frac{4}{15} \qquad \frac{1}{36}$$
$$2, 2, 5 \qquad 2, \underline{3, 3} \qquad 3, 5 \qquad 2, 2, \underline{3, 3}$$

Step 2. Determine the greatest frequency of 2. (underscored above)   2, 2

Step 3. Determine the greatest frequency of 3. (double underscored)   3, 3

Step 4. Determine the greatest frequency of 5.   5

Step 5. Find the LCD:

LCD = (greatest frequency of 2) X (greatest frequency of 3) X (greatest frequency of 5)

LCD = (2 X 2) X (3 X 3) X (5) = 180

The calculated LCD can be proved by dividing it by each of the original denominators to determine if they can be evenly divided into the LCD.

Proof:

$$180 \div 20 = 9$$
$$180 \div 18 = 10$$
$$180 \div 15 = 12$$
$$180 \div 36 = 5$$

Therefore, 180 is the LCD.

Study Problems.

Find the LCD of each of the following groups of fractions.

1. $\dfrac{7}{8}, \dfrac{3}{4}, \dfrac{3}{16}$

2. $\dfrac{1}{15}, \dfrac{3}{10}, \dfrac{4}{5}$

3. $\dfrac{5}{18}, \dfrac{2}{3}, \dfrac{11}{14}, \dfrac{5}{9}$

4. $\dfrac{7}{8}, \dfrac{4}{9}, \dfrac{2}{3}$

5. $\dfrac{3}{8}, \dfrac{3}{5}, \dfrac{17}{20}$

6. $\dfrac{1}{6}, \dfrac{3}{8}, \dfrac{5}{9}, \dfrac{2}{3}$

When the LCD is determined, the next step in the addition and subtraction of fractions with different denominators is to convert the different fractions into equivalent fractions, each with the LCD as its denominator.

Study Example.

Add the fractions: $\frac{1}{2}$, $\frac{2}{3}$, $\frac{3}{4}$, $\frac{5}{6}$, and $\frac{1}{8}$

Step 1. Write prime factors:

$$\frac{1}{2} \qquad \frac{2}{3} \qquad \frac{3}{4} \qquad \frac{5}{6} \qquad \frac{1}{8}$$
$$1, 2 \quad 1, 3 \quad 2, 2 \quad 2, 3 \quad 2, 2, 2 \qquad \text{(primes)}$$

Step 2. Determine frequencies of primes:

$$1, 2, 2, 2, 3$$

Step 3. Multiply frequent primes to get LCD:

$$2 \times 2 \times 2 \times 3 = 24$$

Step 4. Convert to equivalent fractions:

$$\frac{1}{2} = \frac{12}{24}, \quad \frac{2}{3} = \frac{16}{24}, \quad \frac{3}{4} = \frac{18}{24}, \quad \frac{5}{6} = \frac{20}{24}, \quad \frac{1}{8} = \frac{3}{24}$$

Step 5. Add the equivalent fractions:

$$\frac{12}{24} + \frac{16}{24} + \frac{18}{24} + \frac{20}{24} + \frac{3}{24} = \frac{12 + 16 + 18 + 20 + 3}{24}$$

$$= \frac{69}{24} = 2\frac{21}{24} = 2\frac{7}{8}$$

Combined Addition and Subtraction of Fractions

Whenever a mixture of plus and minus fractions appears as a problem in addition, the same procedures apply as when all the fractions have plus signs except that in the step where the equivalent fractions are added, the fractions must each bear their respective signs. Using the Study Example just given, the signs of certain fractions have been changed below to form a problem in addition and subtraction:

$$\frac{1}{2} - \frac{2}{3} + \frac{3}{4} + \frac{5}{6} - \frac{1}{8} =$$

The LCD is 24 and the equivalent fractions are:

$$\frac{12}{24} - \frac{16}{24} + \frac{18}{24} + \frac{20}{24} - \frac{3}{24} = \frac{12 - 16 + 18 + 20 - 3}{24} = \frac{31}{24}$$

Converting the fractional answer to a mixed number:

$$31 \div 24 = 1\frac{7}{24}$$

Study Problems.

Perform the indicated operations.

1. $\frac{1}{2} - \frac{1}{6} + \frac{1}{4} - \frac{1}{8} =$

2. $\frac{1}{2} + \frac{3}{4} - \frac{5}{8} =$

3. $\frac{2}{5} + \frac{3}{4} - \frac{1}{6} =$

4. $\frac{3}{4} + \frac{7}{9} - \frac{2}{9} - \frac{6}{9} =$

5. $\frac{21}{24} + \frac{11}{12} - \frac{3}{8} =$

6. $\frac{1}{3} + \frac{3}{8} - \frac{4}{7} =$

7. $\frac{11}{15} - \frac{2}{9} + \frac{4}{5} - \frac{1}{3} =$

8. $\frac{4}{5} - \frac{3}{4} + \frac{2}{3} - \frac{1}{2} =$

9. $\frac{5}{12} - \frac{4}{5} + \frac{2}{3} =$

10. $\frac{13}{14} + \frac{11}{21} - \frac{9}{14} - \frac{5}{7} =$

11. $\frac{5}{8} + \frac{7}{8} - \frac{3}{4} + \frac{5}{8} - \frac{9}{16} =$

12. A tube has an inside diameter of $\frac{5}{16}$ in. and a wall thickness of $\frac{1}{8}$ in. What is the outside diameter (O.D.)?

13. A bolt is used to fasten two 2 × 6 planks. The 2-in. thickness (nominal) is actually $1\frac{5}{8}$, and the nut for the bolt is $\frac{1}{4}$ in. thick. What is the minimum length of bolt to use if bolts come in length increments of $\frac{1}{4}$ in.

14. A special laminated wooden beam is made up of two pieces of planking $1\frac{5}{8}$ in. thick, and three pieces of 1-in. nominal lumber that is $\frac{13}{16}$ in. thick. What is the actual thickness of the beam?

15. A special jet fuel is made in a laboratory in small quantities for testing. If the formula calls for $\frac{3}{4}$ lb of chemical A, $\frac{1}{8}$ lb of chemical B, and $\frac{2}{3}$ lb of chemical C, what is the total weight of the experimental batch of jet fuel?

16. A carpenter is nailing some $\frac{3}{8}$ in. (actual measurement) plywood onto furring strips that are $\frac{13}{16}$ in. thick. If nails come in $\frac{1}{8}$ in. incremental

lengths, how long a nail must he use to insure maximum penetration but not go completely through the furring strip?

17. A draftsman scale (ruler) has the inches divided into halves, quarters, eighths, and sixteenths. How long a line would result if one-half, three-quarters, five-eighths and eleven-sixteenths were added together?

18. Determine dimensions A, B, C, D, E, and F in Fig. 2-1.

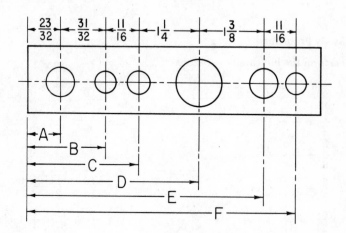

Fig. 2-1. Study problem 18.

19. In Fig. 2-2, determine horizontal dimensions, A, B, C, and D. Also determine vertical dimensions, E, F, G, and H.

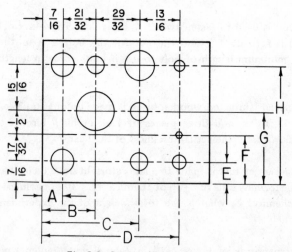

Fig. 2-2. Study problem 19.

20. Determine the overall dimension, A, of the locating pin shown in Fig. 2-3.

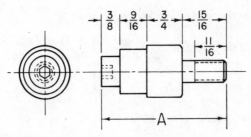

Fig. 2-3. Study problem 20.

Addition and Subtraction of Mixed Numbers

Mixed numbers can be added and subtracted by two methods. In the first method, the mixed numbers are converted into improper fractions which are then added by means of the LCD method.

Study Examples.

1. Add: $8\frac{3}{5}$ and $11\frac{2}{7}$

Step 1. Convert to improper fractions:

$$\frac{(8 \times 5) + 3}{5} = \frac{40 + 3}{5} = \frac{43}{5}$$

$$\frac{(11 \times 7) + 2}{7} = \frac{77 + 2}{7} = \frac{79}{7}$$

Step 2. Add improper fractions: LCD = $5 \times 7 = 35$

$$\frac{43}{5} + \frac{79}{7} = \frac{43 \times 7}{35} + \frac{79 \times 5}{35} = \frac{301 + 395}{35} = \frac{696}{35}$$

Step 3. Reduce improper fraction to a mixed number:

$$696 \div 35 = 19\frac{31}{35}$$

In the alternate method, the whole numbers are added first and then the fractions are added separately by the LCD method. The sum of the whole numbers and the sum of the fractions are then added.

Study Example.

Add: $8\frac{3}{5}$ and $11\frac{2}{7}$

Step 1. Add the whole numbers: $8 + 11 = 19$

Step 2. Add the fractions: $\dfrac{3}{5} + \dfrac{2}{7} = \dfrac{(3 \times 7)}{35} + \dfrac{(2 \times 5)}{35}$

$$= \frac{21}{35} + \frac{10}{35} = \frac{31}{35}$$

Step 3. Add: 19 and $\dfrac{31}{35} = 19\dfrac{31}{35}$

Study Problems.

1. $3\frac{1}{2} + 2\frac{1}{8} - \frac{5}{16} =$ 2. $6\frac{7}{8} + 3\frac{5}{6} =$ 3. $\frac{19}{32} + 6\frac{1}{4} =$

4. Calculate the total length of the detail in Fig. 2-4.

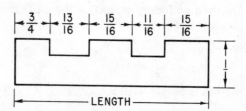

Fig. 2-4. Study problem 4.

5. Determine the length and width of detail in Fig. 2-5.

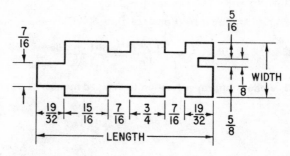

Fig. 2-5. Study problem 5.

6. $10\frac{2}{32} + \frac{7}{16} - 1\frac{1}{8} =$ 7. $12\frac{1}{8} - 2\frac{3}{4} - \frac{5}{6} =$

8. Determine the length and width of the detail in Fig. 2-6.

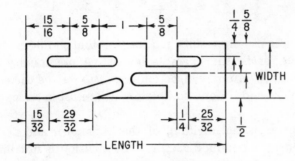

Fig. 2-6 . Study problem 8.

9. Six sheets of steel have the following thicknesses: $\frac{1}{8}, \frac{3}{32}, \frac{3}{16}, \frac{1}{4}, \frac{1}{16}$ and $\frac{11}{64}$. What is the total thickness of the six sheets?

10. In Fig. 2-7, determine dimensions A, B, C, and D.

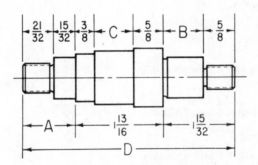

Fig. 2-7. Study problem 10.

11. Eight 2 X 4's, 8 ft long are nailed end to end with each joint having an overlap of $6\frac{3}{8}$ in. What is the total length of the assembly?

12. A rectangular field is 450 ft, 8 in. long by 235 ft, 11 in. wide. How many feet of fencing are needed to enclose the field?

13. $45\frac{3}{5} - 22\frac{7}{10} - 11\frac{14}{15} =$ 14. $22\frac{8}{9} + 11\frac{5}{6} - 8\frac{1}{3} =$

Multiplication of Fractions

The rule for the multiplication of fractions is quite simple.
Multiply all the numerators to obtain the numerator of the answer and multiply all the denominators to obtain the new denominator.

Study Example.

1.
$$\frac{1}{2} \times \frac{2}{3} \times \frac{3}{4} = \frac{1 \times 2 \times 3}{2 \times 3 \times 4} = \frac{6}{24} = \frac{1}{4}$$

The answer usually results in a fraction that should be reduced to lowest terms, as in the example above. A preliminary operation known as "cancellation" can be employed to reduce the complexity of the problem and if done completely will result in an answer in the lowest terms.

Cancellation is the dividing of the numerator and denominator of a fraction by a common factor. This division is not limited to a single fraction. Any numerator and any denominator of a string of fractions may be divided by the common factor. Failure to find and complete all possible cancellations will not cause an error but will require the added operation of reduction to the lowest terms. Cancellation is shown in the following example.

Study Examples.

1.
$$\frac{30}{48} \times \frac{12}{15} \times \frac{4}{6} =$$

Step 1. Cancel 6 into 30 and 6: $\quad \dfrac{\cancel{30}^{5}}{48} \times \dfrac{12}{15} \times \dfrac{4}{\cancel{6}_{1}} =$

Step 2. Cancel 5 into 15 and 5: $\quad \dfrac{\cancel{5}^{1}}{48} \times \dfrac{12}{\cancel{15}_{3}} \times \dfrac{4}{1} =$

Step 3. Cancel 4 into 48 and 4: $\quad \dfrac{1}{\cancel{48}_{12}} \times \dfrac{12}{3} \times \dfrac{\cancel{4}^{1}}{1} =$

Step 4. Cancel 12 into both 12s: $\quad \dfrac{1}{\cancel{12}_{1}} \times \dfrac{\cancel{12}^{1}}{3} \times \dfrac{1}{1} =$

Step 5. Multiply: $\quad \dfrac{1}{1} \times \dfrac{1}{3} \times \dfrac{1}{1} = \dfrac{1}{3}$

2. Cancellation is usually done on the original writing of the problem, making it unnecessary to rewrite after each cancellation.

$$\frac{1}{2} \times \frac{5}{6} \times \frac{24}{35} =$$

Steps: Cancel 2 into 2 and 24, 5 into 5 and 35, and then 6 into 6 and 12 (12 obtained from dividing 24 by 2):

$$\frac{1}{\cancel{2}_1} \times \frac{\cancel{5}^{1}}{\cancel{6}_1} \times \frac{\cancel{24}^{\cancel{12}^{2}}}{\cancel{35}_7} = \frac{2}{7}$$

Study Problems.

Multiply, using cancellation operation.

1. $\dfrac{2}{9} \times \dfrac{3}{8} =$

2. $\dfrac{7}{8} \times \dfrac{3}{21} =$

3. $\dfrac{5}{8} \times \dfrac{3}{10} \times \dfrac{8}{9} =$

4. $\dfrac{22}{27} \times \dfrac{9}{11} \times \dfrac{3}{4} =$

5. $\dfrac{18}{35} \times \dfrac{77}{99} \times \dfrac{55}{66} =$

6. $\dfrac{144}{165} \times \dfrac{15}{24} \times \dfrac{32}{65} \times \dfrac{25}{36} =$

7. $\dfrac{91}{99} \times \dfrac{6}{13} \times \dfrac{22}{49} =$

8. $\dfrac{15}{32} \times \dfrac{16}{45} \times \dfrac{8}{13} \times \dfrac{52}{64} =$

Division of Fractions

The method of dividing fractions is also quite simple as it directly relates to the rule for the multiplication of fractions. The rule of division is:

Invert all fractions immediately following a division sign ($\div$), change the ($\div$) sign to ($\times$) and proceed as in multiplication of fractions.

CAUTION! Cancellation can *not* be done before inversion and sign change.

Study Examples.

1. $\quad \dfrac{5}{6} \div \dfrac{2}{3} = \dfrac{5}{6} \times \dfrac{3}{2} = \dfrac{5}{\cancel{6}_2} \times \dfrac{\cancel{3}^{1}}{2} = \dfrac{5}{2} \times \dfrac{1}{2} = \dfrac{5}{4} = 1\dfrac{1}{4}$

Note that cancellation is done after inversion.

2. $\quad \dfrac{2}{3} \div \dfrac{1}{2} \times \dfrac{1}{4} \div \dfrac{5}{6} = \dfrac{2}{3} \times \dfrac{2}{1} \times \dfrac{1}{4} \times \dfrac{6}{5}$ (inverting)

$\qquad \dfrac{\cancel{2}^{1}}{\cancel{3}_1} \times \dfrac{\cancel{2}^{1}}{1} \times \dfrac{1}{\cancel{4}_1} \times \dfrac{\cancel{6}^{2}}{5} =$

$\qquad \dfrac{1}{1} \times \dfrac{1}{1} \times \dfrac{1}{1} \times \dfrac{2}{5} = \dfrac{2}{5}$

NOTE: The first fraction in a string has no sign of operation ($\times$ or $\div$); such a sign preceding the first fraction would be meaningless and if present, should be ignored.

Study Problems.

1. $\frac{7}{11} \div \frac{14}{22} \times \frac{1}{2} \div \frac{1}{2} =$

2. $\frac{8}{9} \times \frac{3}{4} \div \frac{1}{3} \times \frac{4}{5} \times \frac{1}{8} =$

3. $\frac{21}{32} \times \frac{8}{9} \times \frac{7}{8} \div \frac{3}{4} =$

4. $\frac{25}{36} \div \frac{5}{9} \times \frac{2}{7} \times \frac{3}{4} \div \frac{1}{4} =$

5. $\frac{7}{8} \div \frac{35}{48} \times \frac{6}{7} \div \frac{36}{70} \div \frac{2}{3} =$

6. $\frac{3}{7} \div \frac{5}{34} \div \frac{2}{3} \div \frac{5}{6} \div \frac{45}{21} =$

7. $\frac{94}{100} \times \frac{5}{8} \div \frac{5}{24} \times \frac{2}{15} =$

8. $\frac{3}{4} \div \frac{4}{7} \times \frac{21}{35} \div \frac{3}{4} \times \frac{1}{2} =$

9. $\frac{4}{7} \div \frac{7}{9} \div \frac{3}{14} \times \frac{2}{3} \div \frac{3}{22} =$

10. $\frac{3}{2} \div \frac{6}{5} \times \frac{22}{7} \times \frac{8}{3} \div \frac{11}{12} =$

11. How many holes, spaced on $\frac{9}{16}$ in. centers, can be drilled in the detail shown in Fig. 2-8?

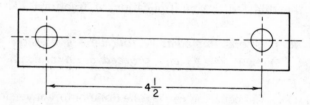

Fig. 2-8. Study problem 11.

12. Determine the number of discs that can be blanked from a 24-ft roll of #18 ga. C.R.S. if each blank is $1\frac{1}{8}$ in. in diameter and there is a $\frac{3}{32}$ in. web between each disc. See Fig. 2-9.

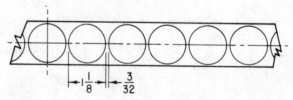

Fig. 2-9. Study problem 12.

13. Face brick is $3\frac{3}{4}$ in. by $2\frac{1}{2}$ in. by 8 in. The mortar joints are $\frac{1}{4}$ in. thick. How many bricks are needed for 200 sq ft of wall?

14. A millwright must use a rough wooden timber measuring 8 by 12 in. (actual). He requires a timber that is $7\frac{13}{16}$ in. thick and he takes equal cuts from each rough side. How thick a piece does he cut off each side?

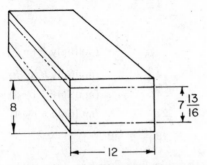

Fig. 2-10. Study problem 14.

Solution: Subtract $7\frac{13}{16}$ from 8:

$$8 = 7\frac{16}{16}$$
$$-\ 7\frac{13}{16}$$
$$\frac{3}{16}$$

Divide $\frac{3}{16}$ by 2: $= \frac{3}{32}$ cut from each side.

15. Masonry walls of cement blocks are made from a standard size block which is $15\frac{5}{8}$ in. long, $7\frac{5}{8}$ in. high, and 8 in. thick. The mortar joint allowance for each block is $\frac{3}{8}$ in. (length and height). How many blocks will be needed to build a wall that is 8 ft high and 160 ft long?

Multiplication and Division of Mixed Numbers

Mixed numbers may be multiplied by two different methods. In the first method, the mixed numbers are arranged one below the other. Multiplication proceeds as shown in the following example:

Study Example.

Multiply: $8\frac{1}{3}$ by $3\frac{3}{5}$

Arrange the mixed numbers one under the other:

$$8\frac{1}{3}$$
$$\times\ 3\frac{3}{5}$$

$$\frac{3}{15} \qquad \text{(multiply } \tfrac{1}{3} \text{ by } \tfrac{3}{5}\text{)}$$
$$4\frac{4}{5} \qquad \text{(multiply 8 by } \tfrac{3}{5}\text{)}$$
$$1 \qquad \text{(multiply } \tfrac{1}{3} \text{ by 3)}$$
$$\underline{24} \qquad \text{(multiply 8 by 3)}$$
$$29\frac{15}{15} = 30$$

The second method is less cumbersome and more practical for the multiplication of three or more mixed numbers. In this method the mixed numbers are converted to improper fractions.

Study Example.

Multiply: $\qquad\qquad 8\frac{1}{3}$ by $3\frac{3}{5}$

Convert to improper fractions: $\quad \dfrac{25}{3} \times \dfrac{18}{5}$

Cancel: $\qquad\quad \dfrac{\cancel{25}^{5}}{\cancel{3}_{1}} \times \dfrac{\cancel{18}^{6}}{\cancel{5}_{1}} = \dfrac{5}{1} \times \dfrac{6}{1} = 30$

The division of mixed numbers *must* be done by conversion into improper fractions:

Study Example.

Divide: $\quad 4\frac{1}{3} \div 3\frac{5}{7}$

Convert to improper fractions: $\dfrac{13}{3} \div \dfrac{26}{7}$

Invert, cancel and multiply:

$$\dfrac{\cancel{13}^{1}}{3} \times \dfrac{7}{\cancel{26}_{2}} = \dfrac{1}{3} \times \dfrac{7}{2} = \dfrac{7}{6} = 1\frac{1}{6}$$

The combined multiplication and division of mixed numbers are accomplished in the same manner as with fractions.

Study Example.

$$13\frac{3}{4} \div 6\frac{7}{8} \times 2\frac{13}{16} \div 4\frac{1}{2} =$$

$$\frac{55}{4} \div \frac{55}{8} \times \frac{45}{16} \div \frac{9}{2} = \qquad \text{(Convert to fractions)}$$

$$\frac{55}{4} \times \frac{8}{55} \times \frac{45}{16} \times \frac{2}{9} = \qquad \text{(Invert and change sign)}$$

$$\frac{\cancel{55}^{1}}{\cancel{4}_{1}} \times \frac{\cancel{8}^{1}}{\cancel{55}_{1}} \times \frac{\cancel{45}^{5}}{\cancel{16}_{4}} \times \frac{\cancel{2}^{1}}{\cancel{9}_{1}} =$$

$$\frac{1}{1} \times \frac{1}{1} \times \frac{5}{4} \times \frac{1}{1} = \frac{5}{4} = 1\frac{1}{4}$$

Study Problems.

1. $6\frac{7}{8} \div 2\frac{3}{4} \times 3\frac{2}{3} =$

2. $33\frac{11}{24} \times 2\frac{5}{8} \div 6\frac{11}{12} =$

3. $14\frac{5}{6} \times 21\frac{15}{18} \div \frac{2}{3} =$

4. $45 \times 2\frac{5}{8} \times \frac{3}{4} \div 5 =$

5. $22\frac{9}{10} \times 14\frac{7}{10} \div \frac{3}{5} =$

6. $\frac{17}{12} \times 15 \div 6\frac{7}{8} + \frac{3}{8} =$

7. $8\frac{13}{25} \div \frac{9}{15} \times \frac{3}{5} \div \frac{19}{25} =$

8. $6\frac{21}{32} \times \frac{17}{24} \div 5\frac{11}{16} \times 4\frac{7}{8} =$

9. $14\frac{3}{7} \times 3\frac{18}{21} \div \frac{11}{14} =$

10. $14\frac{15}{16} \times 8\frac{5}{8} \div 2\frac{1}{2} =$

11. $9\frac{7}{9} \times 31\frac{3}{18} \div 6\frac{25}{27} =$

12. $7\frac{1}{6} \times 2\frac{3}{4} \div \frac{5}{8} \times \frac{3}{4} =$

Complex Fractions

Complex fractions are those in which the numerator or the denominator or both are also fractions. These fractions may be either proper or improper. A complex fraction composed of mixed numbers must have them reduced to improper fractions before solving.

Study Example.

Given the complex fraction: $\dfrac{\frac{3}{5}}{\frac{2}{3}}$

Rewrite as an indicated division: $\frac{3}{5} \div \frac{2}{3}$

Solve by inverting: $\qquad \frac{3}{5} \times \frac{3}{2} = \frac{9}{10}$

This same complex fraction may also be solved by reducing the denominator to 1. Multiply numerator and denominator by the reciprocal of the denominator.

$$\frac{\frac{3}{5}}{\frac{2}{3}} = \frac{\frac{3}{5}}{\frac{2}{3}} \times \frac{\frac{3}{2}}{\frac{3}{2}} - \quad \text{Note that} \quad \frac{\frac{3}{2}}{\frac{3}{2}} \quad \text{equals 1.}$$

$$\text{Since} \ \frac{2}{3} \times \frac{3}{2} = \frac{6}{6} = 1, \text{then} \ \frac{\frac{3}{5} \times \frac{3}{2}}{1} = \frac{\frac{9}{10}}{1} = \frac{9}{10}$$

Study Problems.

1. $\frac{3}{4} \times \frac{12}{9} \times \frac{16}{3} \times \frac{18}{24} =$

2. $3\frac{1}{4} \times 20 \times \frac{5}{32} =$

3. $\frac{13}{20} \div 5\frac{1}{5} =$

4. $5 \times \frac{4}{3} \times \frac{7}{8} \div \frac{5}{6} =$

5. $\frac{6}{10} \times \frac{24}{18} \times \frac{3}{36} \div \frac{36}{18} =$

6. $\frac{32}{49} \div \frac{4}{7} \times \frac{16}{25} \times \frac{1}{2} =$

7. An I-beam weighs 35 lbs per foot. What does a piece 3 ft, 4 in. weigh?

8. How long a piece of $\frac{3}{4}$ in. hex. stock will it require to produce 13 pieces, each $1\frac{7}{8}$ in. long if each cut requires $\frac{1}{32}$ in. waste?

9. A machined detail is shown in Fig. 2-11 dimensioned in the conventional manner. (a) What is the length from a to c? (b) the length from g to d? and (c) the dimension A?

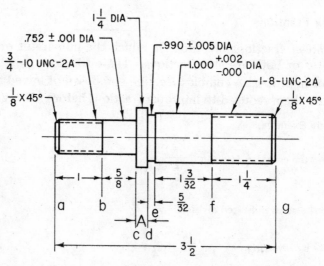

Fig. 2-11. Study problem 9.

10. $\dfrac{3\frac{1}{3} \times 5\frac{3}{8}}{8\frac{1}{8} \times 4\frac{2}{3}}$

11. $\dfrac{\dfrac{\frac{4}{2} \times \frac{1}{3}}{3}}{\dfrac{2}{3} \times \dfrac{3}{4} \times \dfrac{5}{6}}$

12. A hexagon bar of steel is placed between centers on a lathe. The bar is $23\frac{1}{2}$ in. long. If the bar is machined for $\frac{1}{3}$ of its length, how many inches of unmachined hexagon bar remains?

13. A special nonferrous alloy is composed of $\frac{5}{12}$ tin, $\frac{1}{3}$ copper, and $\frac{1}{4}$ zinc. How many pounds of each metal will be used to pour a casting weighing 224 lbs?

14. Multiply the difference of $\frac{6}{15}$ and $\frac{3}{20}$ by their sum.

15. A gasoline storage tank holding $134\frac{2}{3}$ gal is $\frac{3}{8}$ full. How many gallons will it require to fill the tank up?

16. A pipe fitting weighs $\frac{2}{3}$ lbs. How many fittings are in a box weighing 122 lbs if the empty box (tare weight) weighs 32 lbs.

17. Determine the center to center distance of the equally spaced holes in Fig. 2-12.

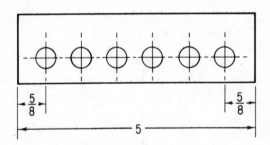

Fig. 2-12. Study problem 17.

18. How many $6\frac{1}{2}$ in. nipples can be cut from a 14 ft length of galvanized pipe? Allow $\frac{1}{16}$ in. per cut.

19. Gasoline is selling for $53\frac{1}{2}$ cents per gallon. How much will it cost to fill a 20 gal gas tank if there are already $6\frac{1}{3}$ gal in the tank?

20. Calculate the weight of 24 brass bars, each $8\frac{1}{2}$ ft long if brass weighs $39\frac{4}{5}$ lbs per foot.

21. $23\frac{4}{6} \div (2\frac{4}{7} + 3\frac{1}{4})$

22. $(10\frac{1}{2} + 3\frac{4}{9}) \div \frac{18}{3} =$

23. A cement contractor has four piles of reinforcing bars. In the first pile there are thirteen $\frac{1}{2}$ in. bars, 14 ft, 6 in. long, weighting $\frac{3}{4}$ lbs per foot. The second pile contains twenty-seven $\frac{1}{2}$ in. bars, $12\frac{1}{2}$ ft long. The third pile has nine $\frac{3}{4}$ in. bars, weighing $1\frac{1}{2}$ lbs per foot, $14\frac{1}{2}$ ft long; and the fourth pile has thirty-five $\frac{3}{4}$ in. bars, 28 ft, 4 in. long. How many pounds of steel does the contractor have?

24. $15\frac{2}{3} \times 8\frac{5}{8} \div 4\frac{3}{4} =$

25. $6\frac{7}{8} + 2\frac{2}{3} - 5\frac{3}{5} =$

CHAPTER 3

Decimals and Decimal Fractions

Decimal notation is a method of writing the fractional portion of whole number in tenths, hundredths, thousandths, ten-thousandths, etc. The word decimal comes from the Latin word for ten: "decem."

Decimal fractions are special forms of common fractions and are used in most of the computational work done in science and engineering. In some industries such as the building trades, details and materials are dimensioned in whole numbers and common fractions. Common fractions, however, are not generally used to specify precise dimensions on machine parts and tooling.

The use of decimals in machine details and in assembly drawings came into general use during World War I. The automobile, aircraft, and machine tool industries pioneered in the use of the decimal system for American manufacturing. Nations using the Metric (SI) System have never had the problems of using common fractions because the Metric System is based on the decimal system.

Decimal Fractions

Decimal fractions are not written as common fractions are since no denominator is shown. In the decimal system, whole numbers are written to the left of a period called the "decimal point" and the decimal fraction is written to the right of the decimal point. For example:

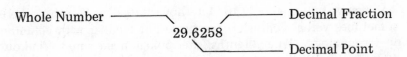

For the time being, we shall concern ourselves only with the decimal fraction. Since no denominator is given, one is assumed. Depending upon the number of digits to the right of the decimal point, the assumed denominator will be 10, 100, 1000, 10,000, etc. For each digit to the right, the denominator increases in the order shown. For example:

$$0.3 = \frac{3}{10} \qquad 0.02 = \frac{2}{100} \qquad 0.252 = \frac{252}{1000} \qquad 0.3946 = \frac{3946}{10,000}$$

A table of decimals is given below to help you become familiar with the system:

TABLE OF DECIMALS

Decimal	Common Fraction	Name
0.1	$= \dfrac{1}{10}$	$=$ one-tenth
0.01	$= \dfrac{1}{100}$	$=$ one-hundredth
0.001	$= \dfrac{1}{1000}$	$=$ one-thousandth
0.0001	$= \dfrac{1}{10,000}$	$=$ one-ten-thousandth
0.00001	$= \dfrac{1}{100,000}$	$=$ one-hundred-thousandth
0.000001	$= \dfrac{1}{1,000,000}$	$=$ one-millionth

Study Examples.

1. The number 123.35 is read: "one hundred twenty-three and thirty-five hundredths."

2. The number 56.462 is read: "fifty-six and four hundred sixty-two-thousandths."

3. The number 9.3895 is read: "nine and three thousand, eight hundred, ninety-five ten-thousandths."

As many shop men refer to "ten-thousandths" as "tenths," there is sometimes verbal confusion for those not familiar with common shop terminology. When "tenths" are spoken, make sure to find out if "ten-thousandths" are meant.

Conversion of Common Fractions to Decimal Fractions

Common fractions with denominators of ten, one hundred, one thousand, ten thousand, etc., are easily converted to decimal fractions by simply writing the numerator behind the decimal point so that there are as many decimal places to the right of the decimal point as there are zeros in the denominator.

Study Examples.

1. Convert $\frac{234}{1000}$ to a decimal fraction: 0.234 (3 zeros, 3 places)

2. Convert $\frac{25}{1000}$ to a decimal fraction: 0.025 (3 zeros, 3 places)

3. Convert $5\frac{196}{10,000}$ to a decimal fraction: 5.0196 (four zeros, four places)

If the common fraction has a denominator that is *not* a ten, one hundred, one thousand, etc., the fraction is converted by dividing the denominator into the numerator.

Study Examples.

1. Convert $\frac{7}{8}$ to a decimal fraction:

$$
\begin{array}{r}
0.875 \\
8\,\overline{\smash{)}\,7.000} \\
6\,4 \\
\hline
60 \\
56 \\
\hline
40 \\
40 \\
\hline
\end{array}
\quad = 0.875
$$

2. Convert $\frac{9}{16}$ to a decimal fraction:

$$
\begin{array}{r}
0.5625 \\
16\,\overline{\smash{)}\,9.0000} \\
8\,0 \\
\hline
1\,00 \\
96 \\
\hline
40 \\
32 \\
\hline
80 \\
80 \\
\hline
\end{array}
\quad = 0.5625
$$

Study Problems.

1. Write the following as spoken dimensions:

    3.46   53.963   0.0693   1237.2562   425.356

2. Convert the following to decimal fractions:

$$\frac{3}{8} \qquad \frac{16}{64} \qquad \frac{4}{100} \qquad \frac{25}{125} \qquad \frac{1}{25} \qquad 5\frac{3}{64}$$

3. Express in numerals: "two hundred and eleven-thousandths."

4. Express in numerals: "two hundred sixteen and one ten-thousandth."

5. Express in numerals: "five hundred twenty-five and three hundred nineteen ten-thousandths."

6. Express in numerals: "three and fifteen-millionths."

Conversion of Decimal Fractions to Common Fractions

Decimal fractions are converted to common fractions by first writing the decimal with its correct ten, hundred, thousand, etc., denominator. Then reduce the fraction, if possible, by dividing both numerator and denominator by a common factor. Since the denominator is a multiple of ten, only multiples of 2 or 5 can be common factors.

Study Examples.

1. Change 0.28 to a common fraction:

$$0.28 = \frac{28}{100}$$

$$\frac{28}{100} \div \frac{4}{4} = \frac{7}{25}$$

2. Change 0.050 to a common fraction:

$$0.050 = \frac{50}{1000} = \frac{5}{100} \div \frac{5}{5} = \frac{1}{20}$$

Study Problems.

Change the following decimal fractions to common fractions:

0.5	0.64	0.875	2.75	3.036	0.375
	0.415	0.145	8.008	3.333	

Fractional Inches Expressed as Decimals

Engineering scales using the fractional inch system are based on a continuing division by 2. Hence, $\frac{1}{2} \div 2 = \frac{1}{4}$, $\frac{1}{4} \div 2 = \frac{1}{8}$, etc. These scales are manufactured to an accuracy of 0.001 inch or less between each division. While these fractional divisions can be *mathematically* expressed as decimal fractions containing from one to six digits, such as; $\frac{1}{2} = 0.5$, $\frac{1}{4} = 0.25$, $\frac{1}{8} = 0.125$, up to $\frac{1}{64} = 0.015625$, tables of decimal equivalents used by engineers, designers, and machinists reflect the implied accuracy inherent in reading these accurate scales. For example, the decimal equivalent of $\frac{1}{2}$ inch and of $\frac{61}{64}$ inch are given as follows in the different tables listed below:

0.5000 and 0.9531   Starrett

0.500   and 0.95313   Illinois Tool Works

0.500   and 0.953125   *Machinery's Handbook*

In these cases, the machinist will use the number of decimal places appropriate to the measurement that he is making and appropriate to the measuring instrument that he is using. For $\frac{61}{64}$ inch he will use 0.953 when using a 0.001 inch micrometer, 0.9531 when using a 0.0001 inch micrometer; and when required by the part accuracy, he will use 0.953125 when using precision gage blocks. However, he could use the values from any of these tables for micrometer measurement, while only *Machinery's Handbook* will provide the answer when he uses precision gage blocks. A certain degree of freedom with respect to the accuracy of answers should be allowed unless specific conditions of a problem impose a limit.

Rounding-Off

Where a figure represents a highly accurate measurement, but is to be used in a computation where such an accurate value is not required, the figure may be "rounded-off." For example, the decimal equivalent of $\frac{43}{64}$ inch is equal to 0.671875 inch. This can be rounded-off to 0.672 inch.

There are definite rules that must be followed in order to round off decimals:

1. When the next digit following the last digit to be retained exceeds 5, the last digit to be retained is increased by 1. Thus, when rounded off to three decimal places 1.2776 will become 1.278; also, 1.27763 will become 1.278.

2. When the next digit following the last digit to be retained is less than 5, the last digit to be retained is left unchanged. Thus, 1.2873 will become 1.287; also 1.28725 will become 1.287.

3. When the digit following the last digit to be retained is exactly 5, the rule specified by ANSI (American National Standards Institute) Z-25-1-1940 (R 1961) "Rules for Rounding Off Numerical Values" should be used. It states that the last digit retained should be the closest *even* value. This rule is applied as follows:

a. If the last digit to be retained is an odd number, increase it to the next largest even number. Thus, 0.1875 will become 0.188.

b. The last digit to be retained is not changed if it is an even number. Thus, 0.0625 will become 0.062.

4. If the next digit beyond the last digit to be retained is a 5 and it is followed by any digits other than 0, the last digit to be retained, whether odd or even, should be increased by 1. Thus, 1.25652 becomes 1.257, when rounded off to three decimal places.

Study Examples.

1. 0.984375 becomes 0.9844

2. 0.984375 becomes 0.984

3. 0.03949 becomes 0.039

4. 0.8295 becomes 0.830

Study Problems.

1. Round-off 0.467494 to three decimal numbers.

2. Round-off 57.8947 to two decimal numbers.

3. Round-off 5.55555 to three decimal numbers.

4. Round-off 0.0059387546 to five decimal numbers.

5. Round-off 0.0059537546 to four decimal numbers.

6. Round-off 4567.535 to no decimal numbers.

7. Round-off 4567.835 to two decimal numbers.

8. Round-off 45.67352 to three decimal numbers.

9. Round-off 82.68499 to two decimal numbers.

10. Round-off 0.7777773 to three decimal numbers.

Addition and Subtraction of Decimal Fractions

Decimal fractions are much easier to add or subtract than common fractions. The basic rule is to be sure to place the decimal points directly below each other when writing down the problem. The addition of two or more decimal fractions is done by placing them in vertical columns and beginning by adding the right column first and then carrying over to the next column to the left and then to the next column until the addition is completed for each column.

Study Examples.

1. 25.327	2. 0.00002	3. 2.635
0.0672	12.001	4.5734
0.0032	1.6003	0.0032
1.6	2.0202	1.4837
26.9974	15.62152	8.6953

Subtraction of decimal fractions is done in similar fashion. The decimal points must be under one another, and subtraction is performed from the right column to the left. If the number to be subtracted is larger than the number above it in the column, ten is "borrowed" from the top number of the immediate left column. This reduces the number by one and must be remembered or noted above the problem.

Study Examples.

		8 1		864
1. 16.3718	2.	9.6273	3. 12.9753
−12.2504		−8.7183	−6.4876
4.1214		0.9090	6.4877

When adding long columns of plus and minus decimal fractions, it is recommended that all the "plus" decimals be added first and set down next to the problem. Then the "minus" decimals are added and set down beneath the plus decimal sum if it is the smaller of the two. If the minus decimal sum is the larger, it is set above the plus decimal sum and the answer will be a minus (negative) number.

Study Examples.

1. Add: 10.362   Step 1. Add + terms:
            −3.792   10.362
            8.22   8.22
            −1.063   18.582

                          Step 2. Add − terms:
                                   −3.793
                                   −1.063
                                   −4.856

Step 3. Add results of Step 1 and Step 2:
$$\begin{array}{r} 18.582 \\ -4.856 \\ \hline 13.726 \end{array}$$

2. Add:
$$\begin{array}{r} 2.9583 \\ -5.8472 \\ 3.0958 \\ -4.0573 \end{array}$$

Step 1. Add ($+$) terms:
$$\begin{array}{r} 2.9583 \\ 3.0958 \\ \hline 6.0541 \end{array}$$

Step 2. Add ($-$) terms:
$$\begin{array}{r} -5.8472 \\ -4.0573 \\ \hline -9.9045 \end{array}$$

Step 3: Add Step 1 and Step 2:
When adding numbers with unlike signs, take the numerical difference between the numbers and prefix with the sign of the larger number.
$$\begin{array}{r} -9.9045 \\ 6.0541 \\ \hline -3.8504 \end{array}$$

Study Problems.

1. Convert to decimal fractions: $\dfrac{3}{4}, \dfrac{5}{8}, \dfrac{3}{16}, \dfrac{1}{4}, \dfrac{11}{32}, \dfrac{7}{8}, \dfrac{21}{64}$

2. Add: $0.7625 + 0.883$

3. Add: $1.234 + 0.567$

4. Add: 1.192 and 0.035

5. Add: 4.62 and 0.0025

6. Subtract: 2.765 from 5.326

7. Subtract: 10.0234 from 11.0000

Perform the indicated operations:

8. $1.2634 + 2.6379 + 1.23 + 0.6324 =$

9. $2.678 - 1.032 - 0.275 - 0.659 =$

10. $0.666 + 0.334 + 1.5 - 2.25 =$

11. $1.7986 + 2.3333 + 3.4567 + 1.0012 =$

12. Evaluate as decimal fractions, rounded-off to five figures:

$$\dfrac{1}{3} \qquad \dfrac{2}{3} \qquad \dfrac{7}{9} \qquad \dfrac{3}{13} \qquad \dfrac{15}{32} \qquad \dfrac{25}{64} \qquad \dfrac{37}{64} \qquad \dfrac{19}{32}$$

Multiplication of Decimals

The method of multiplying decimals is the same as in the multiplication of whole numbers, except for the need to place the decimal point in its proper place. The following examples show how the decimal point is determined.

Study Examples.

1. Multiply: 3.25 by 0.125

> Step 1. Place in regular multiplication form:
>
>   3.25   - (count *two* decimal places)
>   X 0.125   - (count *three* decimal places)
>   1625
>   650   - Total *five* decimal places
>   325
>   0.40625   - (count *five* decimal places from the right to the left)

The rule used above is: *Count the number of decimal places in the multiplicand (the first number: 3.25). Count off the number of decimal places in the multiplier (the second number: 0.125). Add the places (five) and count off in the product (answer) from right to left and place the decimal point ahead of that number.*

2. Multiply: 8.33 by 0.0025

       8.33   Count two decimal places,   2
    X 0.0025   Count four decimal places,   4
       4165   Total = 6
       1666
      20825

               Count off six places, right to left.
    0.020825   A zero is added at the left to make six decimal places.

NOTE: The above answer, rounded-off to five digits, gives 0.02082, as 2 is an even number and rounding off from 5 must give an even number in the fifth place. The six-digit answer, above, rounded-off to three digits, equals 0.021.

Division of Decimals

The division of decimals is done in the same way as the division of whole numbers, but with special preparation of the position of the decimal points in the divisor and in the dividend.

When the divisor (number doing the dividing) contains a decimal fraction such as 8.654, the number must be converted to a whole number; that is, the decimal point must be to the right of the number.

If 8.654 is multiplied by 1000 the answer is a whole number 8654. However, the dividend (number being divided into) must also be multiplied by 1000 so that the problem will not be changed. If the dividend is 92.3647, it becomes 92364.7 after being multiplied by 1000. Below is a detailed method of this division:

Study Example.

Divide 92.3647 by 8.654:

Step 1.   Multiply both divisor   8.654 ⟌ 92.3647
and dividend by 1000.   8654 ⟌ 92364.7
A practical method to move the decimal point is to count the number of digits to the right of the decimal point (three) in the divisor. Then count off the same number to the right of the decimal point in the dividend.

Step 2.   Proceed with division:

$$
\begin{array}{r}
10.673 \\
8654 \overline{\smash{)}92364.700} \\
\underline{8654} \\
58247 \\
\underline{51924} \\
63230 \\
\underline{60578} \\
26520 \\
\underline{25962} \\
558 \text{ remainder}
\end{array}
$$

A second method of moving the decimal point is to move the decimal point in the divisor to the right to make a whole number. A light pencil mark is made tracing the movement as shown in the example below. The decimal point in the dividend is then moved the same number of places using the light line tracer method.

8.654. ⟌ 92.364.7

Study Problems.

Perform the indicated operations:

1. 59375.0 × 0.00482 =

2. 5739 ÷ 0.0946 =

3. 0.4967 × 5.478 =

4. 0.4792 × 0.0468 =

5. 0.00367 × 4973.22 × 0.473 =

6. 0.0045 × 0.00067 =

7. 8.694 ÷ 4.968 × 3.475 =

8. 87.3495 ÷ 0.0045 =

9. 4857 × 0.00369 ÷ 0.0005 =

10. 9.637 × 0.00073 =

11. 49.579 × 3.1416 =

12. 56.0052 ÷ 0.00047 =

13. $0.587 \times 0.472 \times 0.068 =$

14. $23.56 \times 5.73 \times 23 =$

15. $4.6948 \div 0.0055 \times 5.793 =$

16. $0.000056 \times 0.0059 =$

17. $325.4 \times 45.68 \div 929.017 =$

18. $472849 \div 0.0000065 =$

19. $38.75 \times 0.0051 \div 0.075 =$

20. $46.36 \times 0.10405 =$

Measurement Accuracy

It may be said that all figures representing measurements or obtained from computations in which measured values are used are approximations. The more accurate the measuring instrument and the more skillfully it is used, the more precise the resulting measurements will be. Nevertheless they are still approximations and should be recognized as such. Thus, whenever measured values are to be used in mathematical computations, they should be expressed in such a way as to indicate their accuracy. If a measurement is 5 inches to the nearest thousandth of an inch, it should be written as 5.000 inches, not 5 inches. If a measurement is 3.9 inches to the nearest ten-thousandth of an inch, it should be written 3.9000 inches, and not 3.9 inches. In other words, zeros should be inserted, where required, to the number of decimal places which indicate the accuracy of a given figure.

Addition and Subtraction. The accuracy of a sum of figures cannot be greater than that of the least accurate of the figures being added. It may be much less. Take the following example:

$$
\begin{array}{r}
6.34 \\
5.192 \\
3.1 \\
8.9658 \\
\hline
23.5978
\end{array}
$$

Presuming that each figure has been written down correctly to indicate its accuracy, the least accurate figure is 3.1. This figure is accurate only to the nearest tenth. Hence, the answer can be expected to be accurate to the nearest tenth and should be written 23.6. Of course, if the above figures had been written:

$$
\begin{array}{r}
6.3400 \\
5.1920 \\
3.1000 \\
8.9658 \\
\hline
23.5978
\end{array}
$$

indicating that each had been measured to the nearest ten-thousandth, then the answer would be written as 23.5978. It should be recognized, however, that the error in any sum of figures may be greater than the error in any one of them. Thus, in the above example, if each of the figures being added had an error of plus or minus 0.00005, then the resulting errors might be:

+0.00005	+0.00005	−0.00005
+0.00005	−0.00005	−0.00005
+0.00005	−0.00005	−0.00005
+0.00005	+0.00005	−0.00005
+0.00020	0.00000	−0.00020

to give but three possible combinations. In the first case, the correct sum would be 23.5978 − 0.0002 = 23.5976; in the second case it would be 23.5978 + 0.0000 = 23.5978; and in the third case, it would be 23.5978 + 0.0002 = 23.5980. Thus, it can be seen that the sum of a group of figures measured to the nearest ten-thousandth may or may not be correct to the nearest ten-thousandth.

It is frequently assumed, however, that in adding a long column of figures there is a more or less equal distribution of plus and minus errors which tend to balance each other out. This depends, of course, on the type of data that the figures represent. In some cases where a small number of figures of equal accuracy are added, the last place in their sum is rounded off as being inaccurate.

These same general considerations apply in the case of subtraction, except that since usually only two numbers are involved, there is less chance for a balancing out of errors, but also less chance for a large accumulated error.

Multiplication. The accuracy of the product of two figures may be much less than the accuracy of either because of the compounding of error in each; thus in

$$
\begin{array}{r}
12.31 \\
\times 1.05 \\
\hline
6155 \\
12\ 310 \\
\hline
12.9255
\end{array}
$$

the product should be expressed to an accuracy of one decimal place less than that of the least accurate of its factors. Hence, the above answer should be written 12.9.

Division. It might be supposed that in division the error of the quotient would be less than that of either the dividend or the divisor, but such is not the case. As in multiplication, the error in the answer may be much greater than that in either of the factors.

General Rules. Where extremely accurate results are not required, the following rule may be used as a guide in addition and subtraction: *Express the answer to as many decimal places as the least accurate figure being added.*

In multiplication and division, express the answer to one less decimal place than the least accurate of the figures used. In other words, if an answer to three decimal places is desired in an operation requiring multiplication or division, the factors that are used should be accurate to at least four decimal places.

For extreme accuracy in computations, reference should be made to books covering the theory of errors, where equations indicating the probable error in various types of computations can be found.

The Decimal-Inch/Fractional-Inch Measurement System

Measurements of commercial products in the United States are made under two related inch systems: the fractional and the decimal. In the fractional-inch system, a person uses the common yardstick or a 12-inch ruler. Each has the inches subdivided into common fractions based on continued division by 2. Most rules have marks indicating halves ($\frac{1}{2}$), quarters ($\frac{1}{4}$), eighths ($\frac{1}{8}$), sixteenths ($\frac{1}{16}$), and in some cases, thirty-seconds ($\frac{1}{32}$) and sixty-fourths ($\frac{1}{64}$).

The fractional-inch system is used in shop practice and in the construction trades. In tool and die work and in machine shop practice, the precise nature of the work frequently requires that the fraction be converted to a decimal, which is called its decimal equivalent.

A fraction may be converted into its decimal equivalent by dividing the denominator into the numerator. For example:

$$\frac{1}{2} \text{ inch} \quad = \quad 1 \div 2 \quad = \quad 0.500 \text{ inch}$$

$$\frac{3}{8} \text{ inch} \quad = \quad 3 \div 8 \quad = \quad 0.375 \text{ inch}$$

$$\frac{63}{64} \text{ inch} \quad = \quad 63 \div 64 \quad = \quad 0.984375 \text{ inch}$$

$$1\frac{13}{16} \text{ inches} = 1 + (13 \div 16) = 1.8125 \text{ inches}$$

Frequently, the decimal equivalent is a rather large decimal such as 0.984375. This is beyond the machining requirements of the part. In this case, it is rounded off by using the rules given previously in this chapter. For example:

$$\frac{33}{64} \text{ inch} \quad = 0.515625 \text{ inch} \quad = 0.516 \text{ inch}$$

$$\frac{3}{16} \text{ inch} \quad = 0.1875 \text{ inch} \quad = 0.188 \text{ inch}$$

$$1\frac{13}{32} \text{ inches} = 1.40625 \text{ inches} \quad = 1.406 \text{ inches}$$

Tables of decimal equivalents, as given below, are usually available in machine shops and in tool shops so that it is unnecessary to calculate the decimal equivalents. When available, they should be used because they save time and reduce the chance of an error's occurring. However, it is often necessary to round the decimal equivalents off, as explained above.

Table 3-1. Decimal Equivalents of Fractions of an Inch

Fraction	Decimal	Fraction	Decimal	Fraction	Decimal
1/64	0.015 625	11/32	0.343 75	43/64	0.671 875
1/32	0.031 25	23/64	0.359 375	11/16	0.687 5
3/64	0.046 875	3/8	0.375	45/64	0.703 125
1/16	0.062 5	25/64	0.390 625	23/32	0.718 75
5/64	0.078 125	13/32	0.406 25	47/64	0.734 375
3/32	0.093 75	27/64	0.421 875	3/4	0.750
7/64	0.109 375	7/16	0.437 5	49/64	0.765 625
1/8	0.125	29/64	0.453 125	25/32	0.781 25
9/64	0.140 625	15/32	0.468 75	51/64	0.796 875
5/32	0.156 25	31/64	0.484 375	13/16	0.812 5
11/64	0.171 875	1/2	0.500	53/64	0.828 125
3/16	0.187 5	33/64	0.515 625	27/32	0.843 75
13/64	0.203 125	17/32	0.531 25	55/64	0.859 375
7/32	0.218 75	35/64	0.546 875	7/8	0.875
15/64	0.234 375	9/16	0.562 5	57/64	0.890 625
1/4	0.250	37/64	0.578 125	29/32	0.906 25
17/64	0.265 625	19/32	0.593 75	59/64	0.921 875
9/32	0.281 25	39/64	0.609 375	15/16	0.937 5
19/64	0.296 875	5/8	0.625	61/64	0.953 125
5/16	0.312 5	41/64	0.640 625	31/32	0.968 75
21/64	0.328 125	21/32	0.656 25	63/64	0.984 375

When using the fractional-inch system in machine detail dimensioning, the precise dimensions and their tolerances are always given in decimals: that is, in terms of 0.001 inch, 0.0001 inch and finer, but not in terms of 0.01 inch.

The true decimal-inch system is based on dividing the inch into tenths, dividing each tenth into ten parts giving hundredths, dividing each hundredth into tens giving thousandths, dividing each thousandth into ten parts giving ten thousandths, and so on. The decimal divisions lend themselves to more simplified calculations by not relating to common fractions. Because of the dual usage of the two systems by designers and shop men, scales used for measurement are usually in one system or the other; or, as in Fig. 3-1, one side is divided into 32nds and 64ths, while the reverse side has tenths (0.1) and 50ths (0.02).

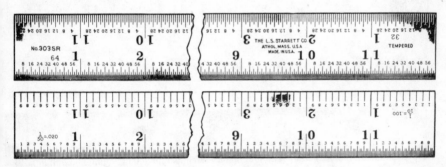

Fig. 3-1. Machinist's scale: 32nds, 64ths, 10ths, 50ths.

In Fig. 3-2, the scale has one edge with tenths, the second edge on the same face with 100ths. On the reverse side it has 32nds and 64ths. On scales and steel rules with both fractional and decimal

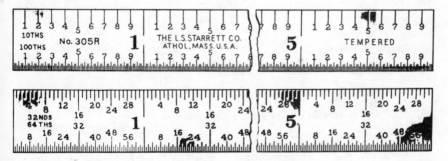

Fig. 3-2. Machinist's scale: 10ths, 100ths, 32nds, 64ths.

divisions, it will be noted that the smallest graduation on the fractional edge is $\frac{1}{64}$ = 0.0156. These divisions are slightly farther apart than the hundredth (0.01) divisions.

This makes the fractional scale somewhat easier to read. However, as both systems are used extensively in industry, a competent workman must be familiar with both systems and be able to use either system.

Figure 3-3 shows a typical Metric-English scale. The smallest metric division on a machinist's steel rule is one-half millimeter (0.5mm). A half millimeter equals 0.0197 inch, which is slightly larger than a sixty-fourth of an inch (0.0156).

Fig. 3-3. Machinist's scale, metric and English units.

Conversion from one system to the other takes some time and effort. Until a person is thoroughly conversant with the metric equivalents, the use of the chart of metric conversions in the appendix of this book is recommended. The Metric System is discussed in full in Chapter 5.

Study Examples.

Conversion of dimensional units.

1. Convert $5\frac{7}{8}$ inch to a decimal fraction.

Divide numerator of fraction by denominator:

$$7 \div 8 = 0.875$$

Add decimal to whole number: 5 + 0.875 = 5.875 inches

2. Convert 3.46875 to fractional notation.

Write decimal as a fraction with proper power of ten as denominator:
$$\frac{46875}{100,000}$$

Find a common factor in both numerator and denominator:
$$\frac{46875}{100,000} = \frac{15 \times 3125}{32 \times 3125}$$

Cancel common factor to give answer: $\frac{15}{32}$

Add fraction to whole number: $3 + \frac{15}{32} = 3\frac{15}{32}$

Study Problems.

Convert the following fractional dimensions to decimals.

1. $3\frac{7}{8}$ 5. $\frac{3}{64}$

2. $2\frac{11}{64}$ 6. $\frac{77}{64}$

3. $9\frac{17}{32}$ 7. $6\frac{15}{64}$

4. $6\frac{3}{8}$ 8. $12\frac{5}{8}$

Convert the following decimal dimensions to fractions.

9. 0.015625   13. 0.40625

10. 3.750   14. 5.96875

11. 2.859375   15. 6.4375

12. 0.8125   16. 0.125

Perform the indicated operations. Give answers in decimals.

17. $8\frac{3}{16} + 0.6875$ 21. $6\frac{15}{32} + 2\frac{7}{64} - 0.90625$

18. $3.625 - 2\frac{7}{32}$ 22. $3\frac{1}{8} - 0.125$

19. $\frac{3}{8} - 0.03125$ 23. $12.653 - \frac{15}{16}$

20. $4.3986 + 2\frac{11}{16} - 0.736$ 24. $\frac{7}{16} - \frac{3}{32}$

25. If, in Fig. 3-4, A,B,C,D,E,F,G, and H are the centers of the respective holes, calculate the distances between these centers: (a) C to E, (b) A to C, (c) A to D, (d) G to A, (e) A to H, (f) G to H.

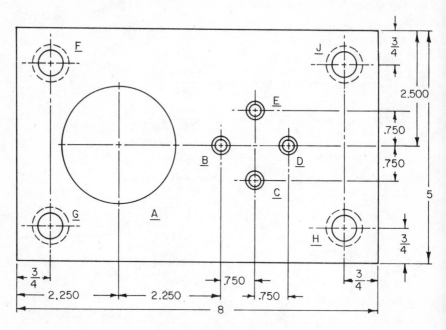

Fig. 3-4. Punch plate detail.

26. In Fig. 3-4, calculate the distances from the left side of the punch plate to the centers of the following holes: (a) B, (b) C and E, (c) D, (d) H and J.

27. In Fig. 3-4, calculate the distances from the upper side of the punch plate to the centers of the following holes: (a) E; (b) A,B, and D; (c) C; (d) G and H.

Decimal Tolerances

Modern mass production depends on interchangeable manufacturing, where parts produced in different factories will all fit together properly at the point of assembly. Unfortunately, it is not possible to make everything to the exact size called for by the dimensions on the part print. Parts can be made to very close dimensions, even to a few millionths of an inch, but the cost of such accuracy for production parts would be prohibitive.

Exact sizes are not needed if the accuracy is related to functional requirements. The degree of accuracy required will determine the amount of variations permitted in the size of the parts. This amount of variation is called the "tolerance," which is the variation "tolerated" by production and assembly requirements.

Drawings of machine parts use decimal fractions in defining dimensional tolerances. Tolerance is the amount by which a given dimension may vary or the difference between the "limits." In Fig. 3-5, the diameter dimension of the hole is written: $\frac{1.251}{1.250}$. These two figures represent the "limits" of the toleranced dimension and tell the machinist that the diameter of the hole may not be machined less than 1.250 inches nor more than 1.251 inches. Likewise, the shaft must not exceed 1.248 inches in diameter nor be smaller than 1.247 inches in diameter.

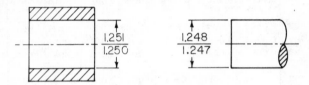

Fig. 3-5. Limit dimensioning.

Frequently, tolerances are expressed by using the plus or minus (±) symbol. The dimensions written after this symbol express the amount that the feature on the part may be more or less than the basic dimension which is given ahead of the plus or minus symbol. For example, in Fig. 3-6 the hole is dimensioned $1.250^{+0.001}_{-0.000}$ and the shaft is dimensioned $1.248^{+0.000}_{-0.001}$ to express the same tolerance as in Fig. 3-5. Typical examples of other tolerances expressed by the plus or minus symbol are: $6.122^{+.002}_{-.003}$, $0.500 \pm.001$, $0.8748 \pm.0002$, and $0.0938^{+.0001}_{-.0002}$. Sometimes the tolerances may be a fractional inch, such as $\pm\frac{1}{16}$ or $\pm\frac{1}{32}$.

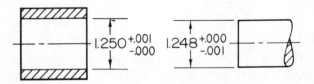

Fig. 3-6. Hole tolerance.

In many designs, no great precision is required and little attention is given to finish or to other than nominal dimensions. The fit of a bolt in a clearance hole is an example of minimal precision. Where more precision is required, the detail drawing will call for the appropriate tolerances. The greater the degree of precision, the finer the tolerance, the higher the cost of machining, and the greater the need for precision measuring instruments.

Table 3-2 illustrates tolerances as related to shop processes.

Table 3-2. Tolerance Ranges for Machining Processes.

Ranges of Sizes From	To & Incl.	Tolerances								
0.000	0.599	0.00015	0.0002	0.0003	0.0005	0.0008	0.0012	0.002	0.003	0.005
0.600	0.999	0.00015	0.00025	0.0004	0.0006	0.001	0.0015	0.0025	0.004	0.006
1.000	1.499	0.0002	0.0003	0.0005	0.0008	0.0012	0.002	0.003	0.005	0.008
1.500	2.799	0.00025	0.0004	0.0006	0.001	0.0015	0.0025	0.004	0.006	0.010
2.800	4.499	0.0003	0.0005	0.0008	0.0012	0.002	0.003	0.005	0.008	0.012
4.500	7.799	0.0004	0.0006	0.001	0.0015	0.0025	0.004	0.006	0.010	0.015
7.800	13.599	0.0005	0.0008	0.0012	0.002	0.003	0.005	0.008	0.012	0.020
13.600	20.999	0.0006	0.001	0.0015	0.0025	0.004	0.006	0.010	0.015	0.025
Lapping & Honing		▨	▨	▨						
Grinding, Diamond Turning & Boring		▨	▨	▨						
Broaching				▨	▨	▨	▨			
Reaming					▨	▨	▨	▨	▨	
Turning, Boring, Slotting, Planing & Shaping							▨	▨	▨	▨
Milling								▨	▨	▨
Drilling								▨	▨	▨

Tolerances calling for $\pm\frac{1}{64}$ may be measured by a rule or a micrometer. A grinding operation with a tolerance of ±0.0002 inch may be checked by a vernier micrometer reading to ten-thousandths of an inch. Lapping, which may require accuracies within a few millionths of an inch, would be checked with gage blocks in conjunction with other precise instruments.

The convention among designers and machinists is that the decimal dimensions with two numbers to the right of the decimal point are "nominal" and usually have no tolerance applied. Dimensions with three or more digits to the right of the decimal point usually require tolerances to be specified, such as: 5.255 ±.002.

A tolerance on a dimension must be properly spoken to avoid misunderstanding in the shop. For example: the dimension $4.375 \, {}^{+.003}_{-.000}$ is read: "four and three-hundred-seventy-five-thousandths plus point zero, zero three, minus nothing." A dimension with a tolerance of ±.001 is read: "plus or minus one-thousandth." Machinists and toolmakers use the word "tenths" to mean 0.0001 inch. It is important that this point be thoroughly understood.

In many tool and die detail drawings, tolerances are often omitted on precise dimensions. A dimension having three decimal places must be made to 0.001 or to 0.0001 inch accuracy depending upon the nature of the work.

Study Examples.

1. A bored hole in a machine detail is to have a diameter between 2.750 and 2.756 in. (a) What are the "limits"? (b) What is the total tolerance? *Answer:* The limits are the maximum and minimum permissible dimensions of the hole. Hence, (a) the limits are 2.750 and 2.756 in. (b) The total tolerance is the total permissible variation in the size of the dimension. Hence, the total tolerance is 0.006 in.

2. In example 1, the basic size of the bored hole is that size from which the limits are obtained. It is sometimes considered as the "ideal" size of the hole. If the ideal size of the hole in example 1 is said to be 2.753 in., what is the "tolerance"? *Answer:* With the basic size of the hole given as 2.753 in. and the limits 2.750 to 2.756 in., the tolerance is ±.003 in. The dimension is written: 2.753 ±.003 in.

3. In the Fig. 3-7, shown below, calculate the limits (maximum and minimum distances) for A. Note that the overall tolerance must not be exceeded.
   Answer: Add the two component dimensions:

   $$\begin{array}{l} 0.750, +0.001, -0.001 \\ \underline{1.125, +0.002, -0.001} \\ 1.875, +0.003, -0.002 \end{array}$$

   Subtract from overall dimensions:

   $$\begin{array}{l} 3.500, +0.004, -0.004 \\ \underline{1.875, +0.003, -0.002} \\ A = 1.625 \ +0.001 \ -0.002 \end{array}$$

   Limits = 1.626, 1.623

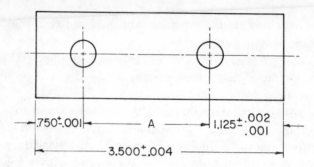

Fig. 3-7. Study example 3.

4. In the Fig. 3-8, shown below, calculate the basic dimension and tolerance of A.

    Answer: Add the component dimension:

$$1.625, +0.002, -0.002$$
$$2.000, +0.000, -0.003$$
$$1.500, +0.003, -0.000$$
$$\overline{5.125, +0.005, -0.005}$$

$$A = 5.125 \pm .005$$

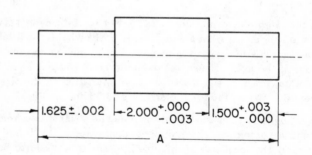

Fig. 3-8. Study example 4.

Study Problems.

1. What are the maximum and minimum distances for the dimension A?

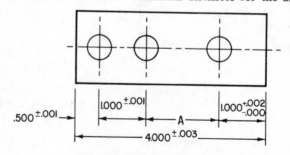

Fig. 3-9. Study problem 1.

2. What is the dimension and tolerance of *A*?

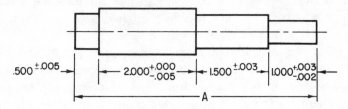

Fig. 3-10. Study problem 2.

3. A shaft is fitted into a sleeve bearing as shown below in Fig. 3-11. The inside diameter of the bearing = 63.50 ±.05 mm. The outside diameter of the shaft = 63.25 + 0.00, −0.01 mm. Find the minimum and maximum clearance between the shaft and the sleeve.

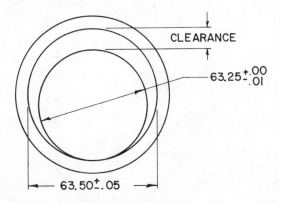

Fig. 3-11. Study problem 3.

Precision Gage Blocks

In modern mass-production industry, it is essential that the parts of an assembly fit into their mating parts with the degree of precision that was engineered into the product. Various types of gages are used to insure this accuracy. There are many kinds of gages, both hand and automated. Some of the more common hand gages are plug gages for hole sizes, thread gages, ring gages for shaft diameter checking, and taper gages. There are many more specialized types.

Gages must be made to the dimensions specified for the part to be checked. Accuracy is achieved in the making of gages by the use of precision gage blocks. Working gages used in toolrooms are made with a gage maker's tolerance and a wear allowance.

Precision gage blocks are small, rectangular steel blocks, heat treated to a high degree of hardness with fine dimensional stability. They are made to have an extremely high degree of dimensional accuracy.

The most accurate blocks in a toolroom are the master blocks used only to check and calibrate the working blocks which are used in the day-to-day checking operations. Gage blocks are assembled in groups, the sum of the individual block dimensions of which equals the required checking dimension. In Fig. 3-12, the use of precision gage blocks is shown in conjunction with a dial test indicator attached to a vernier height-gage in the measurement of a machine detail. The stack of blocks gives the setting used for the dial test indicator.

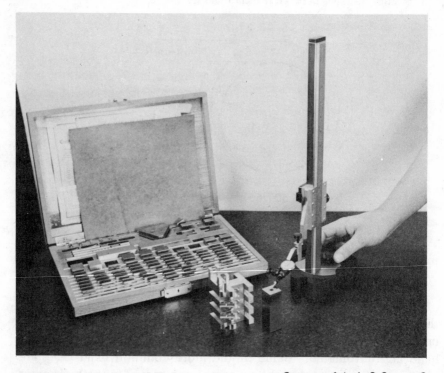

Courtesy of the L. S. Starrett Co.

Fig. 3-12. Using gage block stock to set dial test indicator height.

Another common use of gage blocks is with a sine plate as shown in Fig. 3-13. The stack of blocks gives an accurate vertical measure to set the sine plate to a precise angle.

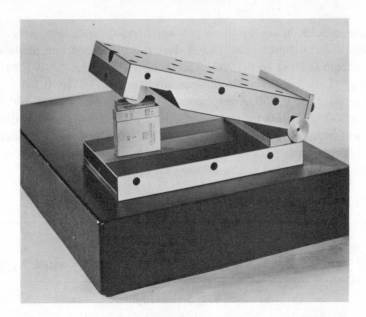

Courtesy of the Brown & Sharpe Manufacturing Co.

Fig. 3-13. Using gage blocks with sine plate to measure angles.

Gage blocks are purchased in sets of 83 pieces, Fig. 3-14. However, they can be purchased in sets of fewer blocks at a lesser cost.

Courtesy of the Brown & Sharpe Manufacturing Co.

Fig. 3-14. A set of precision gage blocks in protective case.

The blocks in an 83-piece set are divided into four series, see Table 3-3. The first series, called the "0.0001 series," consists of 9 blocks from 0.1001 to 0.1009. The second series, the "0.001 series," consists of 49 blocks with dimensions from 0.101 to 0.149. The third series, the "0.050 series," consists of 19 blocks with dimensions from 0.050 to 0.950 by five-hundredths gradations. All the dimensions are in inches. The fourth series, the "1.000 series," consists of four blocks, each increasing by 1.000 inch from 1.000 to 4.000 inches. Also included in the set are two 0.050 wear blocks.

Table 3-3. Sizes for an 83-piece Gage-Block Set

First: .0001 Series (9 Blocks)								
.1001	.1002	.1003	.1004	.1005	.1006	.1007	.1008	.1009
Second: .001 Series (49 Blocks)								
.101	.102	.103	.104	.105	.106	.107	.108	.109
.110	.111	.112	.113	.114	.115	.116	.117	.118
.119	.120	.121	.122	.123	.124	.125	.126	.127
.128	.129	.130	.131	.132	.133	.134	.135	.136
.137	.138	.139	.140	.141	.142	.143	.144	.145
.146	.147	.148	.149					
Third: .050 Series (19 Blocks)								
.050	.100	.150	.200	.250	.300	.350	.400	.450
.500	.550	.600	.650	.700	.750	.800	.850	.900
.950								
Fourth: 1.000 Series (4 Blocks)								
1.000 2.000 3.000 4.000								
Two .050 Wear Blocks								

By combining two or more blocks together, an extremely accurate dimension can be obtained. The blocks are attached to each other by "wringing" their surfaces together. Considerable force is necessary to separate them when their use has been completed, and they are best separated by twisting them apart.

In selecting blocks to obtain a given dimension, the smallest number of blocks that will give the required dimension should be used. This is accomplished by successively eliminating the smallest remaining dimension in order. The assembly of the gage blocks is an excellent example of the addition and subtraction of decimal fractions. Below is an example of the steps taken to set up blocks to give a dimension of 3.6742 inches.

Study Example.

Assemble blocks to give 3.6742 in. See Table 3-3.

Step 1. Eliminate the 0.0002 in. by selecting a 0.1002 block. Subtract 0.1002 from 3.6742:

$$\begin{array}{r} 3.6742 \\ -\ 0.1002 \\ \hline 3.5740 \end{array}$$

Step 2. Eliminate the 0.004 in. by selecting a 0.124 block. Subtract 0.124 from the remainder of step 1:

$$\begin{array}{r} 3.5740 \\ -\ 0.124 \\ \hline 3.450 \end{array}$$

Step 3. Eliminate the 0.450 inch with a 0.450 block. Subtract 0.450 from the remainder of step 2:

$$\begin{array}{r} 3.450 \\ -\ 0.450 \\ \hline 3.000 \end{array}$$

Step 4. Use the 3.000 block for the final step. Assemble and check for required dimension:

$$\begin{array}{r} 0.1002 \\ 0.124 \\ 0.450 \\ 3.000 \\ \hline 3.6742 \text{ in.} \end{array}$$

Study Problems:

Add:	1. 14.16	2. 236.02	3. 102.632
	71.23	62.34	1.023
	45.11	19.10	3.142
	0.015	221.62	10.001

Express as decimals and add:

4. 9 3/4 + 2.19 + 1 1/2 + 3.33 + 2 7/8 =

5. 3 2/3 + 1.33 + 9.60 + 2 1/8 + 3.875 =

6. 10.5 + 1/4 + 5/8 + 1/8 + 1.375 =

7. 19.78 + 1 5/8 + 2 1/3 + 6.75 + 3 2/3 =

Subtract:

8. 0.6239 from 1.0000

9. 10.5 − 0.5275 =

10. 9.375 − 8.125 =

11. 8.23 − 2.72 =

Do indicated operations:

12. 12.36 + 4.68 − 1.23 − 6.74 =

13. 9.62 − 2.76 − 3.67 − 0.23 =

14. 2.35 × 1.25 =

15. 19.10 × 3.1 =

16. 10.62 ÷ 2.31 =

17. 25.15 ÷ 2.6 =

18. 9.12 × 6.3 × 10.2 =

19. Divide (6.73 × 2.69) by 3.66

20. Divide: (2.93 × 15.63 × 19.2) by (33.1 × 1.75)

21. 22.34 × 3.56 ÷ 2.63 + 4.75 − 2.38 × 5.83 = (M.D.A.S.)

22. $4\dfrac{3}{8} + 2\dfrac{5}{16} + 3.587 \times 2\dfrac{7}{8} =$

23. Figure 3-15 shows a typical gaging set-up on a surface plate for the precision measurement of the center distance between two parallel holes. The center distance $D_c = M_1 - M_2 + \dfrac{D_1}{2} - \dfrac{D_2}{2}$. Solve for D_c when $M_1 = 1.125$ in., $M_2 = 0.2188$ in., $D_1 = 0.625$ in. and $D_2 = 0.500$ in. Give answers in thousandths of inches and then convert to millimeters.

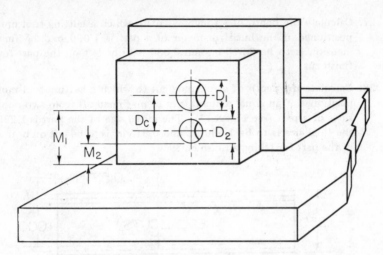

Fig. 3-15. Study problem 23.

24. Figure 3-16 shows a gaging set-up for measuring the center distance of small holes. Gage pins are fitted in the holes and their surfaces measured in relation to the surface plate. Solve for the center distance, D_c. $D_c = M_1 - M_2 - \dfrac{D_1}{2} + \dfrac{D_2}{2}$, when $M_1 = 0.9688$, $M_2 = 0.0781$, $D_1 = 0.4844$ and $D_2 = 0.2813$.

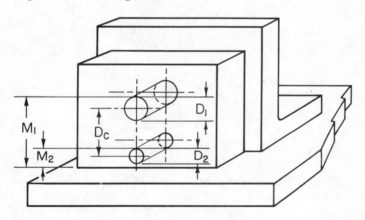

Fig. 3-16. Study problem 26.

25. Determine the gage blocks required to obtain stacks that are equal to M_1 and M_2 in Fig. 3-15.

26. Determine the gage blocks required to obtain stacks that are equal to M_1 and M_2 in Fig. 3-16.

27. Calculate the depth of a roughing cut to which a cutting tool must be positioned if the initial diameter of a part is 1.000 in., the final dimension is to be 0.875 in. and 0.030 is to be left on the part for the finish cut.

28. Calculate the depth of a roughing cut to which a cutting tool must be positioned if an equal amount is to be machined off from two opposite sides of a part (see Fig. 3-17). The initial size of the part is 1.250 in., the final size is to be 1.000 in. and 0.015 in. is to be left on both sides of the part for taking a finish cut.

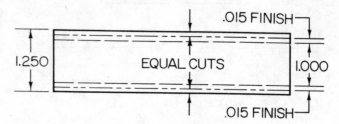

Fig. 3-17. Study problem 28.

CHAPTER 4

Exponents: Powers, Roots, and Percentage

Exponents are often encountered in shop formulas, and their meaning and use should be understood. For example, the following formula is used to calculate the effective outside length of a V-belt:

$$L = 2C + 1.57(D + d) + \frac{(D - d)^2}{4C}$$

where: L = Effective outside length
               C = Center distance between the sheaves
               D = Effective outside diameter of large sheave
               d = Effective diameter of small sheave
               (All dimensions are in inches.)

In this formula, the small 2 above and to the right of $(D - d)$ is the exponent of $(D - d)$. It designates that this expression $(D - d)$ must be multiplied by itself; thus: $(D - d) \times (D - d)$.

Another formula involving an exponent is for the volume of a cube:

$$V = S^3$$

where: V = Volume of the cube
               S = Length of the sides

This formula might have been written as follows:

$$V = S \times S \times S$$

Rather than to repeat the S term over, three times, it is obvious that the notation, S^3, is preferable. Again we see that the exponent, 3, indicates that S must be used three times as a factor.

An exponent, therefore, is simply a notation that shows how many times the term or digit is to be taken as a factor.

There are other terms that are used in conjunction with exponents that must also be learned. Let us refer again to the formula for the volume of a cube. This time, however, let us assume that the length of the sides of the cube (all equal) is four feet.

$$V = S^3$$
$$V = 4^3 = 4 \times 4 \times 4$$
$$V = 64 \text{ cu. ft}$$

As already mentioned, the exponent is 3. The number to which the exponent is attached, 4 in this case, is also given a special name. It is called the "base," or the base of the exponent.

Since $4 \times 4 \times 4 = 64$, the number 4 also has another special meaning. It is the "cube root" of 64. Therefore, it is called a root. Thus, in this example, the number 4 is both the base of the exponent 3 and the root (cube root) of 64.

As one might suspect, the number 64 in the above example is given a special name. It is called a "power." In this case it is the third power of the number 4. Likewise, 256 would be called the fourth power ($4 \times 4 \times 4 \times 4 = 256$) of 4.

We can now see that the terms: exponent, base, root, and power, have special meanings in mathematics and that these meanings should be committed to memory. To help commit these terms to memory, study the example given below:

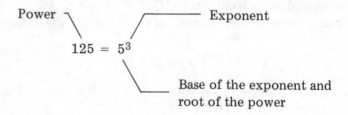

Study Problems.

1. In the problem: $256 = 2^8$,
   a. Check to see if 256 is the correct answer.
   b. Name the exponent, the base, the root, and the power.

2. $5^3 = ?$

3. $4^5 = ?$

4. What number is the fourth power of 10?

5. What number is the cube (third) root of 27?

Powers of Positive and Negative Bases

When the base number is a positive number, all the powers of that base will be positive. When the base number is a negative number, the following rule applies:

Negative base numbers raised to an even power (2,4,6, etc.) give a positive answer. Negative base numbers raised to uneven powers (3,5,7, etc.) result in negative answers.

Study Examples.

1. $(-2)^3 = (-2)(-2)(-2) = -8$

2. $(-2)^5 = (-2)(-2)(-2)(-2)(-2) = -32$

Care must be taken to determine what base an exponent applies to. For example; $(-3)^2$ means that the base is -3 and the action will be: $(-3)(-3) = +9$. However, if the number is written, -3^2, it means that only the number, 3, not the negative sign, is to be squared. Thus: $-3^2 = -(3)^2 = -9$.

Study Examples.

1. $(-4)^2 = (-4)(-4) = +16$

2. $-(3)^2 = -(3)(3) = -9$

3. $(-3)^3 = (-3)(-3)(-3) = -27$

4. $-(3)^3 = -(3)(3)(3) = -27$

In problems involving terms with exponents, the order of operations (M,D,A,S) requires that the powers be raised first, since this action is really a multiplication. As the power is applied to a "grouping symbol," this rule is stated: *The order of operation requires the performing of all operations within grouping symbols first.* While a grouping symbol is not always shown, as in 3^2, 2^3 or 4^2, these operations are to be performed first.

Study Examples.

1. $(-3)^2 + 45 \div 9 = \quad 9 + (45 \div 9)$
$$= \quad 9 + 5$$
$$= 14$$

2. $2^3 + (-3)^3 = 8 + (-27)$
$$= 8 - 27$$
$$= -19$$

3. $3^3 - 4^2 \div 2 = 27 - (16 \div 2)$
$$= 27 - 8$$
$$= 19$$

The Exponent 1

The numeral "1," when used as a factor in multiplication, has no effect on the product. Likewise, when "1" is used as an exponent, it has no effect on the base. For example: 8^1 means 8. Eight is used as a factor only once, or: $8^1 = 8$.

Every digit, term, or expression can be considered as having an exponent of "1" although the digit "1" is never shown. However, the fact that every term has an exponent of "1" if no other is shown is very important and is discussed later in this chapter.

The Exponent Zero

Any value (except 0) raised to the zero power is equal to 1.

Study Examples.

$$V^0 = 1 \qquad 10^0 = 1 \qquad (463,758)^0 = 1$$

The Negative Exponent

A quantity raised to a negative power is equal to the reciprocal of that quantity (one over the quantity) with a positive signed exponent.

Study Examples.

1. $3^{-2} = \dfrac{1}{3^2} = \dfrac{1}{9}$

2. $X^{-4} = \dfrac{1}{X^4}$

3. $10^{-3} = \dfrac{1}{10^3} = \dfrac{1}{1000}$

4. $\dfrac{1}{8^{-2}} = 1 \times 8^2 = 64$

Study Problems.

1. $3^0 =$ 2. $2^{-2} =$

3. $4^3 =$ 7. $(-3)^{-2} =$

4. $(-2)^3 =$ 8. $(-8)^0 =$

5. $-3^2 =$ 9. $2^2 \times 3^{-2} =$

6. $-3^{-3} =$ 10. $(-3)^{-2} \times 2^2 =$

Operations with Exponents

The ability to use exponents properly is very important in the solution of many shop formulas. The following "Rules of Exponents" should be carefully studied so that you can more easily solve the problems which involve exponents.

Like bases with the same or differing exponents may be multiplied or divided.

Rule 1: *When multiplying two or more like bases with exponents, the exponents are added and their sum is placed as the new exponent for the same base.*

Study Examples.

1. $(5^2)(5^3)(5^4) = 5^{2 + 3 + 4} = 5^9$

2. $(2^2)(2)(2^3) = 2^{2 + 1 + 3} = 2^6$

NOTE: As previously explained, a number not having an exponent (2) is considered to have an exponent of one (2^1). Do not overlook this point in future work with similar problems.

Rule 2: *When two values with like bases are divided, the exponent of the divisor is subtracted from the exponent of the dividend.*

Study Examples.

1. $(3)^5 \div (3)^2 = 3^{5 - 2} = 3^3$

2. $(5)^6 \div (5) = 5^{6 - 1} = 5^5$

3. $(9)^3 \div (9)^2 = 9^{3 - 2} = 9$

Rule 3: *When a number with an exponent is in turn raised to a power, the exponents are multiplied and their product placed as the exponent of the number.*

Study Examples.

1. $(5^2)^3 = 5^{2 \times 3} = 5^6$

2. $(3^4)^2 = 3^{4 \times 2} = 3^8$

When the same base numbers are divided and their exponents are the same, a special case develops. Applying Rule 2:

$$5^3 \div 5^3 = 5^{3-3} = 5^0$$

However, a number divided by the same number equals one $(2 \div 2 = 1)$. Therefore:

$$5^3 \div 5^3 = 1; \text{ and } 5^0 \text{ must also equal one.}$$

Rule 4: *Any value (except zero) raised to the zero power is equal to one.*

(This rule was given without proof on page 74.)

When a base number to a power is divided by a base number to a higher power, a negative exponent is generated:

$$4^2 \div 4^5 = 4^{2-5} = 4^{-3}$$

This quotient with a negative exponent is equal to $\frac{1}{4^3}$ because the operation can be written:

$$\frac{4^2}{4^5} = \frac{4 \times 4}{4 \times 4 \times 4 \times 4 \times 4}$$

cancelling:

$$= \frac{1}{4 \times 4 \times 4} = \frac{1}{4^3}$$

therefore:

Rule 5: *A base with a negative exponent can be written as a reciprocal of the base with a positive exponent.*

This rule may be expanded to say that a base to a power may be moved from the numerator to the denominator, or from the denominator to the numerator, by changing the sign of the exponent.

Study Examples.

1. $\dfrac{4^2}{4^5} = 4^2 \times 4^{-5} = 4^{-3} = \dfrac{1}{4^3}$; or, $\dfrac{4^2}{4^5} = \dfrac{1}{4^5 \times 4^{-2}} = \dfrac{1}{4^3}$

2. $\dfrac{5^2}{6^{-3}} = 5^2 \times 6^3$

When numbers are raised to powers, the product increases very rapidly in value to an extremely large number. This is referred to as "exponential expansion." Thus, for example, when 3 is raised to the power of 12, (3^{12}), it will equal 531,450.

Study Problems.

Simplify the following expressions with positive exponents:

1. $(2^2)(2^{-3}) =$

7. $(1^2)(2^{-2})(3^2) =$

2. $3^{-4} =$

8. $(\frac{2}{3})^3 =$

3. $1 \div 4^{-3} =$

9. $(\frac{1}{2})^2(2)^{-2} =$

4. $5^2 \div 5^{-3} =$

10. $(3^2)(2^0)(4^{-2}) =$

5. $2^2 \div 2^{-2} =$

11. $(4^{-1})(4^2) =$

6. $(4^2)(5^0) =$

12. $(5^0)(5^1)(5^2) =$

Solve for numerical answers:

13. $3^4 =$

17. $\dfrac{2^5}{5^2} =$

14. $2^6 =$

18. $4^3 - 3^2 =$

15. $10^9 =$

19. $(5^2)^3 =$

16. $\dfrac{2^3}{4^2} =$

20. $3^2 \times 3^3 =$

Roots

The roots of a number are the equal factors of a power. Powers that have whole numbers as roots are called "perfect powers." The perfect power of 5 is 25; of 10 is 100. The roots of perfect powers are always whole numbers.

The symbol used to denote the extraction of a root is, for example; $\sqrt[3]{}$ as in $\sqrt[3]{27}$ which is read: "cube root of 27 and equals 3. This symbol $\sqrt{}$ is called the "radical," and the number under the bar of the radical is called the "radicand." The small figure in the "v" of the radical is called the index. It is used when roots higher than the square root are to be extracted. When no index is shown, the radical is referred to as the "square root" radical which is $\sqrt{}$.

Study Examples.

1. $\sqrt{36} = 6$ $\sqrt[3]{125} = 5$ $\sqrt{81} = 9$ $\sqrt[3]{-125} = -5$

2. $\sqrt[3]{64} = 4$ $\sqrt[4]{81} = 3$ $\sqrt{400} = 20$ $\sqrt{49} = 7$

When even roots are extracted (2,4,6,.....) they must be signed "±" because either a negative root raised to an even power or a positive root raised to the same power will give a positive product.

$$(+2) \times (+2) = +4 \qquad (-2) \times (-2) = +4$$

hence: $\sqrt{4} = \pm 2$.

When uneven roots are extracted (3,5,7,.....) the root bears the sign of the radicand. There is a single positive cube root for $+27$, which is $+3$ and a single negative cube root for -27, which is -3.

Study Problems.

1. $\sqrt{64} =$ 5. $\sqrt[4]{256} =$

2. $\sqrt{25} =$ 6. $\sqrt[6]{64} =$

3. $\sqrt[3]{64} =$ 7. $\sqrt[3]{-216} =$

4. $\sqrt[3]{-64} =$ 8. $\sqrt{10,000} =$

Extraction of a Square Root

Many problems in mathematics involve squares and square roots. The knowledge of how to extract a square root is very valuable to solve many shop formulas.

There are many methods that can be used to extract the square root of a number. Some of the most common are:

1. Arithmetical, a division-type process;

2. Reference to square root tables;

3. The divide and average method;

4. Slide rule method;

5. Logarithms;

6. Office calculators with a square root key.

Each of these methods has its advantages and its disadvantages. The method used should be based on the accuracy required and the available tables or office machine. Methods 1,2, and 3 will be explained in the following pages.

Arithmetical Method

The extraction of the square root by the arithmetical method must be performed in specific sequential steps. The step is shown in the left column and the work is shown at the right.

Study Example 1.

Extract the square root of 22.5625:

Step 1.   Write the number under the radical sign. Place a new decimal point over the old point.

$$\sqrt{\overset{.}{22.5625}}$$

Step 2.   Mark off groups of two digits from the decimal point to the right and to the left.*

$$\sqrt{\overset{.}{`22.56`25}}$$

Step 3.   Determine the largest perfect square that is in the first group of two to the left, the number 22. This is 16, which is placed under the first pair and its square root, 4, placed above the group.

$$\sqrt{\overset{4.}{`22.56`25}}$$
$$16$$

Step 4.   Subtract the square and bring down the next pair of numbers (the second group).

$$\sqrt{\overset{4.}{22.5625}}$$
$$\underline{16}$$
$$656$$

Step 5.   Form a trial divisor by multiplying the number above the bar, 4, by 2. Place this number, 8, to the left of the difference calculated in step 4 and add a temporary (small) zero after it, as shown in (a), in the right margin. Determine the largest number that can be added to the trial divisor, the sum of which, when multiplied by this number, will give a product equal to or less than the dividend, 656. Find this number, add it to the trial divisor, and multiply as shown in (b). Place the product under the dividend and subtract. Bring down the next pair.

$$\overset{4.}{\sqrt{22.56\ 25}}$$
$$16$$
$$8o \overline{)\ 656} \qquad (a)$$

$$\overset{4.\ 7}{\sqrt{22.56\ 25}}$$
$$16$$
$$87 \overline{)\ 656} \qquad (b)$$
$$\underline{609}$$
$$4725$$

---

*An alternate method of "pointing-off" groups of two is to draw a short bar above or below each group of two numbers.

For example: $\overline{1\ 07\ 32.56\ 25}$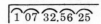

Step 6. Form a second trial divisor by multiplying the number above the bar, 47, by 2. Place this number, 94, to the left of the difference calculated in step 5 and add a temporary (small) zero after it as shown in (a). Determine the largest number, which when added to the trial divisor will give a product equal to or less than the dividend, 4725. See example (b). Place the product under the dividend and subtract. If there is a remainder, add a pair of zeros, bring them down and repeat the process until the required number of significant digits is obtained.

```
        4. 7
      | 22.56 25
        16            (a)
     87 | 656
        609
    94o | 4725
```

```
        4. 7 5
      | '22.56'25
        16            (b)
     87 | 6 56
        6 09
    945 | 47 25
        47 25
            0
```

Study Example 2.

Extract the square root of 107.32:

Step 1. Write the number under the radical sign. Place a new decimal point over the old point.

$$\sqrt{107.32}$$

Step 2. Mark off groups of two digits to the right and to the left of the point. (Note that there is only one digit in the group farthest to the left.)

$$\sqrt{1\,07.32}$$

Step 3. Determine the largest perfect square that is in the first left-hand group, 1. Place the perfect square, 1, under the group and its square root above the group.

```
  1 .
√1 07.32
  1
```

Step 4. Subtract the square and bring down the next pair of numbers (the second group).

```
  1 .
√1 07.32
  1
  0 07
```

Step 5. Form a trial divisor by multiplying the number above the bar, 1, by 2. Place this number, 2, to the left of the difference calculated in step 4 and add a temporary (small) zero after it, as shown in (a). Determine the largest number, which when added to the trial divisor will give a product equal to or less than the dividend, 7. However, since 20 is larger than 7, a zero is placed above the bar in the answer and the next pair of numbers is brought down. See example (b).

```
   1 .
 √1 07.32
   1            (a)
2o|007
```

```
   1 0 .
 √1 07.32
   1            (b)
   7 32
```

Step 6. Form a second trial divisor by multiplying the number above the bar, 10, by 2. Place this number, 20, to the left of the dividend, 732, and add a temporary (small) zero after it as shown in (a). Determine the largest number, which when added to the trial divisor will give a product equal to or less than the dividend, 732. See example (b). Place the product under the dividend and subtract. Bring down the next pair of numbers, 00.

$$
\begin{array}{r}
1\ \ 0.\ \ \ \ \ \ \ \\
\sqrt{01\ 07.32\ 00}\\
1\ \ \ \ \ \ \ \ \ \ \ \ \ \\
20o\ |\ \ 7\ 32\ \ \ \ \ \\
\end{array}
$$ (a)

$$
\begin{array}{r}
1\ 0\ .\ 3\ \ \ \ \ \\
\sqrt{01\ 07.32\ 00}\\
1\ \ \ \ \ \ \ \ \ \ \ \ \ \\
203\ |\ 732\ \ \ \ \ \\
609\ \ \ \ \ \\
\overline{12300}\ \ \\
\end{array}
$$ (b)

Step 7. Form a third trial divisor by multiplying the number above the bar, 103, by 2. Place this number, 206, to the left of the dividend, 12300 and add a temporary (small) zero after it as shown in (a). Determine the largest number, which when added to the trial divisor will give a product equal to or less than the dividend, 12300. See example (b). Place the product under the dividend and subtract. Bring down the next pair of numbers, 00.

$$
\begin{array}{r}
1\ 0\ .\ 3\ \ \ \ \ \\
\sqrt{1\ 07.32\ 00}\\
1\ \ \ \ \ \ \ \ \ \ \ \\
203\ |\ 732\ \ \ \ \\
609\ \ \ \ \\
206o\ |\ 12300\ \ \\
\end{array}
$$ (a)

$$
\begin{array}{r}
1\ 0\ .\ 3\ 5\ \ \ \ \ \ \ \\
\sqrt{01\ 07.32\ 00\ 00}\\
1\ \ \ \ \ \ \ \ \ \ \ \ \ \ \ \\
203\ |\ 732\ \ \ \ \ \ \ \\
609\ \ \ \ \ \ \ \\
2065\ |\ 123\ 00\ \ \ \\
103\ 25\ \ \ \\
\overline{19\ 75\ 00}\ \\
\end{array}
$$ (b)

Step 8. Form a fourth trial divisor by multiplying the number above the bar, 1035, by 2. Place this number, 2070, to the left of the dividend, 197500, and add a temporary (small) zero after it as shown in (a). Determine the largest number, which when added to the trial divisor will give a product equal to or less than the dividend, 197500. See example (b). Place the product under the dividend and subtract. Bring down the next pair of numbers, 00, and repeat the process until the required number of significant digits is obtained.

$$
\begin{array}{r}
1\ 0\ .\ 3\ 5\ \ \ \ \ \\
\sqrt{01\ 07.32\ 00\ 00}\\
1\ \ \ \ \ \ \ \ \ \ \ \ \\
203\ |\ 732\ \ \ \ \\
609\ \ \ \ \\
2065\ |\ 12300\ \ \ \\
10325\ \ \ \\
2070o\ |\ 197500\ \ \\
\end{array}
$$ (a)

$$
\begin{array}{r}
1\ 0\ .3\ 5\ 9\ \ \ \ \ \ \\
\sqrt{1\ 07.32\ 00\ 00}\\
1\ \ \ \ \ \ \ \ \ \ \ \ \\
203\ |\ 732\ \ \ \ \ \\
609\ \ \ \ \ \\
2065\ |\ 12300\ \ \ \\
10325\ \ \ \\
20709\ |\ 197500\ \\
186381\ \\
\overline{11119}\ \\
\end{array}
$$ (b)

Study Problems.

Extract the square root as indicated:

1. $\sqrt{0.361201}$ 4. $\sqrt{64.1601}$

2. $\sqrt{0.040401}$ 5. $\sqrt{7072.81}$

3. $\sqrt{51.076}$ 6. $\sqrt{12345.67}$

Square Root from Tables

The second method of extracting square roots is by using tables. Table 4-1 on pages 83 and 84 lists the squares and square roots of the numbers from 1 to 100. Also listed are the cubes, the cube roots, and the reciprocals ($\frac{1}{N}$).

Using the table, for example, to find the square of 35, locate the number 35 in the left-hand column headed "No." The value of 35^2, namely 1,225, appears immediately to the right in the column headed "Square."

To find the square root of 35, locate 35 in the left-hand column and move to the third column to the right, headed "Square Root." Here we find 5.91608.

Note that any number in the column headed "Square" has its square root in the column headed "No." Thus the square root of 5184 is 72.

In like manner this table may be used to find cubes and cube roots of numbers listed as well as the "Reciprocal" of a number which is the quotient of 1 divided by that number. The reciprocal of 48, from the table, is 0.0208333. This is the quotient of $\frac{1}{48}$. Note that the column at the right of the page is also a "No." column.

While the table in this chapter covers only the numbers 1 to 100, a more complete table covering 0.01 to 2000 may be found in *Machinery's Handbook*.

Study Problems.

Extract the square root of the following using the tables.

1. $\sqrt{39}$.   5. $\sqrt{3.25}$ 9. $\sqrt{36.3}$

2. $\sqrt{45}$.   6. $\sqrt{7.60}$ 10. $\sqrt{49.1}$

3. $\sqrt{81}$ 7. $\sqrt{620}$ 11. $\sqrt{345.25}$

4. $\sqrt{235}$.   8. $\sqrt{0.638}$ 12. $\sqrt{10,000}$

Table 4-1. Powers, Roots, and Reciprocals—1

No.	Square	Cube	Sq. Root	Cube Root	Reciprocal	No.
1	1	1	1.00000	1.00000	1.0000000	1
2	4	8	1.41421	1.25992	0.5000000	2
3	9	27	1.73205	1.44225	0.3333333	3
4	16	64	2.00000	1.58740	0.2500000	4
5	25	125	2.23607	1.70998	0.2000000	5
6	36	216	2.44949	1.81712	0.1666667	6
7	49	343	2.64575	1.91293	0.1428571	7
8	64	512	2.82843	2.00000	0.1250000	8
9	81	729	3.00000	2.08008	0.1111111	9
10	100	1,000	3.16228	2.15443	0.1000000	10
11	121	1,331	3.31662	2.22398	0.0909091	11
12	144	1,728	3.46410	2.28943	0.0833333	12
13	169	2,197	3.60555	2.35133	0.0769231	13
14	196	2,744	3.74166	2.41014	0.0714286	14
15,	225	3,375	3.87298	2.46621	0.0666667	15
16	256	4,096	4.00000	2.51984	0.0625000	16
17	289	4,913	4.12311	2.57128	0.0588235	17
18	324	5,832	4.24264	2.62074	0.0555556	18
19	361	6,859	4.35890	2.66840	0.0526316	19
20	400	8,000	4.47214	2.71442	0.0500000	20
21	441	9,261	4.58258	2.75892	0.0476190	21
22	484	10,648	4.69042	2.80204	0.0454545	22
23	529	12,167	4.79583	2.84387	0.0434783	23
24	576	13,824	4.89898	2.88450	0.0416667	24
25	625	15,625	5.00000	2.92402	0.0400000	25
26	676	17,576	5.09902	2.96250	0.0384615	26
27	729	19,683	5.19615	3.00000	0.0370370	27
28	784	21,952	5.29150	3.03659	0.0357143	28
29	841	24,389	5.38516	3.07232	0.0344828	29
30	900	27,000	5.47723	3.10723	0.0333333	30
31	961	29,791	5.56776	3.14138	0.0322581	31
32	1,024	32,768	5.65685	3.17480	0.0312500	32
33	1,089	35,937	5.74456	3.20753	0.0303030	33
34	1,156	39,304	5.83095	3.23961	0.0294118	34
35	1,225	42,875	5.91608	3.27107	0.0285714	35
36	1,296	46,656	6.00000	3.30193	0.0277778	36
37	1,369	50,653	6.08276	3.33222	0.0270270	37
38	1,444	54,872	6.16441	3.36198	0.0263158	38
39	1,521	59,319	6.24500	3.39121	0.0256410	39
40	1,600	64,000	6.32456	3.41995	0.0250000	40
41	1,681	68,921	6.40312	3.44822	0.0243902	41
42	1,764	74,088	6.48074	3.47603	0.0238095	42
43	1,849	79,507	6.55744	3.50340	0.0232558	43
44	1,936	85,184	6.63325	3.53035	0.0227273	44
45	2,025	91,125	6.70820	3.55689	0.0222222	45
46	2,116	97,336	6.78233	3.58305	0.0217391	46
47	2,209	103,823	6.85565	3.60883	0.0212766	47
48	2,304	110,592	6.92820	3.63424	0.0208333	48
49	2,401	117,649	7.00000	3.65931	0.0204082	49
50	2,500	125,000	7.07107	3.68403	0.0200000	50

Table 4-1. Powers, Roots, and Reciprocals—2

No.	Square	Cube	Sq. Root	Cube Root	Reciprocal	No.
51	2,601	132,651	7.14143	3.70843	0.0196078	51
52	2,704	140,608	7.21110	3.73251	0.0192308	52
53	2,809	148,877	7.28011	3.75629	0.0188679	53
54	2,916	157,464	7.34847	3.77976	0.0185185	54
55	3,025	166,375	7.41620	3.80295	0.0181818	55
56	3,136	175,616	7.48331	3.82586	0.0178571	56
57	3,249	185,193	7.54983	3.84850	0.0175439	57
58	3,364	195,112	7.61577	3.87088	0.0172414	58
59	3,481	205,379	7.68115	3.89300	0.0169492	59
60	3,600	216,000	7.74597	3.91487	0.0166667	60
61	3,721	226,981	7.81025	3.93650	0.0163934	61
62	3,844	238,328	7.87401	3.95789	0.0161290	62
63	3,969	250,047	7.93725	3.97906	0.0158730	63
64	4,096	262,144	8.00000	4.00000	0.0156250	64
65	4,225	274,625	8.06226	4.02073	0.0153846	65
66	4,356	287,496	8.12404	4.04124	0.0151515	66
67	4,489	300,763	8.18535	4.06155	0.0149254	67
68	4,624	314,432	8.24621	4.08166	0.0147059	68
69	4,761	328,509	8.30662	4.10157	0.0144928	69
70	4,900	343,000	8.36660	4.12129	0.0142857	70
71	5,041	357,911	8.42615	4.14082	0.0140845	71
72	5,184	373,248	8.48528	4.16017	0.0138889	72
73	5,329	389,017	8.54400	4.17934	0.0136986	73
74	5,476	405,224	8.60233	4.19834	0.0135135	74
75	5,625	421,875	8.66025	4.21716	0.0133333	75
76	5,776	438,976	8.71780	4.23582	0.0131579	76
77	5,929	456,533	8.77496	4.25432	0.0129870	77
78	6,084	474,552	8.83176	4.27266	0.0128205	78
79	6,241	493,039	8.88819	4.29084	0.0126582	79
80	6,400	512,000	8.94427	4.30887	0.0125000	80
81	6,561	531,441	9.00000	4.32675	0.0123457	81
82	6,724	551,368	9.05539	4.34448	0.0121951	82
83	6,889	571,787	9.11043	4.36207	0.0120482	83
84	7,056	592,704	9.16515	4.37952	0.0119048	84
85	7,225	614,125	9.21954	4.39683	0.0117647	85
86	7,396	636,056	9.27362	4.41400	0.0116279	86
87	7,569	658,503	9.32738	4.43105	0.0114943	87
88	7,744	681,472	9.38083	4.44797	0.0113636	88
89	7,921	704,969	9.43398	4.46475	0.0112360	89
90	8,100	729,000	9.48683	4.48140	0.0111111	90
91	8,281	753,571	9.53939	4.49794	0.0109890	91
92	8,464	778,688	9.59166	4.51436	0.0108696	92
93	8,649	804,357	9.64365	4.53065	0.0107527	93
94	8,836	830,584	9.69536	4.54684	0.0106383	94
95	9,025	857,375	9.74679	4.56290	0.0105263	95
96	9,216	884,736	9.79796	4.57886	0.0104167	96
97	9,409	912,673	9.84886	4.59470	0.0103093	97
98	9,604	941,192	9.89949	4.61044	0.0102041	98
99	9,801	970,299	9.94987	4.62607	0.0101010	99
100	10,000	1,000,000	10.00000	4.64159	0.0100000	100

Square Root of a Fraction

The extraction of the root of a fraction can be done in two ways: First, if the numerator and denominator are perfect squares, the root of fraction can be expressed as the root of the numerator divided by the root of the denominator. Thus:

$$\sqrt{\frac{9}{16}} = \frac{\sqrt{9}}{\sqrt{16}} = \frac{3}{4} = 0.75$$

If the numerator and the denominator are not both perfect squares, the fraction can first be reduced to a decimal and the root then extracted. Thus:

$$\sqrt{\frac{3}{4}} = \sqrt{0.75} = 0.866025$$

Study Problems.

1. $3^4 =$

2. $2^6 =$

3. $10^9 =$

4. $\dfrac{2^3}{4^2} =$

5. $\dfrac{2^5}{5^2}$

6. $4^3 - 3^2 =$

7. $(5^2)^3 =$

8. $3^2 \times 3^3 =$

Square Root by the Divide and Average Method

To compute the square root of a number by the divide and average method, the use of a slide rule or a calculator is suggested. The method consists of dividing the number by an approximate root (from slide rule, tables, or guess) and averaging the product with the approximation. Continue the process as many times as necessary for desired accuracy.

Study Example.

Extract the square root of 682.

Step 1. Approximate the square root: $26 \times 26 = 676$

Step 2. Divide 682 by the approximate root 26:

$$682 \div 26 = 26.23$$

Step 3. Average the approximation (26) and quotient (26.23).

$$\frac{26 + 26.23}{2} = 26.115$$

Step 4. Divide 682 by the Step 3 average:

$$282 \div 26.115 = 26.11526$$

Step 5. Average the approximation 26.115 and 26.11526:

$$\frac{26.115 + 26.11526}{2} = 26.11513$$

Step 6. Divide 682 by the Step 5 average:

$$682 \div 26.11513 = 26.11513$$

therefore: $\sqrt{682} = 26.11513$ (Check in tables.)

Study Problems.

Extract the square root by the divide and average method:

1. $\sqrt{1444}$ 5. $\sqrt{35.4853}$

2. $\sqrt{855,625}$ 6. $\sqrt{739.52}$

3. $\sqrt{1.8769}$ 7. $\sqrt{45.8593}$

4. $\sqrt{0.4587}$ 8. $\sqrt{3.5894}$

Percentage

The word "percent" means by or through a hundred and its notation "%" is a special symbol standing for a fraction. The symbol indicates division by 100. The expression 10% means 10 percent or $\frac{10}{100}$ or 0.10.

The United States monetary system is a decimal system, and a chart of the coins in percentage of a dollar would be as follows:

Dollar	=	$1.00	=	100 cents	=	100%	of a dollar	
Half	=	$ .50	=	50 cents	=	50%	" "	"
Quarter	=	$ .25	=	25 cents	=	25%	" "	"
Dime	=	$ .10	=	10 cents	=	10%	" "	"
Nickle	=	$ .05	=	5 cents	=	5%	" "	"
Penny	=	$ .01	=	1 cent	=	1%	" "	"
Mill*	=	$ .001	=	$\frac{1}{10}$ cent	=	0.1%	" "	"

*The mill is not coined but is used in property tax assessments.

An interest rate of 5 percent means that the annual cost of borrowing one dollar is 5 cents. Credit cards generally cost the user $1\frac{1}{2}$ percent a month interest which is 18 percent per year in interest.

Percentages, decimal fractions, and proper fractions are interrelated mathematically. Table 4-2 gives the equivalents of these three forms.

Table 4-2. Equivalent Percentages, Decimals and Fractions

Per-Cent	Decimal Fraction	Proper Fraction	Per-Cent	Decimal Fraction	Proper Fraction
5%	0.05	$\frac{1}{20}$	40%	0.40	$\frac{40}{100}$
10%	0.10	$\frac{1}{10}$	$43\frac{3}{4}\%$	0.4375	$\frac{7}{16}$
$12\frac{1}{2}\%$	0.125	$\frac{1}{8}$	50%	0.50	$\frac{1}{2}$
$15\frac{5}{8}\%$	0.15625	$\frac{5}{32}$	60%	0.60	$\frac{3}{5}$
$16\frac{2}{3}\%$	0.16667	$\frac{1}{6}$	$62\frac{1}{2}\%$	0.625	$\frac{5}{8}$
$18\frac{3}{4}\%$	0.1875	$\frac{3}{16}$	$66\frac{2}{3}\%$	0.66667	$\frac{2}{3}$
20%	0.20	$\frac{1}{5}$	70%	0.70	$\frac{7}{10}$
$21\frac{7}{8}\%$	0.21875	$\frac{7}{32}$	75%	0.75	$\frac{3}{4}$
25%	0.25	$\frac{1}{4}$	80%	0.80	$\frac{4}{5}$
$28\frac{1}{8}\%$	0.28125	$\frac{9}{32}$	$83\frac{1}{3}\%$	0.83333	$\frac{5}{6}$
$31\frac{1}{4}\%$	0.3125	$\frac{5}{16}$	$87\frac{1}{2}\%$	0.875	$\frac{7}{8}$
$33\frac{1}{3}\%$	0.33333	$\frac{1}{3}$	90%	0.90	$\frac{9}{10}$
$37\frac{1}{2}$	0.375	$\frac{3}{8}$	100%	1.00	1

Many times it is necessary to manipulate fractions to equivalent percentages not shown in Table 4-2. The fractions are first converted into decimal fractions by dividing the numerator by the denominator, such as: $\frac{1}{7} = 7\overline{\smash{)}1.00} = 0.1428$. This decimal fraction is converted into a percentage by moving the decimal point two places to the right: $0.1428 = 14.28$ percent. This last operation is equivalent to multiplying the decimal by 100.

Percentages are converted to fractions in a more complicated manner. The following example will illustrate the steps to be taken:

Convert $6\frac{1}{4}$ percent to a common fraction.

Step 1.   Write $6\frac{1}{4}\%$ as: 6.25%

Step 2.   Convert to a decimal by dividing by 100:

$$\frac{6.25}{100} = 0.0625$$

Step 3.   Express decimal (0.0625) as a fraction with a denominator as a power of ten. *Note*: There will be the same number of zeros in denominator as there are digits to the right of the decimal point.

This gives: $0.0625 = \frac{625}{10,000}$

Step 4.   Extract a common factor from both the numerator and the denominator:

$$\frac{625}{10,000} = \frac{125 \times 5}{125 \times 80} = \frac{5}{80}$$

Percentages greater than 100 percent are converted to common fractions in a similar manner: Convert 125 percent to a common fraction.

Step 1.   Convert to a decimal by dividing by 100:

$$125\% = \frac{125}{100} = 1.25$$

Step 2.   Express decimal as a fraction with the denominator a power of ten:

$$1.25 = 1\frac{25}{100}$$

Step 3. Extract a common factor from both the numerator and the denominator:

$$1\frac{25}{100} = 1\frac{25 \times 1}{25 \times 4} = 1\frac{1}{4}$$

NOTE: percentages over 100 percent always convert to an improper fraction which can be expressed as a mixed number as above.

Study Examples.

1. What is 35% of 358 tons of scrap iron?

   35% = 0.35   (Move decimal point two places to the left.)
   358 X 0.35 = 125.30 tons

2. A builder estimates that he will need 35,000 bricks for a building. How many bricks must be ordered if 9% must be allowed for scrap and loss?

   9% = 0.09
   35,000 X 0.09 = 3150 (extra bricks required)
   35,000 + 3150 = 38,150 bricks to be ordered.

Study Problems.

1. Change 15% to a decimal.

2. Change 0.0725 to a percent.

3. What is 30% of 760 board feet?

4. What percent of 40 inches is 35 inches?

5. 883.20 is 32% of what number?

6. A man earning $12,500 per year pays an income tax of 15%. How much does he pay the government?

7. The walls of a house are to be covered with insulation board. The actual area is 2550 sq ft. How many additional square feet of insulation board must be ordered to provide for 7% waste?

8. A contractor estimates that 10,050 cement blocks are needed for a job. How many additional blocks must he order to provide for 12% scrap and damage?

Simple Interest

The calculation of interest rates, the principal sum at interest, and the amount of interest in dollars can be readily accomplished by using one of the three following formulas:

1. Interest (amount in $) = Principal ($) times Rate (%): $I = PR$.

2. Rate (%) = Interest (in $) divided by Principal ($): $R = \dfrac{I}{P}$.

3. Principal ($) = Interest ($) divided by Rate (%): $P = \dfrac{I}{R}$.

While the interest rate is given as a percent digit in the above formulas, the solution of problems using these formulas requires the percentage (%) to be changed into a decimal fraction. This is done by moving the decimal point two places to the left. For example: $6\% = 0.06$, $5\frac{1}{2} = 5.5\% = 0.055$

Using formula 1, the amount of interest in dollars at a rate of 6 percent applied to a principal of $300 is:

$I = P \times R$, where $P = \$300$ and $R = 6\%$

$I = \$300 \times 0.06 = \$18.00$

When the amount of interest in dollars and the principal in dollars are known, formula 2 is used to calculate the percent rate. Thus, when $I = \$30.$ and $P = \$600$.:

$$R = \frac{I}{P}$$

$$R = \frac{\$30}{\$600} = 0.05 = 5\%$$

Given the interest rate (percent) and the amount of interest (dollars), formula 3 is used to find the principal. Thus when $R = 5$ percent and $I = \$60$:

$$R = \frac{I}{R}$$

$$R = \frac{\$60}{0.05} = \$1200$$

Study Problems.

1. Find the amount of interest paid per year on a loan of $2500 at 6 percent interest.

2. A bank loan at 8 percent cost the borrower $48 per year. How much did he borrow?

3. A loan of $2000 cost a borrower $80. What interest rate was charged?

4. A bank gives 4 percent interest on savings. What would a depositer be required to have in his account to provide him with a monthly income of $100? (Assume annual interest divided by 12.)

5. Import duties are listed as percentages of the value of goods imported. What would be the total cost of a Spanish iron work sold in Spain for $1250 and subject to a duty of 8 percent? What would be the sales price in the United States after a 20 percent retail mark-up?

List Price and Discounts

Wholesale houses and distributors use a system of list price and discount in their catalogs. List price is the fair retail price and is charged to persons buying only a few items at a time. Large quantity buyers are favored by lower prices. The lower prices are obtained by applying a percentage "discount" to the list price. In some cases only one discount is given and the resulting price is known as the "wholesale" price.

Large machine tools are sold at a discount to an agent. The amount of discount from the list or market price would be the agent's gross profit on the sale. Some perishable tools such as drills and reamers have list prices in the manufacturer's catalog followed by a series of percentages, such as:

$$\frac{1}{2}\text{-13 UNC tap}\quad\text{List \$1.65,}\quad\text{less 15\%,}\quad 10\%,\quad 6\%$$

These percentages are discounts and will be deducted from the list price depending on the quantity ordered. In the catalog there is a discount schedule indicating the quantities required in the order to get the discounts. If the schedule indicates that:

1 – 5, list; 6 –20, 15%; 21 –50, 15% and 10%; 51 and over, 15%, 10% and 6%

orders for amounts in the four quantity ranges will get the discounts indicated. An order of 12 will get 15 percent off the list price, an order of 51 or more will get a total multiple discount of 15 percent, 10 percent and 6 percent.

Study Example.

An order called for 60 reamers listing at $10 each. The discounts for this quantity are 15%, 10%, and 6%. How much will the 60 reamers cost the customer?

Solution: In multiple discounts the percentages must be applied in order of statement, one at a time. Each succeeding rate is applied to the new cost after the discount is taken. This is done as follows:

60 reamers @ $10 list totals $600

First discount of 15% = $600 X 15% = $90

Net after first discount: $600 − $90 = $510

Second discount of 10%: $510 X 10% = $51

Net after second discount: $510 − $51 = $459

Third discount of 6%: $459 X 6% = $27.54

Net after third discount: $459 − $27.54 = $431.46

The above complicated computation can be shortened to one calculation on a business machine by multiplying the list price by the difference of each rate from 100%. For example:

$600 X (1.00 − 0.15) X (1.00 − 0.10) X (1.00 − 0.06)
$600 X 0.85 X 0.90 X 0.94 = $431.46

Study Problems.

1. A manufacturer's agent selling machine tools averages a 30% profit on his annual sales. If his annual gross income is $30,000, what was the total price of the machines he sold during that year?

2. A mill supply firm offers multiple discounts of 20%, 15%, 10% and 5% on orders totaling $1,000, $5,000, $7,000, and $10,000, respectively compounded. (An order of $10,000 or over gets all four discounts; $7,000 gets the first three, etc.) How much will the discounted price be on separate orders of: $2,000, $6,000, $8,000, and $12.000?

3. A merchant's invoice from a supplier is for $1550. The supplier offers a $2\frac{1}{2}$% discount if the invoice is paid in 30 days. How much can the merchant save by prompt payment? (This is referred to in business as "discounting one's bills.")

4. Which is the most profitable to a buyer? A single discount of 12%, a multiple of 8% and 4% or a multiple of 7%, 3%, and 2%?

5. A dozen drills list priced at $75 are discounted $15. What percent discount was given?

6. A machinery distributor buys a lathe from a manufacturer for $750 and sells it for the manufacturer's list price of $1250. What was the discount percentage offered by the manufacturer?

7. An ancient Greek mathematician discovered this formula:

$$c^2 = a^2 + b^2$$

This formula is applied to a right (90°) triangle whose sides were labeled as shown below:

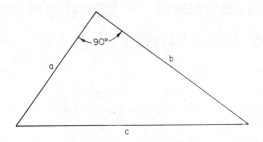

Fig. 4-1. Study problem 7.

How long is c if $a = 30$ ft and $b = 40$ ft?

8. Evaluate: $\sqrt{95.6484}$ $\sqrt{5867.56}$ $\sqrt{701}$.

9. Evaluate: 35^3 47^4 19^5 13^4

10. $30.60 is what percent of $255 ?

11. $21.91 is 7% of what number?

12. $\frac{1}{2}$ is 25% of what number? $\frac{6}{7}$ is $12\frac{1}{2}$% of what number?

13. If a man works 40 hours a week, what percent of the hours in the week does he work?

14. A machine detail has a $\pm.12\frac{1}{2}$% tolerance on its length.   If the base length is 14.625 in., what will be the minimum and maximum allowable lengths?

15. The clearance between a punch and a die for blanking an aluminum part is specified to be 8% of the stock thickness per side. Calculate the clearance required for blanking a sheet of $\frac{1}{16}$-in. thick aluminum. The answer should be given in terms of one-thousandths inch.

16. Babbit metal is used in low-speed, high-pressure bearings.   It is a mixture of the following percentages of metals: tin, 85%; antimony, 10%; copper, 4% and lead, 1%. How many pounds of each metal will be required to rebabbit a bearing using 8.5 pounds of the alloy?

17. Solder used by tinsmiths usually contains 59% tin and 41% lead. Solder used in plumbing has less tin and more lead (35% tin and 65% lead). A tinsmith has 60 pounds of plumber's solder that he wants to convert to tinsmith's solder. What metal must he add and how much of it?

CHAPTER 5

Measurement – English and Metric (SI) Units

Measurement systems have been in use since the dawn of history. They were devised to answer questions relating to: How many? How much? How long? How heavy?

The answers to these common questions involve the use of a "counting" numeral such as: 3, 25, $67\frac{1}{2}$, etc., plus a measurement "unit" such as pounds, feet, acres, gallons, etc.

This chapter will deal with seven general types of measurement: linear or length, two-dimensional or area, three dimensional or volume expressed in "cubic" units, volume expressed in capacity units, weight, pressure and vacuum, and temperatures.

Measurement systems are needed by all people and are used not only by skilled tradesmen but by merchants, bankers, scientists, housewives and children. The English speaking countries of the world have used a measurement system that is centuries old. It is known as the "English" system and while it is complicated and the units are not readily interchangeable, it was the system which fostered the English Industrial Revolution and which later made the United States the "Arsenal of Democracy" in World War II.

The metric system, the other great system of measurement, was first established in France in the years following the French Revolution. In 1954, a modernized metric system was developed which was subsequently named the Système International d'Unités, for which the abbreviation is SI in all languages. There is a worldwide move to adopt this modernized metric system and there is growing interest in it in the United States. Therefore, the rudiments of SI should be learned.

Measures of Extension

Every physical body occupies space. A body has three dimensions; length, width, and thickness. Shop drawings use standard conventions to portray three dimensions on a two-dimensional sheet of paper. Models and patterns are three dimensional for direct use in the shop.

The basic single dimension is the line. A line is defined as the distance between two points in space. In shop practice a line connects two points on a flat surface. The measure of a line is called "linear" or line measure.

Two dimensions are required to define a surface or area. These dimensions; length and width, relate to square or area measurement. When multiplied together, length and width produce what is called a *denominate number* that has the unit of measurement notation of "square," such as square feet.

A solid object is three-dimensional. It has length and width to which is added thickness or height. The measurement of a solid object gives a "cubic" measure. When the three dimensions are multiplied together they produce the unit of measurement referred to as "cubic" or "cube," such as a cubic yard.

Linear Measure

The measure of length in the shop is usually in terms of inches and fractions or decimals of inches. In countries using the metric system the common shop linear dimension is the millimeter (mm). Each measurement system has its subdivisions and multiples to provide for extremely large or small measurements. Technical personnel in shops and offices should be able to use both the English and metric systems.

The SI is based on use of the *meter* for linear measurement. The meter is subdivided into 100 equal parts called *centimeters*. The centimeters are further divided into ten equal parts called *millimeters*. There are 1000 millimeters in a meter. The meter is 39.3700787 inches long. Dividing this inch length by 100 we get the conversion factor of 0.3937 inch equals 1 centimeter (cm). One millimeter will then equal 0.03937 inch.

Knowing that 1 centimeter equals 0.3937 inch, we can get the value of 1 inch in terms of centimeters by dividing 1 centimeter by 0.3937. This gives the value of 1 inch as 2.54 centimeters or 25.4 millimeters.

The conversion of shop drawing dimensions from one system to the other must be made with great accuracy. Many international corporations are using "dual" dimensioning showing both English and metric dimensions on their shop drawings to insure accuracy whether the part is made in a metric system country or in an English system country. Simple measuring tools such as machinists' and draftsmen's scales and tape measures can be obtained with both the inch/foot and metric scales on their faces.

In the metric system, the units are multiples of ten. The metric prefixes and abbreviations for many units are shown in Table 5-1.

Table 5-1. Factors and Prefixes for Forming Decimal Multiples and Sub-multiples of the SI Units

Factor by which the unit is multiplied	Prefix	Symbol	Factor by which the unit is multiplied	Prefix	Symbol
10^{12}	tera	T	10^{-2}	centi	c
10^{9}	giga	G	10^{-3}	milli	m
10^{6}	mega	M	10^{-6}	micro	μ
10^{3}	kilo	k	10^{-9}	nano	n
10^{2}	hecto	h	10^{-12}	pico	p
10	decca	da	10^{-15}	femto	f
10^{-1}	deci	d	10^{-18}	atto	a

Linear Conversion Factors

Conversion of Inches to Millimeters: In construction work it is common practice to use meters and centimeters. The millimeter, however, is the common unit of measurement for all shop drawings in order to prevent any possible error as a result of a misplaced decimal point.

Metric units may be converted to English units and English units to metric units by using Table 5-2. To convert from English to metric, multiply the English unit by the factor shown opposite the English unit in the left column. To convert from metric to English, divide the metric unit by the factor opposite the metric unit in the "metric" column. Abbreviations for metric units are shown in the right-hand column.

Study Examples.

1. How many meters, centimeters, and millimeters are there in 220 yards?

$$220 \text{ yd} \times 0.9144 = 201.168 \text{ m}$$
$$201.168 \text{ m} \times 100 = 20,116.8 \text{ cm}$$
$$201.168 \text{ m} \times 1000 = 201,168 \text{ mm}$$

NOTE: In European countries 201,168 would be written 201 168, that is, a half space would be used instead of a comma to mark off thousands, millions, etc. This practice may be adopted in the United States, if and when the SI metric system becomes standard.

Table 5-2. SI (Metric) Conversion Factors

Multiply	by	To Obtain
Acre	4046.9	meter^2 (m^2)
Cubic foot	0.028317	meter^3 (m^3)
Cubic inch	16.387	centimeter^3 (cm^3)
Degree of angle	17.453	milliradian (mrad)
Degree of angle	0.017453	radian (rad)
Foot	0.3048	meter (m)
Gallon, U.S. liquid	0.0045467	meter^3 (m^3)
Gallon, U.S. liquid	3.7854	liter (1)
Inch	25.400	millimeter (mm)
Microinch	0.0254	micrometer (μ)
Micron	1	micrometer (μ)
Mile, nautical	1.8520	kilometer (km)
Mile, statute	1.6093	kilometer (km)
Minute of angle	0.29089	milliradian (mrad)
Pounds-force	4.4482	newton
Pounds/square inch	6.8948	kilopascal (kPa)
Second of angle	4.8481	microradian (μrad)
Square foot	0.092903	meter^2 (m^2)
Square inch	645.16	millimeter^2 (mm^2)
Square mile	2.5900	kilometer^2 (km^2)
Square yard	0.83613	meter^2 (m^2)
Ton (short, 2000 lb)	907.18	kilogram (kg)
Yard	0.9144	meter (m)

Conversions using Table 5-2 are usually done by means of a calculator. When a calculator is not available, the conversion can be more easily done by using a table similar to Table 5-3.

2. Convert 2.6875 inches to millimeters using Table 5-3.

Step 1. Obtain the metric equivalent of 2 in.

$$2 \text{ in.} = 50.800 \text{ mm}$$

Step 2. Find the metric equivalent of 0.6000 in. (2.6875):

0.6000 in. = 15.24 mm

Step 3. Find the metric equivalent of 0.0800 in. (2.6875):

0.0800 in. = 2.032 mm

Step 4. Find the metric equivalent of 0.0070 in. (2.6875):

0.0070 in. = 0.1778 mm

Step 5. Find the metric equivalent of 0.0005 in. (2.6875):

0.0005 in. = 0.0127 mm

Step 6. To find the answer, add the metric equivalents:

2.0000 in. = 50.8000 mm
0.6000 in. = 15.2400 mm
0.0800 in. = 2.0320 mm
0.0070 in. = 0.1778 mm
0.0005 in. = 0.0127 mm

2.6875 in. = 68.2625 mm

Table 5-3. Inch—Millimeter and Inch—Centimeter Conversion Table
(Based on 1 inch = 25.4 millimeters, exactly)

INCHES TO MILLIMETERS											
in.	mm	in.	mm	in.	mm	in.	mm	in.	mm	in.	mm
10	254.00000	1	25.40000	.1	2.54000	.01	.25400	.001	.02540	.0001	.00254
20	508.00000	2	50.80000	.2	5.08000	.02	.50800	.002	.05080	.0002	.00508
30	762.00000	3	76.20000	.3	7.62000	.03	.76200	.003	.07620	.0003	.00762
40	1,016.00000	4	101.60000	.4	10.16000	.04	1.01600	.004	.10160	.0004	.01016
50	1,270.00000	5	127.00000	.5	12.70000	.05	1.27000	.005	.12700	.0005	.01270
60	1,524.00000	6	152.40000	.6	15.24000	.06	1.52400	.006	.15240	.0006	.01524
70	1,778.00000	7	177.80000	.7	17.78000	.07	1.77800	.007	.17780	.0007	.01778
80	2,032.00000	8	203.20000	.8	20.32000	.08	2.03200	.008	.20320	.0008	.02032
90	2,286.00000	9	228.60000	.9	22.86000	.09	2.28600	.009	.22860	.0009	.02286
100	2,540.00000	10	254.00000	1.0	25.40000	.10	2.54000	.010	.25400	.0010	.02540

MILLIMETERS TO INCHES											
mm	in.	mm	in.	mm.	in.	mm	in.	mm	in.	mm	in.
100	3.93701	10	.39370	1	.03937	.1	.00394	.01	.00039	.001	.00004
200	7.87402	20	.78740	2	.07874	.2	.00787	.02	.00079	.002	.00008
300	11.81102	30	1.18110	3	.11811	.3	.01181	.03	.00118	.003	.00012
400	15.74803	40	1.57480	4	.15748	.4	.01575	.04	.00157	.004	.00016
500	19.68504	50	1.96850	5	.19685	.5	.01969	.05	.00197	.005	.00020
600	23.62205	60	2.36220	6	.23622	.6	.02362	.06	.00236	.006	.00024
700	27.55906	70	2.75591	7	.27559	.7	.02756	.07	.00276	.007	.00028
800	31.49606	80	3.14961	8	.31496	.8	.03150	.08	.00315	.008	.00031
900	35.43307	90	3.54331	9	.35433	.9	.03543	.09	.00354	.009	.00035
1,000	39.37008	100	3.93701	10	.39370	1.0	.03937	.10	.00394	.010	.00039

When the inch unit is a decimal equivalent of a fractional inch, the conversion may be made more rapidly and with less chance for error by using Table 5-4. Table 5-4 provides millimeter equivalents that are precise.

Table 5-4. Millimeter Equivalents of Fractional Inches.

Fraction, Inch	Precise Equivalent	Within 0.002 Inch	Within 0.005 Inch	Within 0.010 Inch
		Millimeters[*]		
1/32	0.793 8	0.8	0.8	1
1/16	1.587 5	1.6	1.5	1.5
3/32	2.381 2	2.4	2.5	2.5
1/8	3.175	3.2	3.2	3
5/32	3.968 8	4	4	4
3/16	4.762 5	4.8	4.8	5
7/32	5.556 2	5.6	5.5	5.5
1/4	6.350	6.3	6.4	6.5
9/32	7.143 8	7.1	7.1	7
5/16	7.937 8	7.9	8	8
11/32	8.731 2	8.7	8.7	8.5
3/8	9.525	9.5	9.5	9.5
13/32	10.318 8	10.3	10.3	10.5
7/16	11.112 5	11.1	11	11
15/32	11.906 2	11.9	12	12
1/2	12.700	12.7	12.7	12.5
17/32	13.493 8	13.5	13.5	13.5
9/16	14.287 5	14.3	14.3	14.5
19/32	15.081 2	15.1	15	15
5/8	15.875	15.9	16	16
21/32	16.668 8	16.7	16.7	16.5
11/16	17.462 5	17.5	17.5	17.5
23/32	17.780	17.8	17.8	18
3/4	19.050	19	19	19
25/32	19.843 8	19.8	19.8	20
13/16	20.637 5	20.6	20.6	20.5
27/32	21.431 2	21.4	21.5	21.5
7/8	22.225	22.2	22.2	22
29/32	23.018 8	23	23	23
15/16	23.812 5	23.8	23.8	24
31/32	24.606 2	24.6	24.5	24.5

[*]NOTE: In European practice a comma is used in place of a decimal point; as 0,793 8.

3. Convert 2.6875 in. to millimeters using Table 5-4. Note that 0.6875 is the decimal equivalent of $\frac{11}{16}$ in. Obtain a precise conversion.

Step 1. From Table 5-4:

$$2 \text{ in.} = 50.8000 \text{ mm}$$

$$\frac{11}{16} \text{ in.} = 17.4625 \text{ mm}$$

$$2\frac{11}{16} \text{ in.} = 68.2625 \text{ mm}$$

This conversion, accurate to 0.005 inch, would be made as follows:

$$2 \text{ in.} = 50.8 \text{ mm}$$

$$\frac{11}{16} \text{ in.} = 17.5 \text{ mm}$$

$$2\frac{11}{16} \text{ in.} = 68.3 \text{ mm}$$

Tolerances must also be converted from inch to millimeter units in shop practice. Although Table 5-3 can be used for this purpose, it is more convenient to use Table 5-5 which lists the normally applied tolerances.

Table 5-5. Equivalent Tolerance Values
(with Appropriate Rounding
of Metric Values)

Inch Tolerance	mm Tolerance
0.0001	0.002 5
0.0002	0.005
0.0003	0.007 5
0.0005	0.013
0.001	0.025
0.002	0.050
0.003	0.075
0.004	0.100
0.005	0.125
0.007	0.18
0.010	0.25
0.015	0.40
0.020	0.50
0.030	0.75
1/16	1.5

4. Convert the following dimensions from inch to millimeter units:

$$2.500^{\pm.001}$$

Step 1. From Table 5-3:

$$2.500 \text{ in.} = 63.500 \text{ mm}$$

Step 2. From Table 5-5:

$$.001 \text{ in.} = 0.025 \text{ mm}$$

Step 3. Add the tolerance to the dimension:

$$
\begin{array}{llll}
2.500 & \text{in.} = 63.500 & \text{mm} \\
\pm.001 & \text{in.} = \quad\quad \pm.025 & \text{mm} \\
\hline
2.500^{\pm.001} & \text{in.} = 63.500^{\pm.025} & \text{mm}
\end{array}
$$

Conversion of Millimeters to Inches. The conversion of millimeters to inches can be handily done by the use of Table 5-3.

Study Example:

1. Determine the inch equivalent of 93.662 mm.

Step 1. Find the inch equivalent of 90 mm:

$$90 \text{ mm} = 3.54331 \text{ in.}$$

Step 2. Find the inch equivalent of 3 mm:

$$3 \text{ mm} = 0.11811 \text{ in.}$$

Step 3. Find the inch equivalent of 0.6 mm:

$$0.6 \text{ mm} = 0.02362 \text{ in.}$$

Step 4. Find the inch equivalent of 0.06 mm:

$$0.06 \text{ mm} = 0.00236 \text{ in.}$$

Step 5. Find the inch equivalent of 0.002:

$$0.002 \text{ mm} = 0.00008$$

Step 6. Add:

$$
\begin{array}{lll}
90 & \text{mm} = 3.54331 & \text{in.} \\
3 & \text{mm} = 0.11811 & \text{in.} \\
0.6 & \text{mm} = 0.02362 & \text{in.} \\
0.06 & \text{mm} = 0.00236 & \text{in.} \\
0.002 & \text{mm} = 0.00008 & \text{in.} \\
\hline
93.662 & \text{mm} = 3.68748 & \text{in.} \\
& = 3.6875 & \text{in. (rounded off)}
\end{array}
$$

Study Problems.

Convert from inches to millimeters:

1. 2.250 in. =   5. 1.53125 in. =

2. 10.3125 in. =   6. 2.90625 in. =

3. 0.9375 in. =   7. 0.09375 in. =

4. 6.4375 in. =   8. 0.6875 in. =

Convert from millimeters to decimal inches:

9. 15.875 mm =   13. 45.836 mm =

10. 24.6062 mm =   14. 4,710,055. mm =

11. 1234. mm =   15. 6.8467 mm =

12. 8.459 mm =   16. 0.00456 mm =

Conversion of Units Used in Building and Construction Work

The construction trades utilize the inch/foot units of the English linear system while the highway construction industry uses yards, rods, and miles. In countries using the metric system, the meter and centimeter units are used predominately in the construction trades with the kilometer added for highway construction work and the millimeter occasionally used in some fine measurements.

In the English system, the yard is subdivided into three equal parts called *feet*. A foot is, in turn, subdivided into twelve equal parts called *inches*. The following table shows some of the common English and metric linear units and their relationships within each system:

Table 5-6. English and Metric Linear Measurement

12 inches (in.) = 1 foot (ft)
3 feet = 1 yard (yd)
1 yard = 36 inches
1 rod = $5\frac{1}{2}$ yards = $16\frac{1}{2}$ feet
1 mile = 320 rods = 1760 yards = 5280 feet
10 millimeters (mm) = 1 centimeter (cm)
100 centimeters (cm) = 1 meter (m)
1000 meters (m) = 1 kilometer (km)
For conversion factors, see pages 484-485

Study Examples.

1. How many feet are there in 50 rods?

$$50 \text{ rods} \times 16\frac{1}{2} \text{ ft per rod} = 825 \text{ ft}$$

2. A 5 meter board is cut into 60-cm lengths. If the saw cut is 2 mm wide, how many pieces can be cut to length and what is the length of the scrap piece?

$$5 \text{ m} = 500 \text{ cm} = 5000 \text{ mm}$$

$$1 \text{ length} = 60 \times 10 = 600 \text{ mm} + 2 \text{ mm} = 602 \text{ mm long}$$

$$5000 \div 602 = 8 \text{ pieces and a scrap piece } 184 \text{ mm long}$$

Study Problems.

1. A rectangular fenced lot is 52 ft, 6 in. by 28 yd. How many feet of fencing were used?

2. How many inches are there in $3\frac{1}{2}$ yd?

3. Change 5 yd, $2\frac{1}{2}$ ft to inches.

4. A ball of chalk cord contains 500 ft of cord. How many yards is this equivalent to?

5. How many pieces $4\frac{1}{2}$ in. long can be cut from a piece of molding that is 6 ft, 8 in. long? Allow $\frac{1}{8}$ in. waste per piece for cutting.

6. Reduce 15,675 cm to kilometers.

7. Change 6,115 m to millimeters.

8. Reduce 532 cm to meters.

9. Change 452 km to meters.

10. A timber is 5.5 m long. How many 150-cm pieces can be cut from it, allowing 3-mm per cut?

Conversion factors to facilitate conversion of metric and English linear units are as follows:

1 inch = 2.54 cm	1 cm = 0.3937+ in.
1 foot = 30.48 cm	1 meter = 39.37+ in.
1 yard = 0.9144 m	1 meter = 1.0936+ yd
1 mile = 1.6+ km	1 kilometer = 0.621+ miles

To facilitate conversion of units of measurement, a table of conversion factors appears in the appendix of this book, pages 484 to 485.

Study Example.

A part is dimensioned as 10 inches long. Find its length in millimeters.

Referring to the table on page 484 under the section headed "LENGTH," find "inch" in the first column, and opposite it "millimetre," in the third column. In-between is the multiplying factor 25.4, thus:

Multiply	By	To Obtain
inch	25.4	millimetre

(Note that the European "re" spelling is used for metric units in this table.)

Hence 10 in. X 25.4 = 254 mm.

Study Problems.

1. How many yards are there in 1 mile?

2. Change 4 yd, 2 ft, 3 in. to centimeters.

3. Reduce 456 in. to meters.

4. Reduce 52.5 ft to centimeters.

5. How many pieces, each measuring 8.3 cm can be cut from a bar 245.6 ft in length? Allow 3 mm per cut.

Nautical Measure

Nautical measure is used on the sea and in the air. The following table gives the common nautical (linear) dimensions:

Table 5-7. Nautical Measure

1 nautical mile = 1.15+ land (statute) miles 1 nautical mile = 6076.11549 feet. One minute of latitude at the equator. 3 nautical miles = 1 league 1 fathom = 6 feet 1 knot = 1 nautical mile per hour = 1.15+ mph
For conversion factors see pages 484-485.

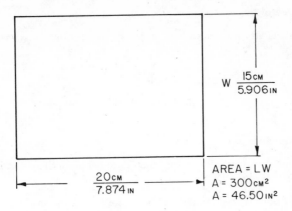

Fig. 5-1. Square measure.

Square Measure

The unit of surface or area measurement is a "square" of the unit used in the linear measurement along two sides, Fig. 5-1. Areas with dimensions in miles are expressed as "square miles," areas with dimensions in feet are expressed in "square feet." Confusion can exist with the terms "two square miles" and "two miles square." A "two-mile square" measures two miles on a side but "two square miles" measure 1.4142 miles on a side ($1.4142 = \sqrt{2}$). In the common English system we have the following square measure:

Table 5-8. English Square (Area) Measurement

1 square foot (sq ft) = 144 square inches (sq in.)
1 square yard (sq yd) = 9 square feet
1 square rod = $30\frac{1}{4}$ square yards
1 acre = 160 square rods
1 acre = 43560 square feet
1 acre = 4840 square yards
1 square mile = 1 section = 640 acres
1 township = 36 sections = 36 square miles
For conversion factors see pages 484-485.

Land measure in links and chains has been superseded by the use of steel tapes, accurately marked off in feet and tenths of feet.

Study Example.

An industrial site is 2 miles long and 0.75 miles wide. How many acres are in the site?

$$2 \times 0.75 = 1.5 \text{ sq miles}$$

$$1.5 \text{ sq miles} \times 640 = 960 \text{ acres}$$

Study Problems.

1. How many square inches are there in a steel plate that is 54 in. wide and 13 ft 8 in. long?

2. The interior measurements of a factory building are 155 ft 8 in. by 368 ft 5 in. How many square yards of flooring are required?

3. A square field containing 8 acres is to be fenced to store material. How many feet of fencing are required?

4. Floor tiles measure 8 in. square. How many will be required to tile an office floor containing 1600 sq yd?

5. A Texas ranch contains 7.5 sections. How many acres does it have?

In the metric system (SI) the square millimeter and the square meter are used. Land areas are measured as ares and hectares, the former being 10 meters square and the latter equal to 100 ares, approximately $2\frac{1}{2}$ acres.

Table 5-9. Metric Square Area Measure

100 square millimeters (mm^2) = 1 square centimeter
100 square centimeters (cm^2) = 1 square decimeter
100 square decimeters (dm^2) = 1 square meter (m^2)
1,000,000 square meters (m^2) = 1 square kilometer (km^2)
100 square meters = 100 centares = 1 are (a)
100 ares = 1 hectare (ha)
100 hectares (ha) = 1 square kilometer (km^2)
For conversion factors see pages 484-485

The conversion of English area measures to metric areas or the conversion of metric areas to English is most easily done by the use of the following factors:

1 square inch = 6.4516 square centimeters
1 square foot = 0.0929 square meters
1 square yard = 0.83613 square meters
1 acre = 40.4686 ares
1 cm^2 = 0.155 square inch
1 m^2 = 10.7639 square feet
1 m^2 = 1.19599 square yards
1 are = 119.599 square yards

Study Example.

Reduce 7,560 m^2 to ares:

$$100 \text{ sq m} = 1 \text{ are}$$

$$7,560 \text{ m}^2 = 75.60 \text{ ares}$$

Study Problems.

1. A sheet of galvanized steel is 150 millimeters square. How many square inches does it contain?

2. How many square meters are contained in 1 acre?

3. A railway right-of-way is 150 ft wide. How many acres of land are there in 56 miles of right-of-way? How many hectares?

4. A rectangular sheet of steel is 150 mm wide and 200 mm long. How many square inches of steel are in the sheet?

5. A sailboat has a sail with 15 m^2 of sailcloth. How many square feet is this?

6. A room is 25 ft by 36 ft with a 8-ft ceiling height. How many square feet of wall space must be painted?

7. If a gallon of paint will cover 425 sq ft, how many gallons will be required to paint the walls in problem 6?

Volume Measure

Volume measure is expressed in cubic units; i.e., cubic inches, cubic centimeters, cubic feet, etc. In Fig. 5-2, the rectangular solid is double dimensioned; that is, both inches and centimeters are shown. Its volume, in cubic units, is the product of its length, width, and height. The unit used to describe volumes is a compound of the word "cubic" and the linear measurement unit.

Measurements of objects that have corner angles of 90 degrees (right-angled or square corners) will give the three linear measurements

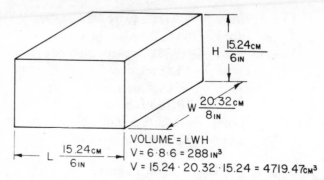

VOLUME = L W H
V = 6·8·6 = 288 ɪɴ³
V = 15.24 · 20.32 · 15.24 = 4719.47ᴄᴍ³

Fig. 5-2. Cubic measure.

necessary to calculate the cubic measure. Bodies such as cones, cylinders, pyramids, spheres, etc., require special formulas to calculate volume. These calculations will be explained in Chapter 14.

The calculation of the cubic displacement of an internal combustion automobile engine requires the measurement of the bore of the cylinders to calculate piston area, the length of stroke of the pistons, the volume of the irregular upper combustion space in the cylinder head, and the number of cylinders in the engine.

The units of cubic measurement in the English system are the same basic units of the linear measure with the addition of the word "cubic." In the English system we have the following relationships:

Table 5-10. English Cubic Measure

1 cubic foot (cu ft) = 1728 cubic inches (cu in.)
1 cubic yard (cu yd) = 27 cubic feet (cu ft)
1 cubic yard (cu yd) = 46,656 cubic inches (cu in.)
1 cord (wood) = 128 cubic feet
For conversion factors, see pages 484-485

In the metric SI system, the cubic units are derived from the basic units by prefixing the word "cubic" as in the English system. The term cubic centimeter is usually written in SI as cm^3 and cubic meter as m^3.

Table 5-11. Metric Cubic Measure

1000 cubic millimeters = 1 cu cm or 1 cm^3
1000 cubic centimeters = 1 cu dm or 1 dm^3
1000 cubic decimeters = 1 cu m or 1 m^3
For conversion factors see pages 484-485

The conversion of English cubic measurements to metric equivalents and metric units to English units may be done by using the following common equivalents:

$$1 \text{ cubic inch} = 16.387 \text{ cubic centimeters}$$
$$1 \text{ cubic foot} = 0.02832 \text{ cubic meter}$$
$$1 \text{ cubic centimeter} = 0.06102 \text{ cubic inch}$$
$$1 \text{ cubic meter} = 61,024 \text{ cubic inches}$$

Study Example.

A stock crib is 20 ft long, 30 ft wide and 12 ft high. How many cubic feet does it contain? How many cubic meters would this be?

$$20 \times 30 \times 12 = 7200 \text{ cu ft}$$

$$7200 \text{ cu ft} \times 0.02832 \frac{\text{m}^3}{\text{cu ft}} = 203.904 \text{ m}^3$$

Study Problems.

1. A box is 3 ft long, 4 ft wide and 2 ft high. What is its volume in cubic inches? In cubic centimeters?

2. A rainfall on a certain day is $3\frac{1}{2}$ in. How many cubic inches will fall on a field of $2\frac{1}{2}$ ha?

3. A storage tank is 8 ft by 20 ft by 6 ft. How many cubic feet are there in the tank? How many cubic meters?

4. The foundation of a building requires 346 cu yd of concrete. This was hauled in concrete buggies holding 9 cu ft each. How many loads were required to pour the foundations?

5. An oil drum holds 1.57 cu ft of oil. How many cubic centimeters would this be?

6. The oil reservoir of a machine that is operated by fluid power is 30 in. long, 14 in. wide, and $2\frac{1}{2}$ ft deep. How many gallons of oil are required to: (a) fill the reservoir, and (b) fill the reservoir to $\frac{7}{8}$ of its depth?

Review Problems.

1. A car is driven 25 mph for $4\frac{1}{2}$ hours. How many miles did it travel? How many kilometers did it travel?

2. A security fence is 12 ft 6 in. high. What is its height in meters?

3. A piece of land is rectangular and is 20 rods long by 15 rods wide. How many acres are in the field and how many meters of fencing will be required to enclose it?

4. A distance runner ran 1,500 m in 5 minutes. This is equivalent to running a mile in how many minutes?

5. A piece of galvanized steel sheet is 150 centimeters square. How many square inches does it contain?

6. A room is 12 ft × 16 ft × 8 ft high. How many cubic feet does it contain? How many square meters of wall does it contain? How many yards of carpet will be required for the floor (wall to wall)?

7. What is the difference in area between 30 inches square and 100 square centimeters? Give answer in square millimeters.

8. A basement for a house is to be 60 ft × 35 ft × 8 ft deep. After 525 cu yd have been removed by the excavator, how many cubic yards remain to be dug?

9. A mountain is 11,367 ft high. What is its height in meters?

10. A rectangular water storage pond is 345 ft long by 675 ft wide. After a heavy rain the water level rose 4 in. How many cubic feet of water entered the pond?

11. A rectangular grain storage bin is 150 ft long, 45 ft wide and 22 ft high. How many cubic yards storage is there? How many cubic meters?

12. A French farm contains 75 hectares. How many acres are there in this farm?

13. A mountain peak is 1,358 m high. How many feet would this be?

14. A ball of chalk line contains 385 ft of line. How many meters is this equivalent to? How many inches?

15. A piece of drill rod is cut into three pieces that measure 3 ft 8 in., 2 ft 11 in., and 4 ft 5 in. The stock length of the rod was 20 ft. What was the length of the remaining piece of rod?

16. Convert the commercial dimension of a 2 × 4 board that is 8 ft long to metric measure.

17. Convert the following metric dimensions to English measure:

$$38 \text{ mm} = \qquad 5 \text{ m} = \qquad 59 \text{ mm} = \qquad 134 \text{ cm} =$$

18. Convert the following English dimensions to metric measure:

$$2\frac{7}{8} \text{ in.} = \qquad 38.258 \text{ in.} = \qquad 2 \text{ ft } 3 \text{ in.} =$$

19. An excavation is 35 ft long, 27 ft wide, and 9 ft deep. How many cubic meters of dirt have been removed?

20. A cubic inch of aluminum alloy is equivalent to how many cubic millimeters?

21. The weight of a copper sheet of B&S gage 16 is 2.355 lb per square foot. What will be the weight of a sheet of copper of the same thickness that measures 2350 mm by 950 mm?

22. The road sign says that it is 45 km to Madrid. How many miles would this be? The sign also gives the speed limit at 55 (kilometers per hour). How fast is this speed in miles per hour?

Board Measure

Although linear units are used in measuring the length of pieces of lumber, the quantity of lumber is measured in *board feet*. A board foot is a piece of lumber that is 12 inches by 12 inches by 1 inch thick. Stock that is less than 1 inch thick is figured as 1 inch. Lumber that is greater than 1 inch is figured by its actual thickness. Hardwood that is between 1 inch and 2 inches is figured to the nearest one-quarter inch of thickness.

To calculate board feet (feet board measure are abbreviated, "fbm") multiply the surface square feet of the board by its thickness in inches using the thickness criteria mentioned in the above paragraph. Hence:

1. board feet $= $ T (inches) $\times$ W (feet) $\times$ L (feet)

2. board feet $= \dfrac{\text{T (inches)} \times \text{W (inches)} \times \text{L (feet)}}{12}$

3. board feet $= \dfrac{\text{T (inches)} \times \text{W (inches)} \times \text{L (inches)}}{144}$

Lumber is priced according to the thousand (M) board feet. The symbol M stands for 1000 and comes from the Latin numerals. Thus: long-leaf yellow pine, 4 by 12 inches in the rough, in lengths up to 20 feet sold for $154.00 per 1000 feet, board measure (fbm). The above quotation means that this lumber sold for $154.00 per thousand board feet. Special processed material such as plywood, hardboard, and other sheet material is sold by the square foot.

Study Examples.

1. A 2 $\times$ 6 that is 12 ft long contains how many board feet?

Formula 1.   board feet $=$ T (inches) $\times$ W (feet) $\times$ L (feet)

board feet $= 2 \times \dfrac{6}{12} \times 12 = 12$ board feet

2. 100 hardwood boards, $1\frac{1}{4}$ in. thick by 8 in. wide and 8 ft long, equal how many board feet?

Formula 2.   board feet $= \dfrac{T \text{ (inches)} \times W \text{ (inches)} \times L \text{ (feet)}}{12}$

$$\text{board feet} = \frac{1.25 \times 8 \times 8}{12} \times 100$$

$$\text{board feet} = \frac{80}{12} \times 100 = 666\frac{2}{3} \text{ board feet}$$

Study Problems.

1. How many board feet are there in each linear foot of a piece of lumber 2 in. thick by 4 in. wide, 6 in. wide, 8 in. wide, and 12 in. wide?

2. No. 1 fir sells for $153.00 per 1000 board feet. How much will 36 pieces of 2 × 12 by 18 ft long cost?

3. A contractor buys 500 Douglas fir planks, 2 × 12 by 20 ft for scaffolding. Find the number of board feet of lumber in one plank and the total number of board feet in the total order.

4. How many board feet are there in 10,000 railroad ties that are 8 by 8 by 8 ft long?

5. Wooden posts were purchased for a fence. The dimensions were 4 by 4 by 12 ft long. How many board feet are in one post, how many in 358 posts?

Angle Measurement

An angle is the inclination of one line with respect to another, or one surface with respect to another. Angular measurement concerns itself with the measurement of the amount of this inclination.

The basic unit of angular measurement is the degree. Quite arbitrarily, a straight line is defined as having 180 degrees, as shown in view A of Fig. 5-3. The angle made when an object, such as the spoke of a wheel, rotates one complete revolution is by definition 360 degrees (see view B). One degree, then, is $\frac{1}{360}$ part of a complete revolution, as shown in view C.

For more precise work, the degree is divided into 60 minutes and the minute, in turn, is divided into 60 seconds. In other words there are 60 minutes in one degree and 60 seconds in one minute. Minutes and seconds here refer to angular measurement, not to time. The abbreviations commonly used are:

$$\text{degrees} = °; \text{such as } 50°$$
$$\text{minutes} = '; \text{such as } 21'$$
$$\text{seconds} = ''; \text{such as } 22''$$

A typical angle might be specified as; $72°\ 18'$ or $14°56'10''$. In summary, the relationship between degrees, minutes, and seconds is as follows:

$$60 \text{ minutes} = 1 \text{ degree}$$
$$60 \text{ seconds} = 1 \text{ minute}$$
$$3600 \text{ seconds} = 1 \text{ degree}$$

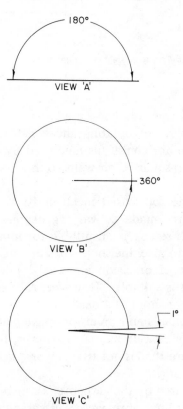

Fig. 5-3. Angular measurement.

When two surfaces or two lines are square with respect to each other, the angle between them is 90 degrees, as shown in Fig. 5-4. Surfaces or lines that are square to each other are said to be perpendicular to each other. The perpendicularity of one surface with respect to another is an important specification which must be measured accurately in many shop applications. Likewise, the perpendicularity of the axis of a hole with respect to the axis of another hole or with respect to a surface is often very important.

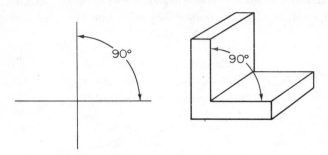

Fig. 5-4. The 90-degree or right angle.

Perpendicularity and squareness are two terms that are used somewhat interchangeably. In building and construction work a plumb line is square to the horizontal (as measured by a level) and builders use the plumb line to square their walls so that they are perpendicular to the floor.

One of the basic tools used in all crafts is a square. Different crafts require squares made to varying degrees of precision. The common carpenter's square is unsuitable for most of the work done in a machine shop. In machine shop work and in toolmaking, squares of different degrees of precision are used. For ordinary work the combination square is suitable. However, for very precise work the hardened steel square should be used. For some classes of toolroom work, cylindrical squares and precision angle plates are used as references for right angle measurements. However, for any class of work in any craft, a square that is not true is worse than useless because it will cause spoiled work.

Angles are measured by the carpenter's square and most frequently by protractors. Ordinary protractors are graduated in degrees. For more precise work in drafting and in the machine shop, vernier protractors, graduated to read in increments of five minutes, are used. In the machine shop other precision instruments are also used to measure angles such as sine bars, dividing heads, angle gage blocks, plug gages, ring gages, and clinometers.

In electrical work and electronic problems, angular measurements are given in decimal degrees rather than in degrees, minutes, and seconds. Decimal degrees may be easily converted into the minute/second system as shown below:

Study Examples.

1. Given the angle measure as $48.793°$, convert to degrees, minutes, and seconds.

   Step 1. Multiply the decimal (0.793) by 60 to convert to decimal minutes:

   $$0.793° \times 60 = 47.580' \, (47' + 0.580')$$

   Step 2. Multiply the minute decimal (0.580) by 60 to convert to decimal seconds:

   $$0.580' \times 60 = 34.8''$$

   Step 3. Round off the $34.8''$ to $35''$

   Step 4. Add:
   $$
   \begin{array}{r}
   48° \\
   +\quad 47' \\
   +\qquad 35'' \\
   \hline
   48° \, 47' \, 35''
   \end{array}
   $$

2. Given the angle in the minute/second system as $33°27'36''$, convert to the decimal angle measurement system.

   Step 1. $36''$ are $\dfrac{36}{60}$ of a minute; therefore convert the fraction to a decimal:

   $$\frac{36}{60} = 0.60'$$

   Step 2. Add $27'$ and $0.60'$ giving: $27.6'$

   Step 3. $27.6'$ are $\dfrac{27.6}{60}$ of a degree; therefore convert the fraction to a decimal:

   $$\frac{27.6}{60} = 0.46°$$

   Step 4. Add:
   $$
   \begin{array}{r}
   33° \\
   +\; 0.46° \\
   \hline
   33.46°
   \end{array}
   $$

3. Addition of the measurements of angles in the decimal system should pose little difficulty as the addition is done in the manner of any decimal numbers. However, the addition of angles in the degrees, minutes, and

seconds system follows a special method shown below. To add $48° 47' 35''$ and $33° 27' 36''$:

Step 1. Arrange the angles, one under the other in columns of degrees, minutes, and seconds:

$$48° \ 47' \ 35''$$
$$\underline{33° \ 27' \ 36''}$$

Step 2. Add each column:

$$48° \ 47' \ 35''$$
$$\underline{33° \ 27' \ 36''}$$
$$81° \ 74' \ 71''$$

Step 3. Convert $74'$ into $1° 14'$

Step 4. Convert $71''$ into $1' 11''$.

Step 5. Add degrees, minutes, and seconds:

$$81°$$
$$1° \ 14'$$
$$\underline{1' \ 11''}$$
$$82° \ 15' \ 11''$$

4. In subtraction of angles measured in the degree, minutes, and seconds system, it is necessary to "borrow" from the next larger division if the minutes and/or seconds in the subtrahend are more than the minutes and/or seconds in the minuend. For example; subtract $33° \ 27' \ 36''$ from $48° \ 19' \ 25''$:

Step 1. Arrange problem in normal subtraction form:

$$48° \ 19' \ 25''$$
$$\underline{-33° \ 27' \ 36''}$$

Step 2. Borrow $1'$ ($60''$) from $19'$ and add to the $25''$:

$$48° \ \ 19' \ \ 25''$$
$$\underline{- \ 1' \ +60''}$$
$$48° \ \ 18' \ \ 85''$$

Step 3. Borrow $1°$ ($60'$) from $48°$ and add to the $18'$:

$$48° \ \ 18' \ 85''$$
$$\underline{- \ 1° \ +60'}$$
$$47° \ \ 78' \ 85''$$

Step 4. Rewrite the problem using the minuend as revised in step 3:

$$
\begin{array}{r}
47^\circ\ 78'\ 85'' \\
-33^\circ\ 27'\ 36'' \\
\hline
14^\circ\ 51'\ 49''
\end{array}
$$

Study Problems.

1. Convert 48.8573° to degrees, minutes, and seconds.

2. Convert $35^\circ\ 25'\ 32''$ to a decimal degree.

3. Add: $28^\circ\ 45'\ 27''$ and $18^\circ\ 25'\ 63''$.

4. Subtract: $95^\circ\ 46'\ 53''$ from $145^\circ\ 23'\ 42''$.

5. Convert: 72.5834° to degrees, minutes, and seconds.

6. Convert: $65^\circ\ 22'\ 47''$ to decimal degrees.

7. Add: $38^\circ\ 22'$ and $45'\ 38''$ and $25^\circ\ 56'\ 33''$.

8. Subtract: $25^\circ\ 58'\ 49''$ from $29^\circ\ 42'\ 38''$.

9. Add: $45^\circ\ 0'\ 22''$ and $59'\ 58''$.

10. Add: $39.5745^\circ + 3^\circ\ 39'\ 22'' + 59.75^\circ$.

Weight Measure

There are many sub-systems of weights in the English system, depending on the material being weighed. The most common weight sub-system is "avoirdupois" using *ounces*, *pounds*, and *tons* as units. This is most widely used in trade and commerce. Jewelers use troy weights for gems and precious metals. These units are *grains*, *carats*, *pennyweights*, *ounces*, and *pounds*. Drugs and medicines are weighed under the "apothecaries'" weight system utilizing *grains*, *scruples*, *drams*, *ounces*, and *pounds*. The ounces and pounds in Troy and apothecaries' systems are the same, but they are not the same weights as the ounces and pounds in the avoirdupois system.

Weight is the result of the force of gravity of the earth on a body. This attractive force "pulls" all objects toward the center of the earth. Science has found that this force varies slightly from sea-level to mountain-tops and in certain land areas because of a bulge at the earth's equator. However, the differences are extremely small and weight systems do not take into account the variance. Weight is measured by balance type scales, which date back many thousands of years, spring balances (fish scales), and the lever-type (no springs) scales. The English avoirdupois weight system is given in Table 5-12.

Table 5-12. English Weight Measure
Advoirdupois weight
16 ounces (oz) = 1 pound (lb) 100 pounds = 1 hundredweight, U.S. (cwt) 112 pounds = 1 British hundredweight 20 U.S. hundredweight = 1 ton (2000 lb) (short ton) 20 British hundredweight = 1 long ton (2240 lb)

The English weight system has two different tons. The U.S.A. "short ton" of 2000 pounds is used in the U.S.A. domestic market whereas the "long" ton of 2240 pounds is used worldwide wherever the English system is used. The British ton is related to an ancient British system of small weights based on the *stone* weighing 14 pounds, 2 stones equaling a *quarter* weighing 28 pounds, four quarters equaling a *hundredweight* weighing 112 pounds. The British ton equals 20 British hundredweight or 2240 pounds. In the U.S.A. a hundredweight weighs 100 pounds.

In the metric SI there are no units of weight, as such. The unit of mass is the kilogram defined as the mass of the international prototype kept in France. In the original metric system, a gram was described as the weight of 1 cubic centimeter of water at 4 degrees Centigrade. For practical shop and general industrial use, the kilogram is taken as the unit of weight. A liter (1000 cc) of water at 4 degrees Centigrade will weigh one kilogram. The following table shows the metric "weight" measures as commonly used in the world.

Table 5-13. Metric Weight Measure
1000 milligrams (mg) = 1 gram (cg) 1000 grams (g) = 1 kilogram (kg) 1000 kilograms (kg) = 1 metric ton (2204.6 lbs)
For conversion factors see pages 484-485

The conversion of English weights to metric weights and from metric to English can be most easily done by the use of the following factors:

$$1 \text{ gram} = 0.03527 \text{ ounce (avoir.)}$$
$$1 \text{ ounce (avoir.)} = 28.35 \text{ grams}$$
$$1 \text{ kilogram} = 2.2046 \text{ pounds} = 35.274 \text{ ounces (avoir.)}$$
$$1 \text{ pound} = 0.4536 \text{ kilogram} = 453.6 \text{ grams}$$
$$1 \text{ metric ton} = 0.9842 \text{ long ton}$$
$$1 \text{ ton (2240 pounds)} = 1.016 \text{ metric tons} = 1.016 \text{ kilograms}$$

Study Problems.

1. Reduce: 90,857 g to kilograms.

2. Reduce: 478 kg to pounds.

3. Reduce: 4586 lb to short tons (2000 lbs).

4. Change 2.3 kg to ounces (avoirdupois).

5. Reduce: 56.7 oz to pounds.

6. Reduce 14 U.S. tons to kilograms.

Liquid and Dry Measurement

Capacity is directly related to cubic measurement. It refers to a confined space or volume of that space. Special names are used in the English system and vary as to liquid and dry measure. Here the names go back many centuries and even the same word varies in meaning as it relates to special trades or professions.

Liquid measure is used in the sale and handling of liquids such as milk, gasoline, water, and beverages. Dry measure is used to measure out grains and vegetables. A third type of capacity measure which is used by druggists and physicians is known as "apothecary" fluid measure.

The English system is illogical in many ways when capacity measure is concerned. While each unit can be defined in terms of other capacity units and in cubic measurement, the units have no common denominator as do the units in the metric SI. Even the English units vary between usage in Great Britain and in the United States. The American gallon contains 231 cubic inches while the British "Imperial" gallon contains 277.42 cubic inches.

In the English system, the capacity units and their relationships are as follows:

Table 5-14. English Capacity Measure

Liquid Measure

2 pints (pts) = 1 quart (qt) (57.75 cu in.)
4 quarts (qt) = 1 gallon (gal) U.S. (231 cu in.)
$31\frac{1}{2}$ gallons = 1 barrel (bbl); 2 bbl = 1 hogshead

$7\frac{1}{2}$ gallons = 1 cubic foot (water at 60°F temperature)
1 gallon water = $8\frac{1}{3}$ pounds at 60°F temperature
1 cubic foot water = 62.4 pounds at 60°F

Dry Measure

2 pints (pt) = 1 quart (qt)
8 quarts (qts) = 1 peck (pk)
4 pecks (pks) = 1 bushel (bu)
1 bushel (bu) = 2150.42 cu in.

For conversion factors see pages 484-485

In SI the units of capacity are derived directly from the cubic measure. The standard metric volume measure is the cubic meter. Another metric capacity unit is the liter. This closely approximates one quart in the English system, since one liter equals 1.057 quarts. The following table of metric capacity measure shows the use of the standard SI prefixes to the base, the meter:

Table 5-15. Metric SI Capacity Measure

1000 cubic millimeters (mm^3) = 1 cubic centimeter (cm^3)
1000 cubic centimeters (cm^3) = 1 cubic decimeter (dm^3)
1000 cubic centimeters (cm^3) = 1 liter (l)
1 cubic decimeter (dm^3) = 1 liter (l)
1000 cubic decimeters (dm^3) = 1 cubic meter (m^3)
1000 liters = 1 cubic meter (m^3)

For conversion factors see pages 484–485

Conversion from English liquid measure to metric measure and from metric liquid measure to the English system is readily accomplished by the use of the following common factors:

1 U.S. quart = 0.946 liter   1 liter = 1.0567 U.S. quarts
1 U.S. gallon = 3.785 liters   1 liter = 0.2642 U.S. gallon

Study Problems. Convert the following:

1. 6 liters to quarts.

2. 45 gallons to liters.

3. 35 gallons to kiloliters.

4. 500 deciliters to pints.

5. 2 barrels to liters.

6. 14,065 milliliters to pints.

Pressure

Liquids, gases, and vapors are all considered to be fluids and they are capable of exerting a pressure. When subjected to a pressure, liquids are considered to be incompressible although a very small amount of compression does take place. On the other hand, gases and vapors are compressed under pressure, and they expand when the pressure is lowered.

Pressure may be defined as the force exerted by the fluid on a unit area. The units in which pressure is expressed and their standard abbreviations are listed below:

pounds per square inch: psi; lb/in^2; lb per sq in.

pounds per square foot: psf; lb/ft^2; lb per sq ft

pascals: Pa. (SI - metric)

The formula for pressure is given below in two forms:

$$P = \frac{F}{A}$$

(5-1)

$$F = PA$$

(5-2)

where: P = Pressure

F = Force

A = Area

When force is in pounds and the area is in square inches, pressure is in "lbs per sq in." Likewise, when force is in pounds and area is in square feet, pressure is in "lbs per sq ft"; and when force is in newtons and the area in square meters, pressure is in pascals, or kg/m^2.

Study Examples.

1. A 4-in. oil-hydraulic cylinder is to be used to lift the tailgate of a truck. It must be able to lift 5000 lb, including the weight of the tailgate. What must the hydraulic pressure of the oil in the cylinder be to lift this weight?

It can be assumed that the diameter of the piston in the cylinder will also be 4 in. The first step in solving this problem is to calculate the cross-sectional area of the cylinder and the piston:

$$\text{Area} = \frac{\pi D^2}{4} = \frac{3.1416(4^2)}{4} = 12.5664 \text{ sq in.}$$

Then applying formula (5-1):

$$P = \frac{F}{A} = \frac{5,000}{12.5664} = 398 \text{ psi}$$

2. The maximum air pressure that can be supplied to a 2-in. diameter pneumatic cylinder is 90 psi. What is the maximum force that can be exerted by the piston rod of this cylinder?

$$\text{Area} = \frac{\pi D^2}{4} = \frac{3.1416(2^2)}{4} = 3.1416 \text{ sq in.}$$

Then applying formula (5-2):

$$F = PA = 90 \times 3.1416 = 283 \text{ lb}$$

The atmosphere exerts a pressure on the surface of the earth. The standard atmospheric pressure is 14.696 pounds per square inch or 101.33 kilopascals (kPa), taken at sea level with the temperature of the air being 32 degrees Fahrenheit, (0 degrees Celsius). For practical purposes, this is often rounded off to 14.7 pounds per square inch and 101.00 kilopascals. The standard pressure is also expressed as 29.9213 inches of mercury, which is equal to 14.696 pounds per square inch. This figure is usually rounded off to 29.92 inches of mercury.

Atmospheric conditions usually cause a slight variation in the actual pressure at sea level. Likewise, the pressure will decrease with distance above sea level. The effect of the actual atmospheric pressure, also called the barometric pressure, is usually ignored in measuring the pressure of hydraulic machinery. However, it cannot be ignored in power plant work and in dealing with the low pressures generated by fans and blowers.

Most ordinary pressure gages measure the difference between the actual pressure in the system being measured and the actual pressure of the atmosphere. This is shown in Fig. 5-5. The pressure measured by the gages is called the "gage pressure," which is measured in terms of pounds per inch gage or "psig." The total pressure that actually exists in the system is called the absolute pressure, which is defined in terms of pounds per square inch absolute or "psia." From Fig. 5-5 the relationship between the absolute and the gage pressure can be seen as follows:

Absolute Pressure = Gage Pressure + Actual Atmospheric Pressure

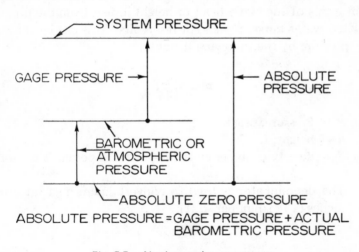

Fig. 5-5. Absolute and gage pressures.

When very low pressures, encountered in dealing with fans and blowers, are to be measured, the actual atmospheric pressure must be measured and added to the gage pressure to obtain the absolute pressure. Where higher pressures are encountered, such as in steam boilers, sufficient accuracy is obtained if the atmospheric pressure is assumed to be 14.7 pounds per square inch. In this case:

$$\text{Absolute Pressure} = \text{Gage Pressure} + 14.7$$

Study Example.

A pressure gage attached to a steam boiler reads 305 psig. What is the absolute pressure of the steam in the boiler?

$$
\begin{aligned}
\text{Absolute Pressure} &= \text{Gage Pressure} + 14.7 \\
&= 305 + 14.7 \\
&= 319.7 \text{ psia}
\end{aligned}
$$

The pressure in boilers is also sometimes specified in terms of atmospheres, where one atmosphere is equal to 14.696 pounds per square inch. The conversion of pressure from atmospheres to pounds per square inch will be treated further on in this section.

In municipal and industrial water systems pressure is usually defined in terms of the static head or height in feet from the use point to elevated water tanks, towers, or reservoirs. This can be converted to gage pressure by the following formula:

$$P = \frac{hd}{144} \tag{5-3}$$

where: P = Pressure, psig
h = height, ft
d = density or lb per cu ft (d = 62.4 lb per cu ft for water).

The denominator is used to convert square feet into square inches (144 sq in. per sq ft).

Study Examples.

1. The water level in a water tower of a small community is maintained at 120 ft above the ground. What is the water pressure at the ground level resulting from this head?

$$P = \frac{hd}{144} = \frac{120 \times 62.4}{144} = 52 \text{ psig} \qquad \text{applying (5-3)}$$

2. A factory requires a minimum water pressure of 100 psig for processing its parts. What head is required to develop this pressure?

$$P = \frac{hd}{144}$$

applying (5-3)

$$100 = \frac{h \times 62.4}{144}$$

$$h = \frac{100 \times 144}{62.4}$$

= 230.77 ft, which for practical purposes is 231 ft

Other measures are also frequently used to measure pressure. Extremely low pressures, such as developed by blowers and fans, are often measured in inches of water (in. H_2O) or centimeters of water (cm H_2O). Low pressures are also frequently measured in inches of mercury (in. Hg) or in centimeters of mercury (cm Hg).

It is necessary to be able to convert from one measure of pressure to another. For this reason the following table of pressure conversion factors is given.

Table 5-16. Pressure Conversion Factors

Multiply	by	to Obtain
Atmospheres	14.7	pounds per square inch
Pounds per square inch	0.0680	atmospheres
Pounds per square inch	27.7	inches of water (H_2O)
Pounds per square inch	2.31	feet of water (H_2O)
Pounds per square inch	70.4	centimeters of water (H_2O)
Pounds per square inch	2.04	inches of mercury (Hg)
Pounds per square inch	5.18	centimeters of mercury (Hg)
Pounds per square inch	6895.	newtons per square meter
Pounds per square inch	6895.	pascals
Inches of water	0.0361	pounds per square inch
Inches of water	249.	pascals
Feet of water	0.433	pounds per square inch
Feet of water	2987.	pascals
Centimeters of water	0.0142	pounds per square inch
Centimeters of water	98.0	pascals
Inches of mercury	0.491	pounds per square inch
Inches of mercury	3377.	pascals
Millimeters of mercury	0.0193	pounds per square inch
Millimeters of mercury	113.	pascals
Pascals	0.000145	pounds per square inch
Pascals	0.00402	inches of water
Pascals	0.000335	feet of water
Pascals	0.0102	centimeters of water
Pascals	0.000296	inches of mercury
Pascals	0.000752	centimeters of mercury

Study Examples.

1. Calculate the head of water at sea level at standard conditions; i.e., at 14.696 psia and $32°$F.

$$\text{ft H}_2\text{O} = \text{lb per sq in.} \times 2.31$$
$$= 14.696 \times 2.31$$
$$= 33.9 \text{ ft}$$

2. A modern super-critical-pressure steam-generating plant operates at a pressure of 340 atmospheres. What is this pressure in terms of lb per sq in. gage?

$$\text{lb per sq in. (abs)} = \text{atmospheres} \times 14.7$$
$$= 340 \times 14.7$$
$$= 4998$$
$$\text{absolute pressure} = \text{gage pressure} + 14.7$$
$$\text{gage pressure} = 4998 - 14.7$$
$$\text{gage pressure} = 4983.3 \text{ psig}$$

NOTE: In this case, because of the very high pressure, slightly more precise results will be obtained if 14.696 is used instead of 14.7, in which case the answer will be 4981.94 psig.

3. Blower discharge pressures are usually measured in inches of water because small differences in pressure can be readily detected. Because of the low pressures involved, the exact atmospheric pressure is usually measured to determine the actual blower discharge pressure. What is the discharge pressure in terms of lb per sq in. abs if the measured discharge pressure of a blower is 55.56 in. H_2O and the barometric (atmospheric) pressure is 29.48 in. Hg?

a. Find the blower discharge pressure in psig.

$$\text{lb per sq in. gage} = \text{in. H}_2\text{O} \times 0.036$$
$$= 55.56 \times 0.036$$
$$= 2$$

b. Find the actual atmospheric pressure in psia.

$$\text{lb per sq in. abs} = \text{in. Hg} \times 0.0491$$
$$= 29.48 \times 0.491$$
$$= 14.47$$

c. Find the absolute pressure of the blower discharge.

$$\text{Absolute pressure} = \text{gage pressure} + \text{actual atmospheric pressure}$$
$$= 2 + 14.47$$
$$= 16.47 \text{ psia}$$

Vacuum

A vacuum is sometimes thought of as being an absence of pressure. Actually a vacuum is any pressure that is less than the prevailing atmospheric or barometric pressure.

Vacuums are used in many industrial applications. In Great Britain vacuum brakes are used on railways. In factories vacuum chucks are used to hold workpieces on machine tool tables to perform machining operations. Vacuum furnaces are used to melt metals. In power plants a vacuum is created in the condenser which appreciably improves the efficiency of the steam turbine, although at the additional cost of pumping the condensate (condensed steam) out of the condenser and pumping cooling water through the condenser.

A vacuum is always measured as the pressure below the actual barometric or atmospheric pressure, as shown in Fig. 5-6. In most cases the vacuum is measured in inches of mercury. Low vacuums, where the pressure is only slightly below the prevailing atmospheric pressure, are measured in terms of inches of water. The actual barometric pressure must always be determined in measuring a vacuum, and this pressure must be corrected to the elevation above sea level of the site of the measurement.

The procedure for calculating the absolute pressure in a vacuum is given below:

absolute pressure = actual corrected atmospheric pressure
  − vacuum gage reading

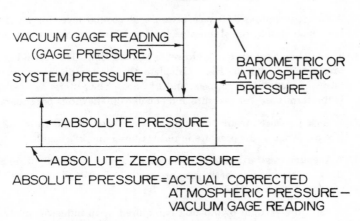

ABSOLUTE PRESSURE = ACTUAL CORRECTED
                    ATMOSPHERIC PRESSURE −
                    VACUUM GAGE READING

Fig. 5-6. Actual barometric and atmospheric vacuums.

Study Example.

A vacuum gage attached to a steam condenser reads 15.20 in. Hg. The corrected barometric pressure reads 29.95 in. Hg. What is the absolute pressure in the condenser in terms of in. Hg and lb per sq in. abs?

$$\text{absolute pressure} = \text{actual corrected atmospheric pressure}$$
$$- \text{ vacuum gage reading}$$
$$= 29.95 - 15.20$$
$$= 14.75 \text{ in. Hg.}$$

$$\text{lb per sq in. abs} = \text{in. Hg} \times 0.491$$
$$= 14.75 \times 0.491$$
$$= 7.24$$

Study Problems.

1. A 3-in. diameter hydraulic cylinder is to be used on a truck tailgate lifting mechanism. If the tailgate is to be required to lift 1400 lbs, including the weight of the tailgate, what should the pressure in the hydraulic system be?

2. A 1-in. diameter pneumatic cylinder is to be used on a tool where it is expected to exert a 50-lb clamping force. Can this cylinder do this job if the air pressure supplied to the cylinder is 90 psi?

3. An oil-hydraulic cylinder is to be attached to a drill press to feed the drill through the workpiece automatically. A force of 1200 lbs will have to be overcome to perform this task. The hydraulic pressure available will be 500 psi. Calculate the diameter of the cylinder required.

4. The pressure on a helium gas tank reads 240 psig. What is the absolute pressure of the helium?

5. A factory requires 150 psig pressure at ground level for its fire lines. How high should the stand-pipe be that is used to obtain this pressure?

6. The discharge pressure of a blower is 102 cm H_2O. The actual barometric pressure is 29.90 in. Hg. Calculate the blower discharge pressure in terms of psig, kPa gage, psia, and kPa abs. The SI term *kPa* (kilopascal) is the term used for pressure and equals 1000 newtons per square meter.

7. A high pressure steam power plant is designed to operate at 50 atmospheres. What is this pressure in terms of psig and kPa gage?

8. A vacuum gage attached to a vacuum furnace reads 28.70 in. Hg, and the corrected atmospheric pressure is 30.11 in. Hg. What is the absolute pressure in terms of psia?

9. A very low vacuum of 0.5 psig is required in an industrial process. The corrected standard atmospheric pressure at the site of this process is 29.90 in. Hg. What should the reading of the vacuum measuring gage be in terms of in. H_2O and in terms of in. Hg?

Temperature Measurement

The measurement of temperature is of great importance in many activities. The scientist must know precise temperature for his experiments and calculations. The power plant operator must know steam temperatures, condenser water temperatures, stack gas temperatures and ambient temperatures to run an efficient operation. The doctor and the nurse must know the patient's temperature for diagnosis and treatment of an illness. You often want to know the temperature inside your house and the outside temperature.

Unfortunately, there are a number of temperature measurement systems in use. The commonly used temperature system in most English speaking countries is the Fahrenheit system. This system places 32 degrees at the point of melting ice and 212 degrees at the boiling point of water. This gives 180 degrees between the two points.

One SI unit for temperature is the kelvin, the scale for which is based on absolute zero (this will be explained in the next section on absolute temperature.) Wide use is made of the Celsius (formerly called Centigrade) scale in engineering, trade, and other nonscientific areas. The relationship between the Celsius temperature t and the kelvin temperature T is expressed by the formula: $t = T - 273.15$. The abbreviation for Celsius is C.

Much confusion exists when people travel from countries using one temperature system to those using the other system. Celsius temperatures for air and water temperatures seem cold for an American traveling in Europe. On the other hand, a European taken sick in the United States will be shocked to learn that he has a fever of 101 degrees—one degree above the boiling point of water in his homeland!

Conversion from one system to the other is easily done by the use of the following conversion formula:

$$\text{Degrees Celsius} = \frac{5}{9}(\text{Degrees Fahrenheit} - 32°)$$

$$\text{Degrees Fahrenheit} = \frac{9}{5}(\text{Degrees Celsius}) + 32°$$

Study Examples.

1. Change $95°F$ to a Celsius reading.

$$C = \frac{5}{9}(F - 32°)$$

$$C = \frac{5}{9}(95° - 32°) = \frac{5}{9}(63°) = 35°$$

2. Change $20°C$ to a Fahrenheit reading.

$$F = \frac{9}{5}(C) + 32°$$

$$F = \frac{9}{5}(20°) + 32° = 36° + 32° = 68°$$

It is well to remember that all metric calculations involving temperatures require that Celsius (Centigrade) degrees be used. All English system calculations require Fahrenheit degrees.

Absolute Temperature

Scientists use a temperature scale based on "absolute" zero. This is based on the fact that as a gas is cooled below $0°$ C it contracts $\frac{1}{273}$ of its volume for every degree it is cooled. In theory, a gas cooled to $-273°$ C would lose all of its volume, all molecular motion would cease and the gas would be without heat. Temperature reckoned from this point, instead of from the zero on an ordinary Celsius (Centigrade) thermometer, is called "absolute temperature." Absolute temperature is measured in the SI on the Kelvin scale and the unit is the kelvin (K). In the English system absolute temperature is measured on the Rankine scale and the unit is the degree Rankine (R).

$$\text{Degrees kelvin (K)} = \text{degrees C} + 273.15$$

$$\text{Degrees Rankine (R)} = \text{degrees F} + 459.7$$

Absolute temperatures are very important to power plant engineers. Except where temperature differences are involved, all measurements and calculations must be made in terms of absolute temperature.

Study Problems.

1. How many liters of gasoline can be put into a service station storage tank in Madrid that measures 4 m in diameter by 7 m long?

2. Convert the following measures:

   5 in. to cm   8 lb to kg

3. How many liters are in a 752 cu in. tank?

4. Convert the speed of sound (1130 ft/sec) to m/sec.

5. A pressure of 50 psi equals how many kg/cm^2?

6. A mountain is 32,756 ft high. How many meters would this be?

7. A 500 cu ft tank contains how many cubic meters?

8. A V-8 engine has a 352 cu in. displacement (cid). What is its metric displacement?

9. A man weighs 198 lb. How many kilograms does he weigh?

10. The average density of the earth is 5.53 gm/cm^3. What is this density in pounds per cubic foot?

11. The acceleration due to gravity is 9.81 m per sec^2. What is this acceleration in the English system?

12. The speedometer in a European automobile reads 125 km per hour. What is this speed in miles per hour (mph)?

13. The density of a substance is 27.68 g per cm^3. What is its density in pounds per cubic inch?

14. The density of gasoline is 0.67 kg per liter. What is this density in pounds per gallon?

15. In 1 kilometer, there are how many inches, centimeters, yards, miles, feet?

16. How many meter sticks, laid end to end, would be needed to enclose a square hectare?

17. What is the normal clinical temperature of a healthy man in Paris, France? (In the United States it is 98.6° F.)

18. How many cubic inches will an eight-liter racing car engine displace?

19. If a flat field measuring 2.5 hectares is evenly covered by water to a depth of 2.5 cm, how many gallons of water are in each hectare?

20. Irrigation men refer to water use in terms of "acre-feet." This is the amount of water that will cover a flat field of 1 acre 1 ft deep. In problem 19, how many acre-feet cover the 2.5 ha field?

21. A machinist milled 0.025 in. from the side of a flat steel bar, 5 in. wide and 20 in. long. How many cubic centimeters of steel did he remove from one side of the bar?

22. The water in a reservoir weighs 1,000,000 tons. How many gallons of water are in the reservoir?

23. In problem 22, how many kiloliters are in the reservoir?

24. Change 50° F into Celsius degrees.

25. Convert 30° C into a Fahrenheit reading.

26. A ship travels at 25 knots. How fast is this in miles per hour; in kilometers per hour?

27. A ranch in Texas is said to contain 200 sq miles. How many acres would this be; how many sections?

28. How much does the water in a full 10,000 gal water tank weigh?

29. An Italian machine detail is 345 mm long and 25 mm in diameter. What are its English system dimensions?

30. The I.D. of a bearing is 1.9685 ±0.0005 in. What is the I.D. and tolerance in the metric system?

31. Convert the following metric dimensions to English units.

    352 mm   0.065 mm   4.58 mm   23.587 mm

32. Convert the following inch dimensions to millimeters.

    $3\frac{17}{32}$ $2\frac{11}{16}$ $5\frac{31}{32}$ $15\frac{3}{8}$ $4\frac{1}{32}$

33. A spherical gas holder has an inner diameter of 95 ft. Convert this dimension to meters.

34. An exterior wall of a factory is to be painted. The wall is 355 ft, 8 in. long and $18\frac{1}{2}$ ft high. The paint will cover 60 sq yd per gallon. How many gallons of paint will be required with 5% extra for loss and spillage?

35. In problem 34, how many liters of paint will be required?

36. If Douglas fir plywood sells at 32 cents per square foot, what will be the price of 23 pieces of regular 4 ft by 8 ft plywood panels?

37. A steel bar is 5.5 in. wide, 0.45 in. thick and 6 ft, 7 in. long. What are its metric dimensions?

38. A German automobile dealer quotes the average mileage at 7 km per liter. Convert this to miles per gallon.

39. An Italian machine tool weighing 500 kg is to be shipped into the United States. The shipping rate quoted was 23 cents per pound. What is the total shipping cost in dollars?

40. An American tourist in France purchases 16 liters of wine. As the duty is figured in gallons by the United States customs, how many gallons does he have to declare?

CHAPTER 6

Fundamentals of Algebra

Algebra is an extension of arithmetic. A knowledge of algebra is important to craftsmen in the industrial workshop, in the building and construction trades, and on the drawing board because the mathematical formulas that are used from time to time are, in reality, algebraic expressions.

Algebra combines the use of the numerals of arithmetic and the letters of the alphabet. The numerals are called *definite* numbers and the letters are called *general* or *literal* (letter) numbers.

In practical shop formulas, the literal numbers are letter symbols for quantities such as horsepower (hp), cutting speed (V), voltage (E), outside diameter of a gear (D_o), etc. Many of these symbols have been standardized by the American National Standards Institute (ANSI) or by trade associations. The multiplication sign ($\times$) used in arithmetic should not be used in algebra and higher mathematics because of the possible confusion of the $\times$ symbol with the literal number x. The symbol ($\cdot$) or ( ) ( ) will be used to denote multiplication in the succeeding pages of this textbook.

All the rules of arithmetic regarding addition, subtraction, multiplication, division, whole numbers, fractions, positive signs, and negative signs are unchanged when applied to algebra. Thus, what we have learned up to this point will be used without change in the study of algebra.

Equations

An equation is a statement of equality between two quantities which are separated by an equal ($=$) sign. The equal sign identifies an equation. Mathematical formulas are equations, and the following rule should be kept in mind:

Always, in an equation or a formula, the terms on one side of the equal sign must be equal to the terms on the other side of the equal sign.

Expressed in a different way, the rule reads:

The value of one side of an equation must always be equal to the value of the other side.

One can say that the two sides of an equation must always "balance," if we think of a balance scale as shown in Fig. 6-1. When both pans are loaded with equal weights, they are level and the scale is in balance.

Fig. 6-1. Balance scale.

Shop formulas are statements in algebraic shorthand. Formulas are solved when known values are assigned to the letters of the formula. As an example, examine an equation that states that the horsepower of one engine is two-thirds as large as that of another engine:

$$hp_1 = \frac{2}{3}hp_2$$

where: hp_1 = Horsepower of the less powerful engine
  hp_2 = Horsepower of the more powerful engine

In order to be an equation, each side of the above equation must be equal to the other. Observe that the horsepower of the more powerful engine (hp_2) must be multiplied by $\frac{2}{3}$ in order to reduce the value of the right-hand side of the equation so that it will equal the value of the left-hand side.

Now if the horsepower of the more powerful engine (hp_2) has a known value, the horsepower of the less powerful engine (hp_1) can have only one value; otherwise the equation will not balance. For each value that may be assigned to hp_2, there is only one value of hp_1 that will maintain the equation in balance. There are equations in which the unknown quantity can have more than one answer, and although they are not too frequently encountered in shop practice, they are treated later in this chapter.

Suppose that the more powerful engine has 300 horsepower, or $hp_2 = 300$. Since they are equal, either hp_2 or 300 can appear in the right-hand side of the equation without changing the value of this side of the equation or unbalancing the entire equation. Thus, we can substitute 300 for hp_2 in the equation:

$$hp_1 = \frac{2}{3} \times 300$$

The value of the unknown term can now be determined by simple arithmetic. Obviously the value of hp_1 is 200 horsepower and if 200 is substituted for hp_1 in the above equation it will balance.

We have seen, then, how we can work with equations by knowing that they must always balance. However, there is more to know, which will be treated in the following pages. Of course, you could have calculated the value of hp_1 without the equation, as this example was intentionally made easy to follow. More complicated equations, such as the one below, are often encountered, which you must be able to work:

$$hp = \frac{2\pi tN}{396,000}$$

where: hp = Horsepower
   t = Torque, inch-pounds
   N = Rotational speed, revs. per minute
   π = 3.1416 (a frequently used number which is always expressed by the symbol, π, called pi.).

This is the basic horsepower formula where the power output of the engine is a rotating shaft. There are three unknown terms in this formula: hp, t and N. If any two of the terms are known, it is possible to calculate the value of the third. Learning to do this is the objective of the text in this chapter.

Axioms of Equality

Formulas and equations are statements of equality and are in arithmetical balance. This balance must be maintained at all times. Many times it is necessary to manipulate a formula in order to solve for a specific unknown term. To do this, the following rules or axioms must be carefully observed.

1. *If an amount is added to one side of an equation or a formula, an equal amount must be added to the other side.*

2. *If an amount is subtracted from one side of an equation or a formula, an equal amount must be subtracted from the other side.*

3. *If one side of an equation or a formula is multiplied by a factor the other side must be multiplied by the same factor.*

4. *If one side of an equation or a formula is divided by a term, the other side must be divided by the same term.*

5. *If one side of an equation or a formula is raised to a power, the other side must be raised to the same power.*

6. *If a root is taken from one side of an equation or a formula, the same root must be taken from the other side.*

These rules of equality are used in the process of manipulation of formulas and equations to move the unknown number to the left of the equal sign and other terms to the right in order to solve for the unknown. Terms may be moved, signs changed, and the unknown quantity placed alone to the left of the equal sign by proper use of these rules. In the following examples, the unknown term is x. Each equation requires the application of the specific rule, as shown, to equate the unknown to the known with the unknown placed on the left of the equal sign:

1.
$$x - 4 = 10$$
$$x - 4 + 4 = 10 + 4$$
$$x = 14$$

    Rule 1, add 4 to each side.

2.
$$x + 7 = 12$$
$$x + 7 - 7 = 12 - 7$$
$$x = 5$$

    Rule 2, subtract 7 from each side.

3. $\dfrac{1}{3}x = 10$ Rule 3, multiply each side by 3. (cancel 3's in left side)

$$3 \cdot \dfrac{1}{3}x = 3 \cdot 10$$

$$x = 30$$

4. $5x = 45$ Rule 4, divide both sides by 5. (cancel 5's in left side)

$$\dfrac{5x}{5} = \dfrac{45}{5}$$

$$x = 9$$

5. $\sqrt{x} = 4$ Rule 5, square both sides.

$$(\sqrt{x})^2 = 4^2$$

$$x = 16$$

6. $x^2 = 25$ Rule 6, extract square root from both sides.

$$\sqrt{x^2} = \sqrt{25}$$

$$x = \pm 5$$ (See page 145)

(NOTE: Powers and roots of literal numbers are treated in greater detail later in the chapter.)

Study Problems.

Solve for the value of x.

1. $x - 10 = 1$ (Solve by addition)

2. $x + 2 = 6$ (Solve by subtraction)

3. $\dfrac{1}{5}x = 2$ (Solve by multiplication)

4. $6x = 42$ (Solve by division)

5. $\sqrt{x} = 3$ (Solve by raising to second power)

6. $x^2 = 144$ (Solve by extracting the square root)

7. $x - 5 = 5$ 8. $\dfrac{1}{4}x = 1$ 9. $10x = 1250$

10. $2x + 4 = 14$

11. $4 - x = 19$ (Hint: Multiply both sides by -1)

12. $40 = 10 - x$ (Hint: Add x to each side and solve)

13. $25 - x = 13$ 14. $3x^2 = 48$

Transposition

Rules 1 and 2 above are sometimes applied in an operation known as *transposition*. If the examples shown for Rules 1 and 2 are examined closely, it will be observed that adding to one side to eliminate the term results in adding the same term to the other side. When subtraction is used, elimination of the term from one side results in subtraction of the same term on the other side. This action of adding a term of opposite sign to one side to eliminate a term on the other side is called "transposition." The rule of transposition can be stated:

Any term may be moved from one side of an equation or formula to the other side (across the equal sign) provided that its sign is changed.

This rule makes it possible to move all unknown terms to the left side of an equation or formula and all the known terms to the right side. This is the conventional placement for solution.

Study Examples.

1. Place the unknown, X, on the left side by transposition in equation:

$$15 = 3 - X$$

Step 1. Move $-X$ to left side and change sign:

$$15 + X = 3$$

Step 2. Move $(+15)$ to right side and change sign:

$$X = 3 - 15$$

Step 3. Add numerical terms:

$$X = -12$$

2. Transpose V_1 to the left side in formula:

$$V_2 = V_1 + at$$

Step 1. Move $+V_1$ to left side and change sign:

$$V_2 - V_1 = at$$

Step 2. Move $+V_2$ to right side and change sign:

$$-V_1 = at - V_2$$

Step 3.   Multiply both sides by (-1)

$$+V_1 = +V_2 - at$$

3. The outside diameter (D_o) of a spur gear is given in a formula as:

$$D_o = D + 2a$$

where: D_o = Outside diameter
           D = Pitch diameter
           a = Addendum

Solve for a:

Step 1.   Transpose $2a$ to the left side and change sign:

$$-2a + D_o = D$$

Step 2.   Transpose D_o to right side and change sign:

$$-2a = D - D_o$$

Step 3.   Divide both sides by (-2) to get $+a$:

$$\frac{-2a}{-2} = \frac{D - D_o}{-2}$$

$$a = \frac{D - D_o}{-2}$$

Step 4.   Multiply the numerator and denominator of the fraction by (-1). Both numeral and literal terms change signs when multiplied by (-1). Note that $\frac{-1}{-1} = 1$; therefore, the value of the fraction is not changed.

$$a = \frac{-1(D - D_o)}{(-1)(-2)}$$

$$a = \frac{D_o - D}{2}$$

4. In electrical circuits, the formula for the sum of the resistances in a series circuit is:

$$R = R_1 + R_2 + R_3$$

Solve for R_3:

Step 1.   Transpose R_3 to the left side and change sign:

$$R - R_3 = R_1 + R_2$$

Step 2.   Transpose R to right side and change sign:

$$-R_3 = R_1 + R_2 - R$$

Step 3. Multiply both sides by (-1) to get a $+R_3$:

$$R_3 = R - R_1 - R_2$$

In study examples 3 and 4 above, transposition produced a negative unknown in the left side of the formula. In example 3 the unknown also had a coefficient other than one (-2). By the use of Axiom 4, the coefficient of (-2) was reduced by dividing both sides of the formula by (-2) which resulted in a $+a$. In example 4, Axiom 3 was used and the negative sign in $(-R_3)$ was changed to a positive by multiplying both sides by -1. These two Axioms are important in transposition since the unknown term should be preceded by a positive $(+)$ sign in the final answer.

Study Problems.

Place X alone in the left side of equation.

1. $6 + X = 10$

2. $X - 10 = 2$

3. $15 = 20 - X$

4. $-8 + X = 4$

5. $8 - X = 4$

6. $8 = X + 4$

7. $5X = 15 + 4X$

8. $\dfrac{X}{3} = 4$

9. $A = 22 - \dfrac{X}{4}$

10. $B = 2C - X$

11. $R = \dfrac{n}{X}$

12. $V = G - \dfrac{X}{2}$

Algebraic Nomenclature

While many of the conventional arithmetic signs and symbols are used in algebra, there are special terms and conventions in use that are described below:

Algebraic Expression. Any combination of numbers and letters related to the four basic mathematical operations: addition, subtraction, multiplication, and division, is referred to as an algebraic expression.   For example, $2a$, $3b - c$, $x^2 + 2x + 1$ are algebraic expressions.

Terms. Each part of an algebraic expression that is separated from other parts of the expression by a plus $(+)$ or minus $(-)$ sign is called a term. For example, in the expression: $3a + 4b - 6c$; $3a$ is a term as is $4b$ and $-6c$. Whenever a minus ($-$) sign precedes a term, it is included as the sign of that term.

Simple Expressions. These are algebraic expressions that consist of only one term. This term may be a single letter such as a or a co-efficient with a letter such as $5b$ or a more complex combination such as:

$$3ab^2c, \text{ or a fraction: } \frac{3wxy}{4abc}$$

Simple expressions are also called *monomials.*

Multiple Expressions. These consist of two or more terms separated by plus ($+$) or minus ($-$) signs. These expressions are usually referred to by special descriptions indicating the number of terms involved (up to three terms). These are:

Binomials. An expression containing two terms. The terms being separated by plus or minus signs. For example:

$$2x - 3b, \text{ or, } x^2 - 4y$$

Trinomials. An expression containing three terms separated by plus or minus signs. For example:

$$X^2 + 2X + 1, \text{ or, } 3ab - 2bc + 5$$

Polynomials. All expressions containing two or more terms are called polynomials. Binomials and trinomials are polynomials.

Product. A product results when two or more terms (factors) are multiplied. In algebra, letters may be multiplied by other letters. This operation is indicated by placing the two letters next to each other. For example, a times b becomes ab, and x times y becomes xy. If literal (letter) terms are multiplied by a numeral, the product is written: $2 \cdot a = 2a$, or; $7 \cdot Z = 7Z$

If there are two or more numerals involved in the factors, they must be multiplied arithmetically and their numerical product used. For example:

$$3a \cdot 4b = 12ab \text{ or } 5cd \cdot 2mn = 10cdmn$$

Study Example.

1. $5 \cdot ab = 5ab$ 2. $5a \cdot b = 5ab$

3. $ab \cdot 5 = 5ab$ 6. $2a \cdot 3b \cdot 4c \cdot 5d = 120abcd$

4. $5a \cdot 2b = 10ab$ 7. $3ab \cdot 4cd = 12abcd$

5. $2 \cdot 5 \cdot ab = 10ab$ 8. $2b \cdot 3x \cdot 5 \cdot c = 30bcx$

Quotient. The result obtained by dividing one number called the *dividend* by another number called the *divisor* is called the quotient. The operational signs used are: $\div$ or a bar ——. In algebra, the dividends and divisors may be monomials or polynomials, such as:

$$ab \div 3c, \text{ also written: } \frac{ab}{3c} \qquad \text{(monomials)}$$

or:

$$\frac{3ab \div 6bc - 9}{3ab + c} \qquad \text{(polynomials)}$$

Study Examples.

1. $2x \div 4y = \dfrac{\overset{}{2x}}{\underset{2}{4y}} = \dfrac{x}{2y}$ 3. $12ax \div 3x = \dfrac{\overset{4}{\cancel{12}a\cancel{x}}}{\cancel{3}\cancel{x}} = 4a$

2. $3a^2 \div 2a = \dfrac{3a^2}{2\cancel{a}} = \dfrac{3a}{2}$ 4. $25abc \div 5ab = \dfrac{\overset{5}{\cancel{25}\cancel{a}\cancel{b}c}}{\cancel{5}\cancel{a}\cancel{b}} = 5c$

Sum or Difference. In arithmetic, specific numbers added to or subtracted from other specific numbers give a sum or a difference, respectively. In algebra, special care must be taken to preserve the identity of the literal (letter) numbers in the process of addition and subtraction. For example, in the problem:

$$2a + 3b + 4a + b + 2c + d$$

the a's can be added together to give $6a$ and the b's added together to give $4b$. The $2c$ and the d cannot be added together or added to the a's or b's. Therefore they remain apart. After the addition the expression becomes:

$$6a + 4b + 2c + d$$

Literal terms must have the same letter to be added or subtracted. Note that the final expression is written in alphabetical sequence (a, b, c, d). This is a basic convention used in algebra.

Study Examples.

1. $2a + 3b + 4a = 6a + 3b$

2. $6x - 2w - 4x = 2x - 2w$

3. $3a + 2b - 2c + a = 4a + 2b - 2c$

4. $7ab + 2z - 7a - 2z = 7ab - 7a$ $(2z - 2z = 0)$

Factors. When two or more numbers are multiplied, each of the numbers (terms) is a factor of the answer, the product. In algebra, both numerals and letters are factors. When a numeral stands before a literal term, the numeral is referred to as the *coefficient*.

Coefficients

The numerical prefix of an algebraic term is called the *numerical coefficient*. For example, in the term $3abc$, 3 is the numerical coefficient. In the term, $35xyz$, 35 is the numerical coefficient. Many times the word "numerical" is dropped and the word "coefficient" is used alone. The term $3abc$ translated into words means: "3 times a times b times c." It can also be written: $3 \cdot a \cdot b \cdot c$ using the symbol ($\cdot$) for the common multiplication sign ($\times$). As stated at the beginning of this chapter, the multiplication symbol ($\times$) is not used in algebra to avoid confusion with X, a literal term.

Coefficients are factors, that is, they are part of the indicated multiplication. In algebra, numerical factors can be arithmetically multiplied but literal terms can only have the multiplication indicated, such as:

$$abc, \text{ or } a \cdot b \cdot c; \text{ or } (a)(b)(c); \quad 3 \cdot 5abc = 15abc$$

Parenthetical Expression. The parenthesis was described in Chapter 1 as a grouping symbol. When an algebraic expression is enclosed by a parenthesis it is known as a parenthetical expression. When a parenthetical expression is immediately preceded by a coefficient, the parenthetical expression is a *factor* and must be multiplied by the coefficient. This is done in the following manner:

$$5(a + b) = 5a + 5b$$
$$3a(b - c) = 3ab - 3ac$$

Study Examples.

1. $2 \cdot 3 \cdot 4 \cdot 5 = 120$

2. $2a \cdot 3bc \cdot 2 = 12abc$

3. $3ab \cdot 2cd = 6abcd$

4. $5xy \cdot 2az = 10axyz$

5. $6(a + b) = 6a + 6b$

6. $3ab(c + d + e) = 3abc + 3abd + 3abe$

Like and Unlike Terms. Terms that differ only in their numerical coefficients are called *like* terms. Like terms such as: $2a + 4a$ may be combined by adding their coefficients: $2a + 4a = 6a$.

Terms that have *unlike* literal numbers cannot be combined other than indicating the arithmetical relationship such as: $3x$ plus $(-4y)$ equals $3x - 4y$.

Therefore the rule can be stated: *Like terms may be combined but unlike terms must remain uncombined.* This rule is sometimes referred to as the "apples and bananas" law; apples can be added to apples and bananas to bananas. If unlike kinds of fruit are added, the result is a salad!

Powers. When the factors in a product are identical, their multiplication is denoted by *exponents* (powers). For example: if $3abc$ is multiplied by $3abc$, the product is:

$$3 \cdot 3 \cdot a \cdot a \cdot b \cdot b \cdot c \cdot c \text{ or;}$$
$$(3abc)^2 = 9a^2b^2c^2$$

This answer is read: "Nine a squared, b squared, c squared." When there are three identical factors, such as $b \cdot b \cdot b$ we get: b^3, which is read "b cubed." Powers higher than cubes are read: fourth power, fifth power, etc.

Study Examples.

1. $a \cdot a \cdot a \cdot a = a^4$ (read: a to the fourth power)

2. $ab \cdot ab \cdot ab = a^3b^3$ or $(ab)^3$

Study Problems.

1. $3a \cdot 2b =$

2. $a \cdot 6b =$

3. $4b \cdot 3 =$

4. $5x + 3 =$

5. $\dfrac{6ab}{2a} =$

6. $\dfrac{12xy}{6y} =$

7. $3b - 2a + b - 3a =$

8. $5x + 3z - y + 2z =$

9. $4h + 2k - f + 3h - k =$

10. $6a - 4b - 6a + 4b =$

11. $6xy \cdot 3z =$

12. $5a \cdot b \cdot 3c =$

13. $a \cdot b \cdot 2d =$

14. $x \cdot z \cdot x \cdot z =$

15. $a \cdot b \cdot c \cdot b =$

16. $3ab \cdot 3ab \cdot 3ab =$

17. $x \cdot xy \cdot xyz \cdot x =$ 　　　19. $3bc(2e - 3g + 2) =$

18. $4x(3y - 2z) =$ 　　　20. $5(3ab - 6bd) =$

21. $12z(xy - 3y + 2x) =$

Roots. The equal factors of a number are called *roots* of that number. For example: 4 is the square root of 16 because $4 \cdot 4 = 16$ or $4^2 = 16$. However, it should be observed that (-4) is also a square root of 16 because $(-4)(-4) = 16$ or $(-4)^2 = 16$. Hence, the square root of a number must always be prefixed by $\pm$ such as: $\sqrt{64} = \pm 8$.

When there are three equal roots, these factors are referred to as *cube* roots; such as: $3 \cdot 3 \cdot 3 = 27$, 3 is the cube root of 27. The cube root of a positive number is only positive and the cube root of a negative number is only negative.

Higher roots are sometimes encountered in algebra such as fourth roots, fifth roots, etc.

The sign $\sqrt{}$ indicates a root is to be taken. The root sign $\sqrt{}$ is called the *radical*. A small number placed in the V of the radical indicates the *index* or the order of the root to be taken. For example: $\sqrt[4]{}$ is the indication of a fourth root. If no index is present, the sign, $\sqrt{}$, indicates that a square root is to be taken. It is important that the upper bar or *vinculum* of the radical be extended over *all* the terms involved in the root extraction.

Study Examples.

1. $\sqrt{100} = \pm 10$
   $(+10)(+10) = +100$
   $(-10)(-10) = +100$

2. $\sqrt[3]{27} = +3; \ (3)(3)(3) = 27$

3. $\sqrt[3]{-27} = -3; \ (-3)(-3)(-3) = -27$

4. $\sqrt[4]{16} = \pm 2$
   $2 \cdot 2 \cdot 2 \cdot 2 = 16$
   $(-2)(-2)(-2)(-2) = +16$

5. $\sqrt[5]{3125} = +5; \ 5 \cdot 5 \cdot 5 \cdot 5 \cdot 5 = +3125$

6. $\sqrt[5]{-3125} = -5$
   $(-5)(-5)(-5)(-5)(-5) = -3125$

Imaginary Numbers. We have seen that when any real number, whether a positive or a negative number, is squared, the product is always a positive number. For this reason, the square root of any positive number is always either a positive or a negative real number.

When a square root term is preceded by a negative number, such as $-\sqrt{16}$, the answer is expressed as follows:

$$-(\pm 4)$$

The principal square root of a number is defined as the positive root of the number. Thus, in the example above, the principal square root would be $-(+4)$ which equals -4.

In some cases, particularly in electrical work, the square root of a negative number, such as $\sqrt{-16}$ does occur. In this case, a new type of number has been invented, called the *imaginary* number. The word imaginary is simply the name of a special kind of number which is, in fact, quite real. In classical mathematics this number is designated as i. However, in electrical work it is called j, or the *j-operator*, because i usually represents current flow. By definition:

$$i^2 = -1 \quad \text{or} \quad j^2 = -1$$

from which:

$$i = \sqrt{-1} \quad \text{or} \quad j = \sqrt{-1}$$

Thus, to find the square root of a negative number, we factor the number under the radical as follows and take the square root as shown below:

Study Examples.

1. $\sqrt{-9} = \sqrt{(9)(-1)} = \sqrt{9}\,\sqrt{-1} = 3i$ or $3j$, since i and $j = \sqrt{-1}$

2. $\sqrt{-25} = \sqrt{(25)(-1)} = \sqrt{25}\,\sqrt{-1} = 5i$ or $5j$

3. $(6i)^2 = 6^2 \cdot i^2 = (36)(-1) = -36$

4. $(10j)^2 = 10^2 \cdot j^2 = (100)(-1) = -100$

5. $\sqrt{-b^2} = \sqrt{(b^2)(-1)} = \sqrt{b^2}\,\sqrt{-1} = bi$ or bj

6. $(bi)^2 = b^2 \cdot i^2 = (b^2)(-1) = -b^2$

Study Problems.

Express each of the following terms of the imaginary unit and simplify:

1. $\sqrt{-4}$ 4. $\sqrt{-9}$ 7. $3\sqrt{-49}$

2. $\sqrt{-25}$ 5. $\sqrt{-16}$ 8. $\sqrt{-4a^2}$

3. $\sqrt{-64}$ 6. $2\sqrt{-9}$ 9. $3\sqrt{-25}$

Exponents:

Exponents are used more often in algebra than in arithmetic. As in arithmetic, an exponent is used to signify the number of times a term or expression is used as a factor in the multiplication. For example:

$$a^4 = a \cdot a \cdot a \cdot a, \text{ and } a^n = a \cdot a \cdot a \cdot \quad (n \text{ times})$$

Laws of Exponents

1. The product of the *multiplication* of the same base numbers equals the base number raised to the power equal to the sum of the exponents of the factors.

   $$a^m \cdot a^n = a^{m+n} \text{ and; } x^3 \cdot x^5 = x^{3+5} = x^8$$
   $$(a + b)^2 \cdot (a + b)^3 = (a + b)^{2+3} = (a + b)^5$$

2. The quotient resulting from the *division* of the same base numbers equals the base number raised to the power equal to the exponent of the denominator subtracted from the exponent of the numerator.

   $$\frac{a^m}{a^n} = a^{m-n} \quad \text{or} \quad \frac{x^6}{x^2} = x^{6-2} = x^4$$

   Note in the case where the exponents of the denominator and numerator are the same:

   $$\frac{x^2}{x^2} = x^{2-2} = x^0 \quad \text{or} \quad \frac{x^2}{x^2} = 1$$

   Therefore a zero exponent equates a base to a value of 1.

3. When a power is raised to a power, the exponents are multiplied.

   $$(a^m)^n = a^{mn} \quad \text{or} \quad (x^2)^3 = x^{2 \cdot 3} = x^6$$
   $$(4x^2z)^3 = 4 \cdot 4 \cdot 4 \cdot x^{2 \cdot 3} \cdot z^{1 \cdot 3} = 64x^6z^3$$

4. The power of a product equals the product of the factors each raised to the power.

   $$(ab)^n = a^n b^n \quad \text{or} \quad (2x)^3 = 2 \cdot 2 \cdot 2 \cdot x^3 = 8x^3$$
   $$(4X^2Z)^3 = 4^3 X^{2 \cdot 3} Z^{1 \cdot 3} = 64X^6Z^3$$

5. The power of a quotient equals the quotient of the numerator and the denominator each raised to the power.

$$\left(\frac{a}{b}\right)^n = \frac{a^n}{b^n}; \quad \left(\frac{w}{4}\right)^3 = \frac{w^3}{64}; \quad \left(\frac{2r^2}{s^3}\right)^3 = \frac{8r^6}{s^9}$$

Note, particularly, that an exponent applies only to the specific quantity to which it is attached. Thus $6a^2b^3$ equals $6 \cdot a \cdot a \cdot b \cdot b \cdot b$, but $(6ab)^3$ equals $6 \cdot 6 \cdot 6 \cdot a \cdot a \cdot a \cdot b \cdot b \cdot b = 216a^3b^3$. Care must be taken with the use of negative $(-)$ signs. For example: $-(x^2)^2 = -x^4$, but $(-x^2)^2 = +x^4$.

Study Examples.

1. $x^5 = x \cdot x \cdot x \cdot x \cdot x$ (read; x to the fifth power)

2. $c^3 = c \cdot c \cdot c$ (read; c cubed)

3. $4^2 = 4 \cdot 4 = 16$ (read; four squared equals sixteen)

4. $3^3 = 3 \cdot 3 \cdot 3 = 27$

5. $8^1 = 8$ (the exponent 1, read as first power, requires no mathematical operation)

6. $(x + y)^2(x + y)^4 = (x + y)^{2+4} = (x + y)^6$

7. $\dfrac{(a + b)^5}{(a + b)^2} = (a + b)^{5-2} = (a + b)^3$

8. $(x^3)^4 = x^{3 \cdot 4} = x^{12}$

9. $(3ab^2)^4 = (3)^4(a)^4(b^2)^4 = 81a^4b^{2 \cdot 4} = 81a^4b^8$

10. $\left(\dfrac{x}{3}\right)^3 = \dfrac{x^3}{3^3} = \dfrac{x^3}{27}$

Study Problems.

Perform the indicated operations.

1. $3(a + b - c) =$

2. $2(3x) =$

3. $x(x - 2) =$

4. $5(b + c) - a =$

5. $z \cdot z \cdot z =$

6. $3 \cdot 3 \cdot 3 \cdot 3 =$

7. $ab \cdot ab =$

8. $-4(x - y + z) =$

9. $-a(-a) =$

10. $x(x) =$

11. $-a(a)(-a) =$

12. $3^2 + 2^3 =$

13. $4^2 \cdot 3^2 =$

14. $5^2 - 25 + 3^3 =$

15. $x^2 \cdot x^4 = 1$

16. $(x^2y)(x^3) =$

17. $(abc^2)(ab)(ac) =$

23. $\dfrac{W^5}{W^5} =$

18. $(x^2y^3)^2 =$

24. $\dfrac{x^2y^5}{xy^2} =$

19. $(-w)^2 =$

25. $(d^5)^2 =$

20. $(-w)^3 =$

26. $-(c^6)^2 =$

21. $\dfrac{a^4}{a^2} =$

27. $\left(\dfrac{a^2}{2b}\right)\left(\dfrac{b^4c}{3ad^2}\right) =$

22. $\dfrac{a^2}{a^4} =$

28. $(5a^2x^3)^2 \div (3b^2xy^3)^3 =$

29. $\dfrac{(pq^2)^3(p^2q^3)^2}{(p^5y^2)^2} =$

Fractional Exponents. The five Laws of Exponents have been presented with the exponents as positive whole numbers. The term a^n means that n is a positive integer (whole number). The question then arises: What to do with an exponent such as $\frac{1}{2}$ as in $a^{\frac{1}{2}}$?

If $a^{\frac{1}{2}}$ is squared: $(a^{\frac{1}{2}})^2 = a^{\frac{1}{2} \cdot 2} = a^1 = a$.

The term a can also be generated by: $\sqrt{a}\sqrt{a} = a$. From this we see that the term $a^{\frac{1}{2}}$ has the same meaning as $\sqrt{a}$. Thus we can develop a rule for fractional exponents:

For any base with a fractional exponent, the numerator of the exponent denotes the power to which the base is to be raised and the denominator of the exponent denotes the root to be extracted.

In engineering calculations, terms having a decimal exponent are frequently encountered. Decimal exponents behave according to the five laws of exponents. However, the treatment of this subject is beyond the scope of this chapter.

Study Examples.

1. $27^{\frac{2}{3}} = \sqrt[3]{27^2} = \sqrt[3]{729} = 9$

   or: $27^{\frac{2}{3}} = \sqrt[3]{27^2} = 3^2 = 9$

2. $(-8)^{\frac{2}{3}} = \sqrt[3]{(-8)^2} = \sqrt[3]{64} = 4$

Zero and Negative Exponents. Many times when the powers of like bases are divided, the resulting power becomes 0 or a negative number. For example:

$$\frac{a^3}{a^3} = a^{3-3} = a^0 \quad \text{and:} \quad \frac{x^2}{x^5} = x^{2-5} = x^{-3}$$

In the case of the zero exponent, it is apparent that the numerator and the denominator of the original term are identical. Thus, if a^3 is divided by a^3, the result should be 1. ($5 \div 5 = 1$, $8 \div 8 = 1$, etc.). From this we may conclude that:

Any quantity with an exponent of 0 is equal to unity (1).

$$x^0 = 1, \quad (4z^3)^0 = 1, \quad (-w)^0 = 1$$

When the division of powers with like bases results in a negative exponent, it is necessary to express it as a positive exponent. The rule states:

A quantity with a negative exponent is equal to the reciprocal of the same quantity with a positive exponent.

If we take x^{-m} and multiply it by unity, or $\frac{x^m}{x^m}$, the following results:

$$x^{-m} = \frac{x^m \cdot x^{-m}}{x^m} = \frac{x^{m-m}}{x^m} = \frac{x^0}{x^m} = \frac{1}{x^m}$$

Thus:

$$x^{-m} = \frac{1}{x^m}$$

Study Examples.

1. $b^{-3} = \dfrac{1}{b^3}$

4. $\dfrac{1}{x^{-m}} = \dfrac{x^m}{x^{-m} \cdot x^m} = \dfrac{x^m}{x^0} = x^m$

2. $(ab^2)^{-5} = \dfrac{1}{(ab^2)^5} = \dfrac{1}{a^5 b^{10}}$

5. $\dfrac{2a^{-2}b}{b^3} = \dfrac{2}{a^2 b^3 b^{-1}} = \dfrac{2}{a^2 b^2}$

3. $x^2 y^{-3} = \dfrac{x^2}{y^3}$

6. $\dfrac{ab}{a^{-1}b^3} = \dfrac{a \cdot a}{b^3 \cdot b^{-1}} = \dfrac{a^2}{b^2}$

Study Problems.

Express the following in equivalent radical form:

1. $w^{\frac{1}{2}}$, $z^{\frac{2}{3}}$, $a^{\frac{4}{3}}$, $q^{0.4}$, $x^{-\frac{1}{2}}$, $p^{2.5}$, $r^{-\frac{2}{3}}$

2. $(x - y)^{\frac{1}{2}}$, $(x + y)^{\frac{1}{4}}$, $z^{\frac{6}{5}}$, $y^{\frac{2}{3}}$, $x^{\frac{3}{2}}$

3. $3b^{\frac{1}{2}}$, $(3b)^{\frac{1}{2}}$, $(x^2 - y^2)^{\frac{1}{2}}$, $(-8y)^{\frac{1}{3}}$, $2w^{\frac{3}{4}}$

4. $4p^{\frac{2}{3}}$, $(4p)^{\frac{2}{3}}$, $5(2x - 3y)^{\frac{1}{2}}$, $5w^{1.5}$, $2z^{3.5}$

5. $-3x^{\frac{1}{2}}$, $(-4z)^{\frac{1}{2}}$, $-8a^{\frac{1}{3}}$, $8b^{\frac{2}{3}}$, $(8n)^{\frac{2}{3}}$

Multiply by laws of exponents:

6. $x^{\frac{1}{2}} \cdot x^{\frac{1}{4}}$ 7. $w^2 \cdot w^{\frac{1}{4}}$ 8. $z^{\frac{1}{2}} \cdot z^3 \cdot z^{\frac{2}{3}}$

Divide by laws of exponents:

9. $z^{\frac{2}{3}} \div z^{\frac{1}{4}}$ 10. $x^{\frac{3}{4}} \div x^{\frac{1}{2}}$ 11. $\dfrac{q^{\frac{1}{2}}}{q^{\frac{1}{4}}}$

Perform the indicated operations:

12. $x^{-\frac{1}{2}} \cdot x^{\frac{1}{2}}$ 15. $\left(\dfrac{k^{\frac{1}{4}}}{k^{\frac{1}{2}}}\right)^4$

13. $(w^{\frac{1}{2}})^0$ 16. $\left(\dfrac{c^2}{d^3}\right)^{\frac{5}{6}}$

14. $w^{0.5} \div w^{0.25}$ 17. $\dfrac{(pd^2q \cdot p^3dq^2)}{p^{\frac{2}{3}}d^{\frac{5}{2}}q^2}$

Algebraic Addition and Subtraction

While the basic rules of addition and subtraction as used in arith-metic apply to algebra, some differences exist in their use since both numerals *and* letters appear in algebra. Numerals have specific quan-tity values but literal terms (letters) do not.

As mentioned in the preceding section, a numerical coefficient can stand before a single letter, $3a$; a series of letters, $3abc$; or a grouping symbol, $3(a + b)$. All of these expressions are called terms and the literal part of the term is a distinct entity, much like a family name or the name of an object.

A mechanic, for instance, may have a kit of tools containing three wrenches, two hammers, three screwdrivers and two pliers. Each of these tools is an entity, a name of an object, and cannot be added or subtracted from a different type of tool or entity. The algebraic expression for this mechanic's tool kit could be:

$$3W + 2H + 3SD + 2P$$

If another mechanic had a tool kit with two wrenches, four hammers, five screwdrivers and three pliers, his kit would be expressed as:

$$2W + 4H + 5SD + 3P$$

If the two tool kits were put together we could add up the different types of tools as follows:

First tool kit:	$3W + 2H + 3SD + 2P$
Second tool kit:	$2W + 4H + 5SD + 3P$
Tools in new kit:	$5W + 6H + 8SD + 5P$

Notice that we still keep separate the different types of tools and add only "like" tools.

Handling Monomials and Polynomials. The addition and subtraction of algebraic expressions must follow two rules:

1. *Remove parentheses.*
2. *Combine like terms.*

Study Examples.

1. Simplify the expression: $3x + 2y - (x - 2y)$

    Step 1. Remove parentheses: Multiply the terms in the parentheses by (-1):

$$3x + 2y - 1(x) - 1(-2y)$$
$$3x + 2y - x + 2y$$

    Step 2. Combine like terms:

$$(3x - x) + (2y + 2y)$$
$$2x + 4y$$

2. Subtract: $6A - 12B - 3A - 8B$

     Step 1.   Identify like terms and group together:

$$(6A - 3A) + (-12B - 8B)$$

           Note: Place a plus sign (+) between groups to keep the expression together.

     Step 2.   Add like terms in each group:

$$(3A) + (-20B)$$

     Step 3.   Remove the parentheses by multiplying each term by its unwritten coefficient, 1:

$$3A - 20B$$

Study Problems.

Add the following: (Answers to be in alphabetical order.)

1. $2a + 3b + a + c + 4b + 2c =$

2. $5x + 3xy + 2y + xy + 2x =$

3. $a + bc + c + 3a + 5c =$

4. $xy + yz + 2x + 3xy =$

5. $5p + 3q + 4pq + 5p + pq =$

6. $8vw + v + 3w + 2v + vw + 3v =$

7. $3xyz + 2xy + 4x + 5y + xyz + 2x =$

8. $4B + 8F + 4C + B + BF + 3B + 2B =$

Add and subtract the following:

9. $5x + 4y - 2x + 3y =$

10. $3a + b - 5a + 2b =$

11. $7w - 3b - 2w + 5b - w =$

12. $3p - 2q + 5q - p =$

13. $3x + xy - y + 3xy + 2y =$

14. $5A + 8B - 3A + AB - 5B + 2AB =$

15. $2xy - 3y + 6x + 4y + 5xy - 6y + 10x =$

16. $9r - 2t + 3w + 5t - 5r + 2w - 2t - r =$

Simplify:

17. $(3x - 2) - (5x + 2) =$

18. $(a - 6c) - (3a + 2b) =$

19. $(3x + y) - (4x - 2y) + (9x - 2y) =$

20. $(3x + y - z) - (3x - 2z) - (2x - 6y) =$

Multiplication of Monomials. The multiplication of monomials follows the law of exponents. When a^2 is multiplied by a^3 we get: $a^2a^3 = a^{2+3} = a^5$. In the case of monomials with numerical coefficients, the coefficients are multiplied. For example:

$$2a^2 \text{ times } 6a^4 = 2 \cdot 6 \cdot a^{2+4} = 12a^6.$$

In monomials with like and unlike terms, each like term is combined while the unlike terms remain uncombined. For example:

$$4a^2bx^3 \text{ times } 3ax$$
$$4 \cdot 3 \cdot a^{2+1} \cdot b \cdot x^{3+1} = 12a^3bx^4$$

When three or more monomials are multiplied, the same basic rule holds true:

1. *Multiply the coefficients.*
2. *Add exponents of like literal factors.*

Study Example.

$$(-3x^3y^2)(3yz^2)(-xyz)$$

Step 1. Multiply coefficients:

$$(-3)(3)(-1) = +9$$

Step 2. Add exponents of like terms:

$$9 \cdot x^{3+1}y^{2+1+1}z^{2+1} = 9x^4y^4z^3$$

Study Problems.

1. $(-3uv^2)(-4u^2w)(uw)(-4vw) =$

2. $(6pqr)(2qrs)(-3)(-2prs)(q^2s^2) =$

3. $(-3ab)(-3ab)(-3ab) =$

4. $(3a^3b)(-4a^2b^2)(-2ab^3)(6bc) =$

5. $(-5x^2y^3)(2y^2z^3)(-3z) =$

6. $(22acd)(2bce)(ade)(3de) =$

Multiplication of Polynomials. Polynomials are multiplied in much the same method as used in arithmetic. Each term in the multiplicand polynomial is multiplied in turn by each term in the multiplier. The conventional manner of operation is to work the multiplications from left to right, although some persons work from right to left, as in arithmetic. Examples of both methods are given below. Care must be used in setting down the individual products in their proper vertical columns of like terms.

Study Examples.

1. Multiply:

$$(2x - 2) \text{ by } (x + 3)$$

Step 1.   Arrange problem vertically: $\begin{aligned} 2x - 2 \\ x + 3 \end{aligned}$

Step 2.   Multiply $2x - 2$ by x: $2x^2 - 2x$

Step 3.   Multiply $2x - 2$ by 3: $+ 6x - 6$

Step 4.   Add like terms: $2x^2 + 4x - 6$

2. Multiply: $(2x - 2)$ by $(x + 3)$, working from right to left.

Arrange problem vertically: $\begin{aligned} 2x - 2 \\ x + 3 \end{aligned}$

Multiply $(2x - 2)$ by 3: $6x - 6$

Multiply $(2x - 2)$ by x: $2x^2 - 2x$

Add like terms: $2x^2 + 4x - 6$

3. Multiply: $(4x^2 - 3x + 3)$ by $(4x - 3)$

Arrange problem vertically: $\begin{aligned} 4x^2 - 3x + 3 \\ 4x - 3 \end{aligned}$

Multiply $(4x^2 - 3x + 3)$ by $4x$: $16x^3 - 12x^2 + 12x$

Multiply $(4x^2 - 3x + 3)$ by 3: $- 12x^2 + 9x - 9$

Add like terms: $16x^3 - 24x^2 + 21x - 9$

Study Problems.

Multiply:

1. $(2b - 7)(b + 2) =$

2. $(7p + 3q)(3p - 2q) =$

3. $(x + 3)(5x - 2) =$

4. $(2x + 2)(3x - 3) =$

5. $5x(x^2 + 3x) =$

6. $2x^2(x - 1) =$

7. $4a(a^2 + 5a + 4) =$

8. $6b^2(7b + 2) =$

9. $(4a + 3)(6a^2 + 5a + 4) =$ 11. $(a^2 - ab + b^2)(a^2 + ab + b^2) =$

10. $(2x - 2)(7x^2 + 3x + 2) =$ 12. $(9 + z^3)(z^2 - 5) =$

Division of Monomials. The division of monomials involves writing the problem as a fraction, cancellation of common factors, and, in the case of factors with exponents, subtraction of the exponent of the denominator from the exponent of the numerator of all like bases. For example:

$$\frac{4x^4}{2x^2} = \frac{2 \cdot 2 \cdot x \cdot x \cdot x \cdot x}{2 \cdot x \cdot x} = 2x^2$$

or:

$$= \frac{2 \cdot 2 \cdot x^4}{2 \cdot x^2} = 2 \cdot x^{4-2} = 2x^2$$

In more complex monomials with coefficients, the procedure is to divide the coefficient of the numerator by the coefficient of the denominator, noting their signs. Then the law of exponents is applied. For example:

$$\frac{21x^3y^2}{-3xy} = \frac{21 \cdot x^{3-1}y^{2-1}}{-3} = -7x^2y$$

Study Example.

Divide: $-27a^4b^3$ by $-9a^2b$

1. Set up as fraction: $\dfrac{-27a^4b^3}{-9a^2b}$

2. Divide coefficients: $\dfrac{-27}{-9} = +3$

3. Subtract exponents: $+3 \cdot a^{4-2}b^{3-1} = 3a^2b^2$

Study Problems.

1. $20a \div (-4)$

2. $21y^4 \div y$

3. $-56x^3y^4 \div (-7xy^2)$

4. $35cd \div 7cd$

5. $\dfrac{-32rst^2}{4rs}$

6. $\dfrac{128a^3bc^2}{-8abc}$

7. $\dfrac{6(-4a^3b^2)}{-12a}$

8. $\dfrac{2a(-3ab^2)}{ab}$

Division of Polynomials. Polynomials, also called multinomials, are divided in a manner similar to long division in arithmetic. The operation is best described by close observance of the following example:

Divide: $(3x^2 + 6x - 9)$ by $(x + 3)$

Write problem as a long division:

$$x + 3 \,\overline{)\, 3x^2 + 6x - 9}$$

Divide first term of the trinomial, $3x^2$, by first term of divisor, x:

$$\begin{array}{r} 3x \\ x + 3 \,\overline{)\, 3x^2 + 6x - 9} \end{array}$$

Multiply divisor, $x + 3$, by result of second step:

$$\begin{array}{r} 3x \\ x + 3 \,\overline{)\, 3x^2 + 6x - 9} \\ 3x^2 + 9x \end{array}$$

Subtract product of third step and bring down the third term of trinomial:

$$\begin{array}{r} 3x \\ x + 3 \,\overline{)\, 3x^2 + 6x - 9} \\ 3x^2 + 9x \\ \hline -3x - 9 \end{array}$$

Divide first term of the difference $(-3x - 9)$ by the first term of the divisor:

$$\begin{array}{r} 3x - 3 \\ x + 3 \,\overline{)\, 3x^2 + 6x - 9} \\ 3x^2 + 9x \\ \hline -3x - 9 \end{array}$$

Multiply divisor, $x + 3$, by the result of fifth step, -3:

$$\begin{array}{r} 3x - 3 \\ x + 3 \,\overline{)\, 3x^2 + 6x - 9} \\ 3x^2 + 9x \\ \hline -3x - 9 \\ -3x - 9 \end{array}$$

Subtract the product of the sixth step $(-3x - 9)$ from the previous difference $(-3x - 9)$:

$$\begin{array}{r} 3x - 3 \\ x + 3 \,\overline{)\, 3x^2 + 6x - 9} \\ 3x^2 + 9x \\ \hline -3x - 9 \\ -3x - 9 \\ \hline 0 \end{array}$$

To check the answer: Multiply the answer by the divisor:

Arrange problem vertically:

$$\begin{array}{r} 3x - 3 \\ x + 3 \\ \hline \end{array}$$

Multiply by x:
Multiply by $(+3)$:
Add like terms:

$$\begin{array}{r} 3x^2 - 3x \\ + 9x - 9 \\ \hline 3x^2 + 6x - 9 \end{array}$$

Basic rules for division of polynomials:

1. *Arrange terms of both dividend and divisor in the descending powers of the common literal term:* $x^2 - 2x + 1$.

2. *Divide first term of dividend by first term of the divisor to obtain first term of quotient (answer).*

3. *Multiply divisor by first term of quotient, place product under like terms in dividend and subtract to obtain a new sub-dividend.*

4. *Repeat above steps until a remainder or zero is obtained.*

Study Example.

Divide:

$$(4y^3 + 6y^2 + 1) \text{ by } (2y - 1)$$

$$
\begin{array}{r}
2y^2 + 4y + 2 \\
2y - 1 \overline{\big)\, 4y^3 + 6y^2 + 1} \quad \text{(Leave } y \text{ space open)}\\
\underline{4y^3 - 2y^2} \\
+ 8y^2 \\
\underline{+ 8y^2 - 4y} \\
+ 4y + 1 \\
\underline{+ 4y - 2} \\
+ 3 \quad \text{Remainder}
\end{array}
$$

Express the answer as: $2y^2 + 4y + 2, R \quad (+3)$

or: $2y^2 + 4y + 2 + \dfrac{3}{2y - 1}$

Study Problems.

Divide as indicated:

1. $x^5 \div x^3$

2. $32a^4 \div (-8a^3)$

3. $(5a + 15c) \div 5$

4. $(P + Prt) \div P$

5. $(45x^2 - 15) \div 15$

6. $(x^3 - 7x^2 + 10x) \div (-x)$

7. $(d^2 + 9d + 14) \div (d + 7)$

8. $(x^2 - 3x - 10) \div (x - 5)$

9. $(15a^2b^2 - 8ab + 1) \div (3ab - 1)$

10. $(9y^2 + 24y + 16) \div (3y + 4)$

Applications

The purpose of learning to perform the operations treated before in this book is to be able to apply these operations to the solution of more complex formulas and equations. The following examples will illustrate how this is done.

Study Examples.

1. Solve for a in the equation given below:

$$5(a - b) = a + b + c$$
$$5a - 5b = a + b + c$$
$$5a - a = 5b + b + c$$
$$4a = 6b + c$$
$$a = \frac{6b + c}{4}$$

2. Solve for a in the equation below:

$$\frac{20a^2}{4a} = \frac{8a^2 - 2}{2a - 1}$$

Step 1.   Perform: $(8a^2 - 2) \div (2a - 1)$

$$
\begin{array}{r}
4a + 2 \\
2a - 1 \overline{\smash{\big)}\ 8a^2 \qquad\quad - 2} \\
\underline{8a^2 - 4a} \\
+ 4a - 2 \\
\underline{+ 4a - 2}
\end{array}
$$

thus: $\dfrac{20a^2}{4a} = 4a + 2$

Step 2.   Cancel:

$$\frac{\overset{5}{\cancel{20}}\overset{1}{a^2}}{\underset{11}{\cancel{4a}}} = 5a = 4a + 2$$

Step 3.   Transpose and solve for a:

$$5a - 4a = 2$$
$$a = 2$$

3. The tool life, T (min), that can be obtained when turning a certain steel at a cutting speed of V feet per minute (fpm) can be expressed by the following formula, where C is a constant that depends upon the type of steel being turned and the cutting tool material. Calculate the tool life when $V = 100$ fpm and $C = 200$, if $VT^{1/5} = C$.

$$VT^{\frac{1}{5}} = C$$

$$T^{\frac{1}{5}} = \frac{C}{V}$$

$$T^{\frac{1}{5} \cdot 5} = \left(\frac{C}{V}\right)^5$$

$$T = \left(\frac{200}{100}\right)^5 = 2^5$$

$$T = 32 \text{ min}$$

Study Problems.

1. $125a^3 = \dfrac{5a}{x^2}$ (solve for a):

2. $3(x + 2y) = 2(5x - y)$ (solve for x):

3. $10(abc)^2 = 1440$ (solve for a):

4. $11 = \sqrt{\dfrac{T}{3}}$ (solve for T):

5. $3j^2 = 5E - 13$ (solve for E):

6. $15 - (2j^2) - 5V = 3$ (solve for V):

7. $\left(\dfrac{a}{b}\right)^3 = 125b^3$ (solve for a):

8. $\dfrac{(xy)^2}{x} = 36y^4 + y^2$ (solve for x):

9. $3P + 4H + 10 = 6P - 8H + 4$ (solve for P):

10. $3x^2 + 4xy - (2x + y)^2 = 11y^2 - (5x^2 + y^2)$ (solve for x):

11. $(x + 2)(x - 2) = 0$ (solve for x):

12. $\dfrac{90a^3}{3a^2} = \dfrac{18a^2 - 8}{3a - 2}$

13. The Taylor tool life equation is $VT^n = C$, where V is the cutting speed in feet per minute (fpm), T is the tool life in minutes, and C is a dimensionless constant that is dependent on the work material and on the cutting tool material. The following values of V, n, and C apply for a certain grade of steel and the cutting tool material given:

	V	n	C
High Speed Steel	90	1/5	180
Cemented Carbide	200	1/3	600
Ceramic	800	1/2	4000

Calculate the tool life (T) that will be obtained when machining at the above conditions for all of the tool materials listed above, using the Taylor equation.

14. Calculate the belt length, L, required for a V-belt drive, using the formula given below:

$$L = 2C + 1.57(D + d) + \frac{(D - d)^2}{4C}$$

where: $C = 24$ in., center distance
$D = 10.6$ in., large sheave diameter
$d = 5$ in., small sheave diameter

15. Calculate the center distance, C, for a V-belt drive, using the formula given below:

$$C = \frac{1}{2}[L - 1.57(D + d) - h(D - d)]$$

where: $L = 74.8$ in., length of V-belt
$D = 10.6$ in., large sheave diameter
$d = 5$ in., small sheave diameter
$h = 1.05$, belt length correction factor

Quadratic Equations

When binomials are multiplied together, as described earlier in the chapter, the product can result in an expression with the unknown literal term being a square (x^2). For example:

$$(2x - 2)(x + 3) = 2x^2 + 4x - 6$$

Some shop formulas have the unknown with an exponent of "2" as in x^2. These equations are known as *quadratic* equations or *second-degree* equations. Examples of quadratic equations are:

$$2x^2 - 8 = 0, \quad 14x^2 = 56, \quad 7x^2 - 34x + 24 = 0$$

All quadratic equations with one unknown can be solved by the use of the *quadratic formula*. This formula is stated:

$$x = \frac{-b \pm \sqrt{b^2 - 4ac}}{2a} \tag{1}$$

and refers to a quadratic equation written in a "standard" form, which is:

$$ax^2 + bx + c = 0 \tag{2}$$

The letters a, b and c are the numerical factors of the terms in the equation and *include* the *sign* of that term.

Study Example.

1. Solve: $x^2 + 6x = -8$

Step 1. Rearrange the terms into the standard form:

$$ax^2 + bx + c = 0$$
$$+1x^2 + 6x + 8 = 0$$

Step 2. Assign values for a, b, and c:

$$a = +1, \quad b = +6, \quad \text{and} \quad c = +8$$

Step 3. State the quadratic formula:

$$x = \frac{-b \pm \sqrt{b^2 - 4ac}}{2a}$$

Step 4. Substitute values, including signs, for a, b and c:

$$x = \frac{-(+6) \pm \sqrt{(6)(6) - (4)(8)}}{2(+1)}$$

Step 5. Do indicated operations and solve for x:

$$x = \frac{-6 \pm \sqrt{36 - 32}}{2}$$

$$x = \frac{-6 \pm \sqrt{4}}{2}$$

$$x = \frac{-6 \pm 2}{2}$$

$$x = \frac{-6 + 2}{2} \quad \text{and} \quad \frac{-6 - 2}{2}$$

$$x = \frac{-4}{2} \quad \text{and} \quad \frac{-8}{2}$$

$$x = -2 \quad \text{and} \quad -4$$

Note that quadratic equations always have two values for the unknown. This is similar to the fact that the square roots of positive integers always have two roots: $\sqrt{4} = \pm 2$.

Step 6. Check the answer by substituting the two values of the unknown, x, in the original equation:

$$x^2 + 6x = -8$$

$$(-2)^2 + 6(-2) = -8 \qquad (-4)^2 + 6(-4) = -8$$
$$4 + (-12) = -8 \qquad 16 - 24 = -8$$
$$-8 = -8 \qquad\qquad -8 = -8$$

In some cases the x term is missing in the equation as in: $3x^2 - 12 = 0$. This can be solved by the quadratic formula. However, it can also be solved more easily by transposing the constant term to the right-hand side and solving as follows:

$$3x^2 - 12 = 0$$

$$3x^2 = +12 \qquad \text{(Transposed)}$$

$$\frac{3x^2}{3} = \frac{12}{3} \qquad \text{(Divide each side by 3)}$$

$$x^2 = 4$$

$$\sqrt{x^2} = \sqrt{4} \qquad \text{(Extract square root from each side of equation)}$$

$$x = \pm 2$$

The above quadratic equation can also be solved by using the quadratic formula as follows:

Step 1. Arrange the equation in standard form, leaving blank space for the nonexisting x term:

$$ax^2 + bx + c = 0$$
$$3x^2 \ \cdots \ -12 = 0$$

Step 2. Assign values for a, b, and c:

$$a = +3, \quad b = 0 \quad \text{and} \quad c = (-12)$$

Step 3. State the quadratic formula:

$$x = \frac{-b \pm \sqrt{b^2 - 4ac}}{2a}$$

Step 4. Substitute values for a, b, and c including signs:

$$x = \frac{-(0) \pm \sqrt{(0)^2 - 4(+3)(-12)}}{2(+3)}$$

$$x = \frac{\pm \sqrt{+144}}{+6}$$

$$x = \frac{\pm 12}{6}$$

$$x = \pm 2$$

The foregoing examples have had perfect squares under the radical which provided whole numbers as answers. In many formulas, the answers will be in decimals, repeating decimals or irrational numbers.

Study Example.

Solve: $3x^2 - 4x = 5$

Step 1. Arrange the terms in the standard equation form:

$$ax^2 + bx + c = 0$$
$$3x^2 - 4x - 5 = 0$$

Step 2. Assign values for a, b, and c:

$$a = +3, \quad b = (-4) \quad \text{and} \quad c = (-5)$$

Step 3. State the quadratic formula:

$$x = \frac{-b \pm \sqrt{b^2 - 4ac}}{2a}$$

Step 4. Substitute values of a, b, and c including signs:

$$x = \frac{-(-4) \pm \sqrt{(-4)^2 - 4(+3)(-5)}}{2(+3)}$$

$$x = \frac{+4 \pm \sqrt{16 + 60}}{6}$$

$$x = \frac{+4 \pm \sqrt{76}}{6}$$

$$x = \frac{+4 + \sqrt{76}}{6} \quad \text{and} \quad \frac{+4 - \sqrt{76}}{6}$$

$$x = \frac{4 + 8.7178}{6} \quad \text{and} \quad \frac{4 - 8.7178}{6}$$

$$x = \frac{12.7178}{6} \quad \text{and} \quad \frac{-4.7178}{6}$$

$$x = +2.1196.. \quad \text{and} \quad -0.7863..$$

Rounding off: $x = 2.12$ and -0.786

Study Problems.

1. $x^2 - 6x + 8 = 0$

2. $x^2 = -12 - 7x$

3. $x^2 + x - 6 = 0$

4. $x^2 + 2x = 8$

5. $16x^2 - 16x = -3$

6. $2x^2 = 18$

7. $5x^2 = 1 + 4x$

8. $3x^2 = x + 2$

9. $3x^2 + 14x = 5$

10. $5x^2 - 3x = 5$

11. $6x^2 - 4x = 5$

12. $x^2 + 7 = 9x$

CHAPTER 7

Ratio and Proportion – Gear Ratios

A ratio is a mathematical way of making a comparison. Comparisons are being made continually in our daily lives. In the home, the shop, and the office we compare costs, weights, quality, quantity, and other elements of our activities.

Comparisons are made by stating a relationship of one item to another. If one casting weighs 150 pounds and another weighs 200 pounds, we say that the first casting weighs $\frac{150}{200}$, or $\frac{3}{4}$ as much as the second. We can also say that the second casting weighs $\frac{200}{150}$, or $\frac{4}{3}$ as much as the first casting.

If a person pays $1000 in federal income tax and $200 in state tax, the taxes are compared by saying that the federal and the state taxes are in a ratio of 1000 to 200, or 5 to 1.

The rule of Ratios is:

*The ratio of one quantity to another **like** quantity is the quotient obtained by dividing the first quantity by the second quantity.*

Ratios can express: *greater than*, *less than*, or *equal to* statements. Ratios express relationships in multiples of one another, such as "twice as heavy," and in percentages such as "100 percent heavier." The two items being compared must be the same kind, or like items. Pulley diameters are compared to pulley diameters and units of length to the same units of length. We cannot compare yards to pounds, rpm's to dollars or taps to drills. Comparison as a ratio is only possible when *like* quantities are compared.

A ratio is a fraction. All the rules applying to fractions apply to ratios. A ratio can be expressed as a proper fraction, such as $\frac{2}{3}$; or as an improper fraction, such as $\frac{7}{5}$; as a division, such as $2 \div 3$ or $7 \div 5$, or with the symbol ":" such as 2:3, 7:5. A ratio is always written in the same sequence as the relationship is stated verbally.

For example, the statement, "Motor A is running at twice the speed of motor B," is written as the ratio 2 to 1, or $\frac{2}{1}$. In the case of a ratio of an automobile rear axle, the statement will usually be: "The axle has a 3.57 ratio." This means that the ratio is 3.57 to 1 or $\frac{3.57}{1}$, and that the input rpm is 3.57 rpm faster than the output rpm of the rear wheels.

Study Examples.

Writing ratios:

1. The ratio of 4 to 2 is $\frac{4}{2}$ and equals 2, or 2 to 1.

2. The ratio of 3 to 12 is $\frac{3}{12}$ and equals $\frac{1}{4}$, or 1 to 4.

3. The ratio of 8 to $\frac{1}{2}$ is $\frac{8}{\frac{1}{2}}$ and equals 16, or 16 to 1.

4. The ratio of $\frac{1}{2}$ to $\frac{1}{4}$ is $\frac{\frac{1}{2}}{\frac{1}{4}}$ and equals 2, or 2 to 1.

Study Problems.

1. Write the ratio of 27 to 3. What does it equal?

2. Write the ratio of 4 to 16. What does it equal?

3. Write the ratio of 16 to 48. What does it equal?

4. What is the ratio of 1 to 9? What does it equal?

5. What is the ratio of 15 to 3? What does it equal?

6. Write the ratio of 12 to $\frac{1}{2}$. What does it equal?

7. What is the ratio of 3 feet to 6 inches? (Convert units.)

8. State the ratio of three-quarters to one-half.

9. What is the ratio of 5 quarts to 4 pints? (Convert units.)

10. What is the ratio of 1 foot to 1 mile? (Convert.)

11. State the ratio of 2 miles to 25 kilometers. (Convert.)

12. Write the ratio of 8 ounces to 1 kilogram. (Convert.)

Direct and Indirect Proportions

A proportion is an expression of equality between two ratios. If two ratios are numerically equal they can be written as an equality. For example, the ratios of $\frac{1}{2}$ and $\frac{4}{8}$ are numerically equal and can be written: $\frac{1}{2} = \frac{4}{8}$.

Proportions are valuable tools in the calculation of costs, speeds, quantities, etc., when a known ratio can be equated to a second ratio, one of whose terms is known. For example, the statement, "the diameter of a pulley in a belt transmission system is three times the diameter of the second pulley," gives us the first ratio, three to one, or $\frac{3}{1}$, sometimes written as 3:1. If the diameter of the second pulley is known, the first pulley's diameter will be three times the diameter of the second pulley. The proportion is:

$$\frac{3}{1} = \frac{\text{Diameter of the first pulley}}{\text{Diameter of the second pulley}}$$

If the second pulley is 15 inches in diameter, then the proportion will read:

$$\frac{3}{1} = \frac{\text{Dia. of the first pulley}}{15 \text{ in.}}$$

This proportion is readily solved by applying an arithmetical operation called *cross-multiplication.*

When a proportion is written as equated ratios (fractions), an imaginary line is drawn from the numerator of the left fraction to the denominator of the right fraction; for example:

$$\frac{3}{1} \diagdown \frac{\text{Dia. first pulley}}{15 \text{ in.}}$$

Then a second imaginary line is drawn from the denominator of the left fraction to the numerator of the right fraction; for example:

$$\frac{3}{1} \diagup \frac{\text{Dia. first pulley}}{15 \text{ in.}}$$

Then the connected numbers at the end of the line connecting the unknown term, "Dia. first pulley" in the example, are multiplied:

$$1 \times \text{ Dia. first pulley}$$

Then the connected numbers at the end of the other line are multiplied:

$$3 \times 15 \text{ in.}$$

Giving the equality: 1. Dia. first pulley $= 3 \cdot 15$ in.

or   Dia. of first pulley $= 45$ in.

Study Examples.

Solving proportions:

1. One foot has 12 inches. How many inches are in 5 feet? Let $X =$ the unknown number of inches in 5 feet. State known ratio of feet to inches:

$$1 \text{ ft} : 12 \text{ in., i.e., } \frac{1 \text{ ft.}}{12 \text{ in.}}$$

State unknown ratio:

$$5 \text{ ft} : X \text{ in., i.e., } \frac{5 \text{ ft}}{X \text{ in.}}$$

Equate the ratios: $\dfrac{1}{12} = \dfrac{5}{X}$

Solve by cross multiplication:

$$1 \cdot X = 12 \cdot 5$$
$$X = 60 \text{ in.}$$

Problems involving descriptive terms such as: inches, pounds, dollars, miles, etc., must have the correct descriptive term in the answer to the problem to give a *correct* answer. For example: $50, 60 in., 10 lbs, $6\frac{1}{2}$ miles.

2. In study example 1 the ratios are direct and the proportion is likewise direct and written in the order of the terms (feet : inches = feet : inches). Ratios are directly proportional when an increase in one denomination causes an increase in the other. A problem in purchasing tools will illustrate a "direct" proportion: Eight taps cost $12. How much will 20 taps cost?

Step 1. Write ratio of tap quantities: $\dfrac{8}{20}$

Step 2. Write ratio of tap costs: $\dfrac{12}{X}$

Step 3. Equate the two ratios: $\dfrac{8}{20} = \dfrac{12}{X}$

Step 4. Solve by cross-multiplication: $8X = 20 \cdot 12$

$8X = \$240.$

Cost of 20 taps: $X = \$30.$

3. Ratios are *indirectly* or *inversely* proportional when an increase in one ratio causes a decrease in the other. Problems involving belt pulleys and gearing utilize inverse ratios in the solution of speed and diameter problems. For example, an electric motor running at 1750 rpm has a 5 in. diameter pulley. What size pulley is required on the driven shaft to turn it at 875 rpm?

The ratio of speeds is: $\dfrac{1750}{875} \quad \dfrac{\text{(motor speed)}}{\text{(driven shaft speed)}}$

The ratio of pulley diameters is: $\dfrac{5}{X} \quad \dfrac{\text{(motor pulley)}}{\text{(shaft pulley)}}$

Since an increase in the driven pulley diameter will decrease its speed, the proportion must be written inversely; that is, the speed of the motor is to the speed of the shaft as the diameter of the shaft pulley is to the diameter of the motor pulley. This is written:

$$\frac{1750}{875} = \frac{X}{5}$$

Step 1. Cross-multiply:

$875 \cdot X = 5 \cdot 1750$

$875X = 8750$

$X = 10$ in. (dia. of driven pulley)

4. Gear ratios are related to the number of teeth on each of the gears in a matching pair. In Fig. 7-1, gear A has 36 teeth and gear B has 22 teeth. One tooth of gear A must contact one tooth of gear B as they rotate in mesh. When gear A makes one revolution, 36 teeth on gear B will have been contacted and gear B will have made more than one revolution. A handy rule for speeds of meshing gears is:

The gear with the lesser number of teeth makes more revolutions (turns faster) than the gear with the greater number of teeth.

Gear A has 36 teeth and gear B has 22 teeth. How many revolutions will gear B make if gear A makes 5 turns?

Step 1. Write ratio of teeth: $\dfrac{36}{22}$

Step 2. Write ratio of revolutions: $\dfrac{5}{X}$

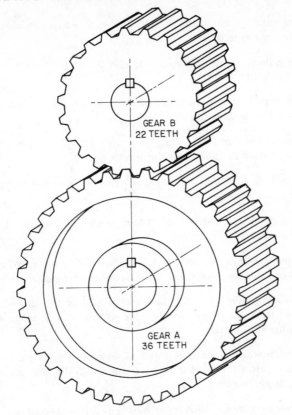

Fig. 7-1. Study problem 4.

Step 3.   Equate ratios but *invert* the second ratio:

$$\frac{36}{22} = \frac{X}{5}$$

Step 4.   Cross-multiply:

$$22X = 5 \cdot 36$$
$$22X = 180$$
$$X = 8.18 \text{ revs.}$$

Study Problems.

1. What is the ratio of 50 to 10? What is the inverse ratio?

2. Solve for the unknown X in the proportion: $\dfrac{4}{32} = \dfrac{X}{16}$

3. Solve for the unknown Z in: $\dfrac{3}{\frac{1}{2}} = \dfrac{18}{Z}$

4. Two numbers are in the ratio of 3 to 4. The lesser number is 60. What is the greater number?

5. The ratio of teeth in gears A and B is $\frac{3}{5}$. How many teeth has gear B if gear A has 15 teeth?

6. Five taper reamers cost $12.00. What will be the cost of a dozen? (Solve by a proportion).

7. A machinist turns 56 pieces in 8 hrs. How long does it take him to turn 14 pieces? (Show your proportion).

8. When gear A in Fig. 7-1 makes $8\frac{1}{2}$ revolutions, how many revolutions will gear B make?

9. A length of railroad rail, 30 ft long weighs 90 lbs per yd. How much will a piece 9 ft 8 in. weigh?

10. A spur gear has 73 teeth, and its pinion has 52 teeth. When the pinion makes 5.3 turns, how many turns does the spur gear make?

11. An automobile has a rear axle with a 3.52 ratio. In high gear (direct drive) how many rpm do the rear wheels turn at when the motor is running at 3800 rpm?

12. A pump which discharges 4 gals of water per minute can fill a tank in 24 hrs. How long will it take a pump to fill the tank when discharging 12 gals per minute?

13. An electric motor running at 800 rpm drives through a speed reducer whose output shaft turns at 125 rpm. What is the reduction ratio of the speed reducer?

14. A pattern for a casting weighs 5.25 lbs. What will the casting weigh if the pattern wood weighs 31.5 lbs per cu ft and the casting metal weighs 425 lbs per cu ft?

15. A grinding-wheel pulley is 5 in. in diameter. The driving motor is running at 1600 rpm and has a 3 in. pulley. What is the speed of the grinding wheel in rpm?

16. Railroad grades (slopes) are measured as the number of feet of rise per 100 ft horizontally. A 2-ft rise in a distance of 100 ft on the "level" is designated as a 2% grade. What is the rise of a 1% grade over a distance of 1 mile (5280 ft on the level)?

17. The volume of a quantity of gas is inversely proportional to the pressure upon it. If the volume of a quantity of gas is 740 cu ft when under a pressure of 16 psia, how many cubic feet will there be when the pressure rises to 30 psia?

Mean Proportional

In a common proportion, when three of the four terms are known, the fourth can be calculated by cross-multiplication. A special type of proportion arises when the second and third terms are equal. (The denominator of the first fraction and the numerator of the second fraction.) When this occurs, the second and third terms are referred to as the "mean proportional."

Study Example.

A typical problem involving mean proportionals would ask for the mean proportional between one number and another. For example, what is the mean proportional between 27 and 3?

Step 1. Write the proportion with the second and third terms designated by X (the mean proportionals).

$$\frac{27}{X} = \frac{X}{3}$$

Step 2. Solve by cross-multiplication:

$$X \cdot X = 27 \cdot 3$$
$$X^2 = 81$$

Step 3. Extract the square root from each side:

$$\sqrt{X^2} = \sqrt{81}$$
$$X = \pm 9$$

Therefore, +9 or −9 are the "mean proportionals."

To understand the meaning of a mean proportional, let us substitute for X in the original proportion, $X = +9$:

$$\frac{27}{9} = \frac{9}{3}$$

Then, reducing each fraction by dividing by the denominator:

$$3 = 3$$

Obviously, a similar result will be obtained when −9 is substituted for X:

$$-3 = -3$$

Study Problems.

1. What is the mean proportional between 16 and 4?

2. The mean proportional between 16 and X is 12. What is the value of X?

3. Find the mean proportional between 2 and 18.

4. Find the mean proportional between 20 and 5.

5. What is the mean proportional between $\frac{3}{4}$ and $\frac{4}{27}$?

6. A motor running at 675 rpm has a pinion gear with 12 teeth. The number of teeth in the driven gear equals its rpm. What is this number? (Use the mean proportional.)

7. Two buildings have equally pitched roofs as shown in Fig. 7-2. An equally pitched roof has both sides at the same angle with the horizontal. The span of the first roof is 160 ft and the vertical height of the second roof is 10 ft. What is the span of the second roof if its span equals the vertical height of the first roof?

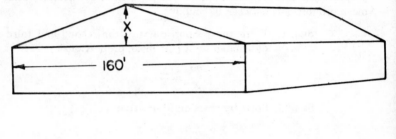

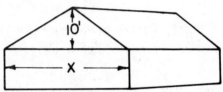

Fig. 7-2. Study problem 7.

8. Two spur gears are in mesh. One gear has 25 teeth and rotates at 81 rpm. How many teeth does the other gear have if its number of teeth and rpm are the same?

9. Two auto engines have equal ratios of horsepower to cubic inch displacement. One has a displacement of 288 cu in. while the second has a rated horsepower of 200. If the horsepower of the first engine equals the displacement of the second engine, what is the displacement of the second engine?

Compound Ratios and Proportions

A compound ratio is the product of two or more simple ratios. In many shop problems, more than one condition exists which affects

their solutions. Two conditions exist which can change the value of
a number:

1. If a number is multiplied by a ratio (fraction) which is less than
    one, the value of that number is decreased; for example:

$$35 \cdot \frac{5}{7} = 25 \quad \text{(decrease)}$$

2. If a number is multiplied by a ratio whose value is greater than
    one, the value of the number is increased; for example:

$$35 \cdot \frac{7}{5} = 49 \quad \text{(increase)}$$

In solving proportions with a series of ratios (compounding), it is
absolutely necessary to put down the individual ratios as stated in the
problem, all in proper order. A problem in compound proportion
usually compares one situation or condition with another. As an example of solving a compound proportion, let us consider a problem
involving two shipments of steel bars.

Study Example.

1. A shipment of 8 steel bars; 6 ft long by 4 in. wide and 1 in. thick weighs
    652.8 lbs. How much will a shipment weigh that consists of 7 steel bars;
    8 ft long by 3 in. wide by 2 in. thick?

    Step 1. Write the ratio of the unknown quantity (weight of the
             second shipment) to the known weight of the first
             shipment:

$$\frac{X \text{ lbs}}{652.8 \text{ lbs}}$$

    Step 2. Write the ratio of the number of bars in second shipment
             to number of bars in the first shipment:

$$\frac{7 \text{ pieces}}{8 \text{ pieces}}$$

    Step 3. Write the ratio of the length of the bars in the second
             shipment to length of bars in first shipment:

$$\frac{8 \text{ ft}}{6 \text{ ft}}$$

    Step 4. Write the ratio of the width of the bars in the second
             shipment to width of bars in first shipment:

$$\frac{3 \text{ in.}}{4 \text{ in.}}$$

Step 5. Write the ratio of the thickness of the bars in the second shipment to thickness of bars in the first shipment:

$$\frac{2 \text{ in.}}{1 \text{ in.}}$$

Step 6. Write the proportion as the ratio of the weights equated to the product of the ratios describing the bars. Note that the terms of the second shipment are the *numerators* of each ratio while the terms of the first shipment are the *denominators*.

$$\frac{X}{652.8 \text{ lbs}} = \frac{7 \text{ pcs}}{8 \text{ pcs}} \cdot \frac{8 \text{ ft}}{6 \text{ ft}} \cdot \frac{3 \text{ in.}}{4 \text{ in.}} \cdot \frac{2 \text{ in.}}{1 \text{ in.}}$$

Step 7. Write as a single ratio equated to a ratio:

$$\frac{X}{652.8 \text{ lbs}} = \frac{7 \cdot 8 \cdot 3 \cdot 2}{8 \cdot 6 \cdot 4 \cdot 1}$$

Step 8. Cancel terms:

$$\frac{X}{652.8 \text{ lbs}} = \frac{7 \cdot \cancel{8} \cdot \cancel{3} \cdot \cancel{2}}{\cancel{8} \cdot \underset{2}{\cancel{6}} \cdot \underset{2}{\cancel{4}} \cdot 1}$$

Step 9. Multiply factors:

$$\frac{X}{652.8 \text{ lbs}} = \frac{7}{4}$$

Step 10. Cross-multiply:

$$4 \cdot X = 7 \cdot 652.8 \text{ lbs}$$

Step 11. Solve for X:

$$4X = 4569.6 \text{ lbs}$$
$$X = 1142.4 \text{ lbs}$$

Study Problems.

1. Fifteen men working 8 hrs per day assemble 72 motors in 6 days. How many motors would 20 men assemble, working 10 hrs per day for 5 days?

2. Five boxcars can be unloaded by three men driving fork trucks in 10 hrs. How many boxcars can be unloaded by four men on four fork trucks working 6 hrs?

3. Four men can carpet a house with 4000 sq ft of floor in 8 hrs. How many hours will it take six men to carpet 5000 sq ft?

4. Ten lengths of 6 in. channel iron @ 13 lbs per ft, 8 ft long, weigh 1040 lbs. How much will 16 lengths of 6 in. channel @ 10.5 lbs per ft, 10 ft long, weigh?

5. Three pumps, each running at 250 gal per min, can fill a reservoir in 6 hrs. How long will it take two pumps, running at 300 gal per min, to fill the reservoir to 75 percent capacity?

6. An automobile final assembly line produces 464 cars during an 8-hr shift. The line is manned by 232 men. How many cars can be produced in 10 hrs by 200 men?

7. A painting contractor quoted $1400.00 to paint a house exterior based on six painters at $5.55 per hr working three 8-hr days. What would his quote be if he used nine painters at $5.25 per hr working two 8-hr days?

8. A cement block wall for a small factory can be laid-up by 7 block layers in five 8-hr days. How many 10-hr days would be required to complete 60% of the construction with a 5-man crew?

Mixture Proportions

A special type of problem involving ratio and proportion is the "mixture" problem. For example, many commercial alloys are composed of two or more metals. The proportions of these metals determine the physical characteristics of the alloy, and care must be used to insure that exact amounts of each alloy metal be added to the mixture. Alloy quantities are expressed either as percentages of the whole or as "parts." If percentages are used, the separate percentages must add up to 100% (the whole amount). For example:

Tinsmith solder is 59% tin, 41% lead. How many pounds of each would be required to make 400 pounds of solder?

Write a proportion for the tin in the mixture:

$$\frac{59}{X} = \frac{100}{400}$$
$$100X = 23,600$$
$$X = 236 \text{ lbs of tin in 400 lbs of solder.}$$

Write a proportion for the lead in the mixture:

$$\frac{41}{X} = \frac{100}{400}$$
$$100X = 16,400$$
$$X = 164 \text{ lbs of lead in 400 lbs of solder.}$$

Study Example.

1. When quantities are expressed as "parts," the whole amount is the sum of the "parts." For example, an alloy is composed of five parts copper, three parts lead, and one part tin. How many pounds of each metal will be needed to cast a 240 lb part?

Step 1.   Add parts to obtain whole:

$$5 + 3 + 1 = 9$$

Step 2.   Form proportions:

$$\frac{5}{9} = \frac{X}{240} \quad \text{(copper proportion)}$$

$$\frac{3}{9} = \frac{Y}{240} \quad \text{(lead proportion)}$$

$$\frac{1}{9} = \frac{Z}{240} \quad \text{(tin proportion)}$$

Step 3.   Cross-multiply and solve each proportion:

$$9X = 5 \cdot 240 \quad \text{(copper)}$$
$$9X = 1200$$
$$X = 133.33 \text{ lbs copper}$$

$$9Y = 3 \cdot 240 \quad \text{(lead)}$$
$$9Y = 720$$
$$Y = 80 \text{ lbs lead}$$

$$9Z = 1 \cdot 240 \quad \text{(tin)}$$
$$9Z = 240$$
$$Z = 26.67 \text{ lbs tin}$$

Step 4.   Add the weights of each metal:

$$
\begin{array}{r}
133.33 \text{ lbs of copper} \\
80.00 \text{ lbs of lead} \\
\underline{26.67 \text{ lbs of tin}} \\
240.00 \text{ lbs of alloy}
\end{array}
$$

Study Problems.

1. A solder is composed of 42% lead and 58% tin. How many pounds of solder can be made from 40 lbs of tin and how many pounds of lead will be used?

2. A tool steel has 1 part molybdenum, 12 parts chrome, 1 part carbon and the balance iron to make 100 parts. How much of each alloy is required to make 95 lbs of tool steel?

3. In 50 oz of a jeweler's alloy there are 18 oz of pure gold. How much of a baser metal must be added to make an alloy that will be 28% pure gold?

4. Bell metal is 25 parts copper to 12 parts tin. How ma
are there in a bell weighing 935 lbs?

5. A casting has 4 parts copper, 3 parts lead and 2 p
pounds of each metal are there in a casting weighin∕

6. How many pounds of lead must be added to change 84∪
which is 44% tin and 56% lead, to a mixture which is 35% ப
65% lead?

7. A bearing metal (Babbit) has 84% tin, 10% antimony, 5% copper, and
1% lead. Find the weight of each metal in a bearing weighing $25\frac{1}{2}$ lbs.

Spur Gears

The most common type of gear in use today is the "spur" gear
whose teeth are cut perpendicular to the face of the gear (parallel to
the axis of the shaft). Spur gears are used in many machines and
mechanisms such as transmissions. Figure 7-3 shows a spur gear.

Fig. 7-3. Spur gear.

Gears are used to transmit power where positive ratios between
the driving and the driven shafts must be maintained. Wheels in contact, Fig. 7-4, will transmit motion and a small amount of power; but
the surface friction is not sufficient to prevent the wheels from slipping upon each other as the power transmitted is increased. A gear
can be viewed as a "non-skid" wheel.

In Fig. 7-5, the principle definitions of gear teeth are shown. The
"pitch circle" corresponds to the circumference of a smooth-faced
wheel in contact with another smooth-faced wheel. The teeth are
formed on either side of the pitch circle with the tooth part that ex-
tends above the pitch circle called the *addendum* and the part below
the pitch circle called the *dedendum*.

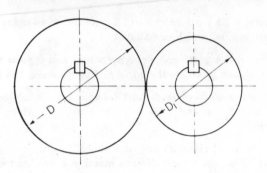

Courtesy of Philadelphia Gear

Fig. 7-4. Wheels in contact-pitch diameters.

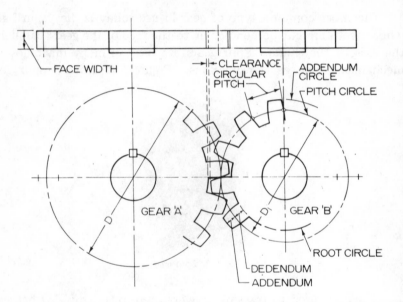

Fig. 7-5. Spur gear nomenclature.

Meshing gears must provide a constant speed relationship as the teeth engage one another. This is accomplished by the use of a special curved surface on the face of the teeth. This curve is known as an *involute* curve and is used on spur and helical gear teeth. Tooth sizes vary, and Fig. 7-6 shows the different sizes of gear teeth in common use cut to the involute profile. The size designation shown is in the "diametral pitch" system. This system is applied to most of the cut gearing that is produced in the United States. Notice that the size of the tooth gets bigger as the diametral pitch number becomes smaller.

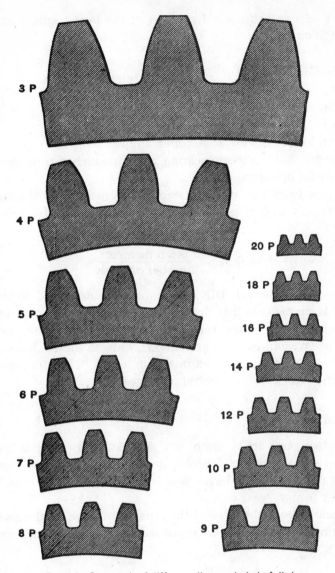

Fig. 7-6. Gear teeth of different diametral pitch, full size.

Diametral pitch is the ratio of the number of teeth to the number of inches of pitch diameter (diameter of the pitch circle). This equals the number of gear teeth per inch of pitch diameter. If gear teeth are larger than about one diametral pitch, it is common practice to use the "circular pitch" system. The circular pitch is the distance from the center of one tooth to the center of the next tooth, measured on the pitch circle.

$$\text{Diametral pitch } (P) = \frac{\text{number of teeth}}{\text{diameter of pitch circle}} = \frac{N}{D}$$

$$\text{Circular pitch } (p) = \frac{\text{circumference of pitch circle}}{\text{number of teeth}} = \pi \frac{D}{N}$$

Circular pitch is also sometimes applied to smaller gears, cast gearing, and usually in worm gearing, although such gearing may also be designed for diametral pitch.

Countries using the metric system (SI) generally use the *module* to indicate the size of gear teeth. It is defined by the following formula:

$$\text{module} = \frac{\text{pitch diameter}}{\text{number of teeth}}$$

For many gears (German DIN), the module is also equal to the addendum. It can be seen that the module is simply the inverse ratio of the diametral pitch. However, in the module system, the pitch diameter is always expressed in millimeters, while in the diametral pitch system, the pitch diameter is expressed in inches. Therefore, module gears will not work with diametral pitch gears.

Gear Ratio

When two gears are in mesh, one gear is the driving gear and the other gear is the driven or follower gear. Usually the smaller gear is called the pinion and the larger gear is simply called the "gear." Together they form a "gear train."

The gear ratio is the ratio of the number of teeth on the gear and on the pinion. This ratio is expressed by the following formula:

$$m_G = \frac{N_G}{N_P}$$

where: m_G = Gear ratio
  N_G = Number of teeth on the gear
  N_P = Number of teeth on the pinion

Since the gear always has more teeth than the pinion, the gear ratio is always a number larger than unity (1). Sometimes the ratio of the speed of the pinion and the gear is called the "speed ratio." This ratio is related to the gear ratio as follows:

$$m_G = \frac{N_G}{N_P} = \frac{\omega_P}{\omega_G}$$

where: ω_P = Speed of the pinion, rpm
         ω_G = Speed of the gear, rpm

Study Example.

1. The gear ratio of a pair of spur gears must be $1\frac{2}{3}$ to 1. The pinion has 18 teeth and rotates at 100 rpm. How many teeth must the gear have and how fast must it rotate?

   Step 1. In the formula:

   $$m_G = \frac{N_G}{N_P}$$

   substitute the known values and solve:

   $$1\frac{2}{3} = \frac{N_G}{18}$$

   $$N_G = 1\frac{2}{3} \cdot 18 = 30 \text{ teeth}$$

   Step 2. In the formula:

   $$m_G = \frac{\omega_P}{\omega_G}$$

   substitute the known values and solve:

   $$1\frac{2}{3} = \frac{100}{\omega_G}$$

   $$\omega_G = \frac{100}{1\frac{2}{3}} = 60 \text{ rpm}$$

The following alternate method could be used to obtain speed of the gear:

$$\frac{N_G}{N_P} = \frac{\omega_P}{\omega_G}$$

substitute the known values:

$$\frac{30}{18} = \frac{100}{\omega_G}$$

$$\omega_G = \frac{18 \cdot 100}{30} = 60 \text{ rpm}$$

Study Problems.

1. The gear ratio of a pair of spur gears is to be 5 to 1. The pinion has 22 teeth and rotates at 250 rpm. How many teeth must the gear have and how fast must it rotate?

2. A gear and pinion in mesh have 32 teeth on the gear and 12 teeth on the pinion. What is the gear ratio?

3. A gear with 42 teeth rotates at 150 rpm. Its pinion rotates at 450 rpm. How many teeth are on the pinion?

4. A spur gear set has a speed ratio of 6 to 1. The gear has 48 teeth and rotates at 460 rpm. How many teeth are on the pinion and how fast does it rotate?

5. A gear rotates at 345 rpm. Its matching pinion rotates at 1380 rpm. What is the speed ratio? If the pinion has 15 teeth, how many teeth has the gear?

Sometimes it is more convenient to express the gear ratio in terms of driving and follower gears. In this event, the formula is written as:

$$\frac{\omega_F}{\omega_D} = \frac{N_D}{N_F}$$

where: ω_F = Speed of the follower gear, rpm
    ω_G = Speed of the driving gear, rpm
    N_D = Number of teeth on the driving gear
    N_F = Number of teeth on the follower gear

Study Example.

1. Two shafts are connected by spur gears. The driving gear has 20 teeth and is fastened to a shaft that rotates at 200 rpm. The follower gear has 32 teeth. Calculate the speed of the follower gear.

Step 1.   Use the formula:

$$\frac{\omega_F}{\omega_D} = \frac{N_D}{N_F}$$

and substitute known values:

$$\frac{\omega_F}{200} = \frac{20}{32}$$

$$\omega_F = \frac{200 \cdot 20}{32} = 125 \text{ rpm}$$

Idler gears that are positioned between the driving and the driven gears do not affect the gear ratio or the speed ratio of the gears. As shown in Fig. 7-7, idler gears serve only to change the direction of rotation of the following gear.

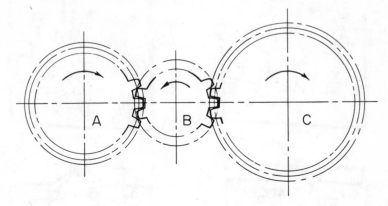

Fig. 7-7. Idler gear used to change direction of the follower gear.

Study Problems.

1. A set of spur gears has 22 teeth on the driving gear and 32 teeth on the follower gear. If the driver turns at 436 rpm, how fast does the follower turn?

2. Gear A in Fig. 7-7 has 12 teeth, gear B has 10 teeth, and gear C has 28 teeth. How fast must gear A turn to turn gear C at 738 rpm?

3. Gear A in Fig. 7-7 rotates at 20 rpm. Gear B turns at 55 rpm. If gear A has 22 teeth, how many teeth are there on gear B?

4. In Fig. 7-7, gear A has 19 teeth, gear B has 12 teeth, and gear C has 28 teeth. If gear B rotates at 1725 rpm, how fast do gears A and C turn?

5. Gear A in Fig. 7-7 has 45 teeth and turns at 1725 rpm. Gear C has 65 teeth. How fast does it turn?

Compound Gear Trains

Compound gears are used extensively in machines where larger gear ratios are required. A typical example is the manual shift transmission gear box used in automobiles (see Fig. 7-8).

A compound gear train is shown in Fig. 7-9. The load is applied at the driving gear D_1 and discharged at the follower gear F_2, after passing through the compounding gears, F_1 and D_2. The two

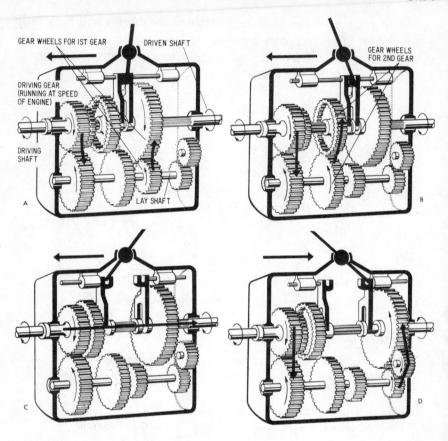

Fig. 7-8. Automotive manual gear shift; three speeds forward and reverse gear. A. Gearbox with 1st gear engaged. B. 2nd gear. C. 3rd gear (direct connection between driving shaft and driven shaft. D. Reverse gear (direction of rotation of driven shaft is reversed).

compound gears (F_1 and D_2) rotate together at the same speed, usually by being keyed to the same shaft. Moreover, they always have a different number of teeth. For this reason, they *do* affect the speed ratio between gears D_1 and F_2. In this respect they are unlike idler gears.

In the compound gear ratio formula below, the subscripts refer to the gears in Fig. 7-9.

$$\frac{\omega_{F2}}{\omega_{D1}} = \frac{\text{Product of driving gears}}{\text{Product of follower gears}} = \frac{N_{D1} \cdot N_{D2}}{N_{F1} \cdot N_{F2}}$$

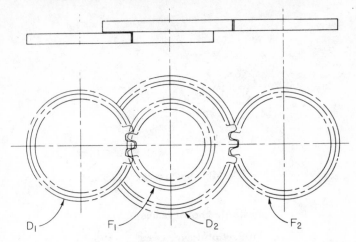

Fig. 7-9. Study example 1.

where: ω_{D1} = Speed of outside driving gear, rpm
ω_{F2} = Speed of outside follower gear, rpm
N_{D1} = Number of teeth on outside driving gear
N_{D2} = Number of teeth on compound driving gear
N_{F1} = Number of teeth on compound follower gear
N_{F2} = Number of teeth on outside follower gear

Study Examples.

1. If, in Fig. 7-9, gear D_1 has 18 teeth and operates at 180 rpm, gear F_1 has 32 teeth and gear D_2 has 20 teeth; if gear F_2 has 39 teeth, how fast does it rotate?

Step 1. Substitute the known values in the formula on page 186.

$$\frac{\omega_{F2}}{180} = \frac{18 \cdot 20}{32 \cdot 39}$$

$$\omega_{F2} = \frac{180 \cdot 18 \cdot 20}{32 \cdot 39} = 51.9 \text{ rpm}$$

2. A shaft that is rotating at 750 rpm is to be connected to another shaft that is to rotate at 150 rpm by means of a compound gear train. (See Fig. 7-10.) The smallest pinion that can be used in this gear train must not have less than 18 teeth. Determine the number of teeth on the driving gears and the follower gears.

Step. 1. In the formula:

$$\frac{\text{Product of driving gears}}{\text{Product of follower gears}} = \frac{\omega_{F2}}{\omega_{D1}}$$

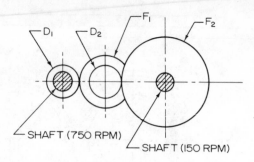

Fig. 7-10. Study example 2.

substitute known values:

$$\frac{\text{Product of driving gears}}{\text{Product of follower gears}} = \frac{150}{750}$$

Step 2. Reduce fraction to its lowest terms:

$$\frac{\text{Product of driving gears}}{\text{Product of follower gears}} = \frac{3}{15}$$

Step 3. Change the fraction again to obtain two terms for both the numerator and denominator:

$$\frac{\text{Product of driving gears}}{\text{Product of follower gears}} = \frac{3 \cdot 1}{5 \cdot 3}$$

Step 4. Expand the fraction by multiplying all of the terms in numerator and denominator by numbers that are equivalent to unity:

$$\frac{\text{Product of driving gears}}{\text{Product of follower gears}} = \frac{(3 \cdot 6)(1 \cdot 20)}{(5 \cdot 6)(3 \cdot 20)}$$

(Note that $\frac{6}{6} = 1$, and $\frac{20}{20} = 1$)

Step 5. Multiply in each parentheses:

$$\frac{\text{Product of driving gears}}{\text{Product of follower gears}} = \frac{18 \cdot 20}{30 \cdot 60}$$

The number of teeth in the driving gears = 18 and 20
The number of teeth in the follower gears = 30 and 60

The gears may be arranged in four different ways. However, they must be set up in the gear train so that the driving gears act as drivers and the follower gears act as followers. Referring again to Fig. 7-10, the correct speed ratio will be obtained when the gears are arranged in any one of the four arrangements shown in the following table.

GEAR	GEAR ARRANGEMENT			
	1	2	3	4
N_{D1}	18	18	20	20
N_{D2}	20	20	18	18
N_{F1}	30	60	30	60
N_{F2}	60	30	60	30

Of course, other gear combinations that satisfy the $\frac{3}{15}$ ratio could have been calculated and used. For most problems in calculating the gears, as above, there are many possible answers. In practice, however, this is usually narrowed down by the requirements of the available center distance and the permissible diametral pitch. As will be seen later, the available change gears impose a limit to the number of possible answers in the case of lathe gearing for thread cutting and for helical milling.

The compound gear formula can be used to extend gear ratios as shown in the following Fig. 7-11.

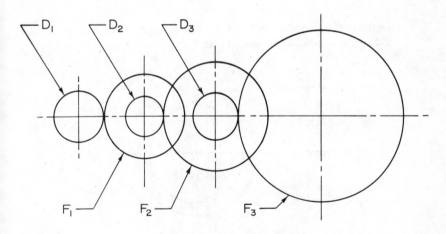

Fig. 7-11. A series of compound gears.

by extending the formula as follows:

$$\frac{\omega_{F3}}{\omega_{D1}} = \frac{\text{Product of driving gears}}{\text{Product of follower gears}} = \frac{N_{D1} \cdot N_{D2} \cdot N_{D3}}{N_{F1} \cdot N_{F2} \cdot N_{F3}}$$

Study Problems.

1. A series of compound gears, as shown in Fig. 7-11, have the following number of teeth: $D1 = 12$, $D2 = 10$, $D3 = 11$, $F1 = 22$, $F2 = 29$, $F3 = 40$. If gear $D1$ turns at 1725 rpm, how fast does $F2$ turn?; how fast does $F3$ turn?

2. In the gear train shown in Fig. 7-12, if gear $D1$ turns at 50 rpm, what is the speed of gear $F2$?

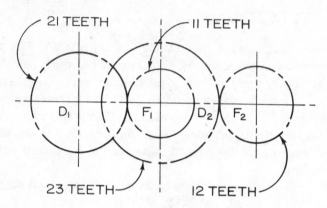

Fig. 7-12. Study problem 2.

3. In the gear train shown in Fig. 7-13, find the number of teeth on the gear $D2$ to give a speed of 70 rpm to gear $F2$.

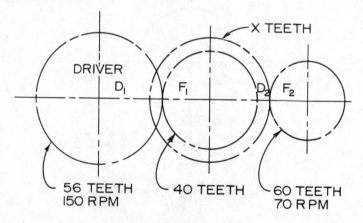

Fig. 7-13. Study problem 3.

Note: If an idler gear, such as shown in Fig. 7-7, is included in a compound train, it must be ignored and not included in the formula as in the case of any other idler gear. Actually, the idler would be both a follower and a driving gear with a ratio of one and, therefore, it would not affect the working of the formula.

Screw Threads

There are two basic types of screw threads: 1, those used primarily for holding parts together, called locking or holding threads; and, 2, those used to provide motion to a component, called "power-transmission" threads. Most common threads are holding threads. They have a 60-degree thread form as shown in View A, Fig. 7-14. These threads are sometimes used to transmit motion. However, for this purpose they are inefficient and limited in that they can transmit only relatively light loads.

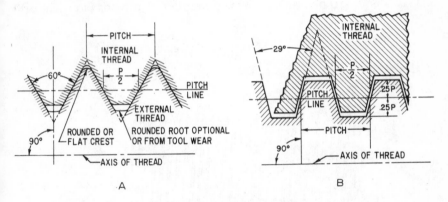

Fig. 7-14. A. American National Standard Unified screw thread. B. American National Standard General Purpose Acme screw thread

Most power transmitting threads used today are of the Acme type. An American National Standard General Purpose Acme thread is shown in View B in Fig. 7-14. The sides of this thread are at an angle of 29 degrees (included angle). Detailed information on the many types of screw threads used in this country can be found in handbooks such as *Machinery's Handbook*.

Screws are normally cut right-handed; i.e., when the screw is turned clockwise, it will travel *away* from the turning force. A left-handed thread requires a counterclockwise rotation to move away from the turning force.

Most threads are cut with a single thread; i.e., a single groove winds its way around the body of the screw. For special purposes, however, double, triple, and even quadruple thread screws are used. These have, respectively: two, three, or four separate grooves wound around the screw.

Threads are usually specified by their nominal outside diameter, the number of threads per inch, the thread series, and the thread class. For example, $\frac{1}{2}$-13-UNC-2 is the specification for a $\frac{1}{2}$-inch diameter, 13 threads per inch, Unified National Coarse thread with a class 2 fit. Further details on thread specifications can be found in *Machinery's Handbook*.

Two important thread terms are shown in Fig. 7-15. The pitch is the distance from a point on the screw thread profile to the corresponding point on the next thread profile, measured parallel to the axis. It is equal to one divided by the number of threads per inch, or:

$$\text{Pitch} = \frac{1}{\text{No. threads per in.}}$$

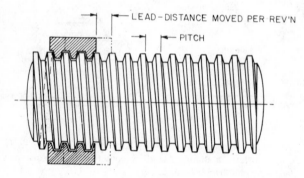

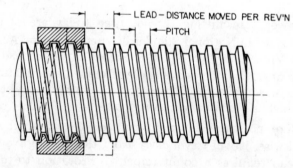

Fig. 7-15. Upper: lead and pitch on a single thread screw. Lower: lead and pitch on a double thread screw.

The lead is the distance that the thread advances axially in one turn. On a single thread screw it is equal to the pitch; on a double thread screw it is twice the pitch; on a triple thread screw it is three times the pitch, etc.

Thread Translation Ratios

The movement of the jaw on a vise or the carriage on a lathe is termed "translation." Formulas for solving problems involving thread-caused translation are as follows:

$$L = n_T p, \qquad S = \frac{n_R n_T}{t_i} \qquad V = N_T L \qquad t_i = \frac{n_R n_T}{S}$$

$$L = \frac{S}{n_R} \qquad v = N_T n_T p \qquad p = \frac{1}{t_i}$$

$$S = n_R L \qquad V = \frac{N_T L}{12} \qquad t_i = \frac{1}{P}$$

where: N_T = No. revolutions per minute (of the thread)
   L = Lead of screw
   n_T = No. of threads (single or multiple)
   t_i = No. of threads per inch
   p = Pitch of thread
   n_R = No. revolutions that thread is rotated
   v = Linear velocity of member actuated by thread, in. per min.
   V = Linear velocity of member actuated by thread, ft per min.
   S = Distance moved by thread or by member actuated by thread, per revolution, in.

Study Examples.

1. How far will a triple-threaded screw with 12 threads per inch move a slide with $3\frac{1}{2}$ turns?

   Step 1. Determine lead of screw with formula: $L = n_T p$

$$\text{Pitch} = \frac{1}{t_i} = \frac{1}{12}$$

$$L = n_T p$$

$$L = 3 \cdot \frac{1}{12} = \frac{1}{4}$$

Step 2.   Solve for travel with formula:

$$S = n_R L$$

$$S = 3\frac{1}{2} \cdot \frac{1}{4}$$

$$S = \frac{7}{8} \text{ in.}$$

2. A lead screw with a single thread travels along the slide 2 inches when turned 14 times. How many threads per inch are on the lead screw?

Step 1.   Solve for t_i in formula:

$$t_i = \frac{N_R\, n_T}{S}$$

Step 2.   Substitute known values:

$$t_i = \frac{14 \cdot 1}{2}$$

$$t_i = \frac{14}{2} = 7 \text{ threads per in.}$$

Study Problems.

1. Determine the lead of a screw that is triple threaded with 7 threads per inch.

2. A screw has a lead of 0.375 in. How far will it travel into a threaded hole if it is turned six times?

3. A double-threaded screw is rotated nine times and travels 6 in. into a threaded hole. How many threads per inch are on the screw?

4. What is the velocity of a machine slide if the screw is double-threaded, has six threads per inch, and is turned at 10 rpm.

5. A double-threaded screw in a vise is turned six times. The vise jaw moves 3 in. How many threads per inch are on the screw?

6. A milling-machine table is actuated by a single-threaded lead screw having 6 threads per inch. If the table is to move at a rate of 16 in. per minute, at what rpm must the lead screw be turning?

Lathe Change Gears

In cutting a screw thread in a lathe, the stock on which the thread is being cut is rotated at a surface speed suitable for the cutting tool and the material being cut. While the work is making one revolution, the tool must be fed along parallel to the axis of the work a distance equal to the lead of the screw thread which is to be cut.

Most modern lathes have a feed-thread gear box where the required number of threads per inch can be obtained by simply moving a few levers and handles. Older lathes and many smaller lathes do not have this attachment; the gears connecting the lathe spindle and the lead screw must be manually changed to obtain gear ratios to cut the required threads per inch on the workpiece. A set of "change-gears" comes with the older lathes and it is an advantage to be able to calculate the proper change-gears to use for cutting a required thread. A typical change-gear set for a lathe together with their placement on the lathe is shown in Fig. 7-16.

Courtesy of South Bend Lathe

Fig. 7-16. Lathe showing change gear set with a list of change gears available.

In calculating change gears, the lead number of the lathe must first be obtained. The lead number, also called the lathe screw constant, is defined as the number of threads per inch cut by a lathe when gears of the same size are placed on the lead screw and the lathe stud. The gears for any thread are then calculated by the following formula:

$$\frac{\text{Lead number of lathe}}{\text{Threads per inch to be cut}} = \frac{\text{Number of teeth on spindle-stud gear}}{\text{Number of teeth on lead-screw gear}}$$

$$= \frac{\text{Driving gear}}{\text{Driven gear}}$$

Study Examples.

Available change gear sizes must be known.

1. A $\frac{3}{4}$-10 UNC-2A thread is to be cut on a lathe having a lead number equal to 8. Calculate the change gears required. See Fig. 7-16 for list of change gears available.

   Step 1. Substitute known values in above formula:

   $$\frac{\text{Lead number of lathe}}{\text{Threads per inch to be cut}} = \frac{8}{10} = \frac{\text{Number of teeth on driving gear}}{\text{Number of teeth on driven gear}}$$

   Step 2. Multiply ratio by $\frac{4}{4}$ (1) to get size gears in set:

   $$\frac{8 \cdot 4}{10 \cdot 4} = \frac{32}{40} = \frac{\text{Number of teeth in driving gear}}{\text{Number of teeth in driven gear}}$$

Sometimes a set of two gears cannot be used because the calculated gears to cut the desired threads are not available. In this case it is necessary to use four gears, or compound gearing. The calculation of compound gearing is shown in the following example:

2. A $\frac{5}{16}$-24 UNF-2A thread is to be cut on a lathe that has a lead number equal to 8. Calculate the size of the change gears required to cut this thread by using a compound-gear ratio.

   Step 1. $\dfrac{\text{Lead number of lathe}}{\text{Threads per inch to be cut}} = \dfrac{8}{24}$

   Step 2. Factor numerator and denominator to form two ratios:

   $$\frac{8}{24} = \frac{2 \cdot 4}{4 \cdot 6}$$

   Step 3. Multiply each ratio by a unity (1) ratio, say $\frac{10}{10}$ and $\frac{8}{8}$:

   $$\frac{(2 \cdot 10)(4 \cdot 8)}{(4 \cdot 10)(6 \cdot 8)} = \frac{20 \cdot 32}{40 \cdot 48} = \frac{\text{Driving gears}}{\text{Driven gears}}$$

   Driving gears have 20 teeth and 32 teeth.
   Driven gears have 40 teeth and 48 teeth.

Study Problems.

1. A screw cutting lathe has a lead number equal to 8. Calculate the size of change gears required to cut a $\frac{7}{8}$-9 UNC-2A thread.

2. It is desired to cut a number of $1\frac{1}{2}$-6 UNC-2A threads on a lathe with a lead number of 8. Calculate the size of change gears required.

3. On a lathe with a lead number of 6, calculate the size of compounded change gears to cut a $\frac{1}{2}$-28 UNEF-2A thread.

4. Calculate the size of compounded change gearing to cut $\frac{3}{8}$-16 UNF-3A threads on a screw cutting lathe with a lead number of 8.

Helical Milling

A helical surface is produced on a milling machine by rotating the workpiece while at the same time feeding it in the direction of the axis of rotation. This motion is accomplished by using a universal dividing head which is driven from the table feed screw through change gears.

The lead of the milling machine must be known or calculated before the change gears required to cut the helix can be calculated. The lead of a milling machine is the ratio of the Dividing Head Ratio to the Number of Threads per Inch of Lead Screw, or as a formula:

$$\text{Lead of a milling machine} = \frac{\text{Dividing head ratio}}{\text{No. threads per inch on lead screw}}$$

Most dividing heads have a 40 to 1 ratio. Figure 7-17 and Fig. 7-18 show two views of the change-gears in a milling machine.

In Fig. 7-18 gears D and B are driving gears and C and A are driven gears. Gears B and C are compound gears, since they are located on the same shaft and rotate together at the same speed. Because these gears have different numbers of teeth, they affect the speed ratio between gears A and D. Single idler gears, on the other hand, affect only the direction of rotation and not the speed ratio of the driving and driven gears.

Variations in speed ratios are required to rotate the dividing head in cutting different helical shapes. This is accomplished by using "change gears" available in a set. Proper change gears are determined from the lead of the helix to be cut and the lead of the machine. The calculation is made by using the following formula:

$$\frac{\text{Driven gears}}{\text{Driving gears}} = \frac{\text{Lead of helix to be cut}}{\text{Lead of machine}}$$

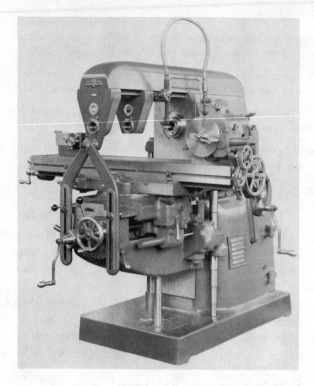

Courtesy of Cincinnati Milacron

Fig. 7-17. Older model of a milling machine showing exposed change gears.

The standard change gears available for the dividing head in Fig. 7-18 have the following number of teeth: 17, 18, 19, 20, 21, 22, 24 (2 gears), 27, 30, 33, 36, 39, 42, 45, 48, 51, 55, 60.

Study Example.

1. A helix with a lead of 18 in. is to be cut on a milling machine with a lead screw with 4 threads per inch and a dividing head ratio of 40 to 1. Calculate the required change gears using the ones supplied with the dividing head.

> Step 1.   As the lead screw has 4 threads per inch, the lead of the milling machine would be:
>
> $$\frac{40}{4} = 10$$
>
> Step 2.   Apply formula:

Courtesy of Cincinnati Milacron

Fig. 7-18. Newer model milling machine with enclosed change gears. A. driven gear; B. driving gear; C. driven gear; D. leadscrew.

$$\frac{\text{Driven gears}}{\text{Driving gears}} = \frac{\text{Lead of helix to be cut}}{\text{Lead of machine}}$$

$$= \frac{18}{10} = \frac{3 \cdot 6}{5 \cdot 2}$$

Step 3. Multiply ratios by unity factors to get teeth:

$$\frac{\text{Driven gears}}{\text{Driving gears}} = \frac{(3 \cdot 9)(6 \cdot 10)}{(5 \cdot 9)(2 \cdot 10)} = \frac{27 \cdot 60}{45 \cdot 20}$$

Driven gears = 27 teeth and 60 teeth
Driving gears = 45 teeth and 20 teeth

Referring to Fig. 7-18, the driving gears (45 and 20) are placed in positions D and B. It does not matter which gear is on D or which gear is on B.

The driven gears (27 and 60) are placed in positions C and A. Again, it does not matter which of the driven gears is in position C or A. For example, the correct lead (18 inches) would be cut if the gears were positioned in any of the following sequences:

	Sequence No.			
	1	2	3	4
Gear and Position	No. of Teeth			
Driving Gear D	45	45	20	20
Driving Gear B	20	20	45	45
Driven Gear C	27	60	27	60
Driven Gear A	60	27	60	27

Study Problems.

1. Calculate the change gears needed to cut a helix with a lead of 40 in. The dividing head ratio is 40 to 1 and the lead screw has 4 threads per inch. Use gears supplied with dividing head as listed above.

2. A milling machine with a dividing head ratio of 40 to 1 and a lead screw with 5 threads per inch is to be used to cut a helix with a lead of 24 in. Calculate the change gears needed using gears supplied with mill.

3. A helix with a lead of 48 in. is to be cut on a mill with a 4 threads per inch lead screw and a dividing head ratio of 40 to 1. Calculate the change gears used.

4. The lead screw on a milling machine has 4 threads per inch, a dividing head ratio of 40 to 1 and is to be set up to cut a helix with a lead of 35 in. Calculate the change gears required.

Helical and Bevel Gears

Helical gears have involute tooth profiles as do spur gears. This shape provides the constant speed relationship required in precision machines. However, a helical gear has its teeth cut at an angle with the shaft or centerline of the gear. Helical gears may be used to connect parallel or non-parallel shafts.

Figure 7-19A shows a helical gear. The contact between meshing teeth is very smooth and almost noiseless. Figure 7-19B shows a special type of helical gear called the "herringbone." This is a double helical gear with the teeth opposed. This construction eliminates the end thrust factor.

Courtesy of Philadelphia Gear

Fig. 7-19. A. Helical gear. B. Herringbone gear.

When a machine designer needs to connect two shafts that are in the same plane and intersect at an angle, a bevel gear can be used. A bevel gear set is shown in Fig. 7-20. The tooth form is involute but tapers from the outer to the inner edges. The speed of a simple bevel gear train is calculated in the same manner as for spur gears; that is, using the inverse proportion of the number of teeth on each gear. This is stated in the formula:

$$\frac{\omega_F}{\omega_D} = \frac{N_D}{N_F}$$

Fig. 7-20. A set of bevel gears.

where: ω_F = Speed of the follower gear, rpm
ω_D = Speed of the driving gear, rpm
N_D = Number of teeth on driving gear
N_F = Number of teeth on follower gear

Worm Gears and Worm Gear Ratios

When high reduction rates are required, the worm and worm gear are used, singly or in multiples. A typical worm gear set is shown in Fig. 7-21. This method of power transmission is nonreversible in most applications, with the worm driving the worm gear. The worms may be single, double, triple, or four threaded and are available from stock. Right-handed worms are carried in dealers' stock but left-handed worms must be specially ordered. A right-handed worm is defined as one in which the worm thread has a right-hand helix, such as a right-hand screw thread. A single-thread worm will turn the worm gear the distance of one tooth each time the worm makes one turn. A double-threaded worm moves the gear two teeth, a triple-threaded worm three teeth, and a four-threaded worm four teeth. The same inverse proportion relationship applies to worm gearing as to spur gearing. The number of teeth on a worm is its number of threads. A double-threaded worm meshed with a 40-tooth gear will have to be rotated 20 times to turn the gear one turn. Figure 7-22 illustrates single, double, and triple-threaded worms.

Courtesy of Boston Gear

Fig. 7-21. A worm gear set.

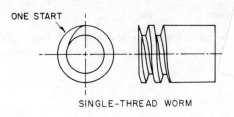

SINGLE-THREAD WORM

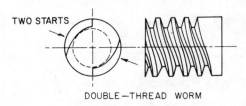

DOUBLE—THREAD WORM

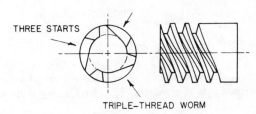

TRIPLE—THREAD WORM

Fig. 7-22. Single, double and triple-thread worms.

If the angle that the worm thread makes with the centerline of the worm is sufficiently small, the worm can be driven by its gear. This type of high reduction, yet reversible, gearing is used in the rear drive axle systems of some heavy-duty trucks.

Calculation of speed ratios of worm gear sets relates to the threading of the worm. A single-thread worm is calculated as a one-tooth gear, a double-thread worm as a two-tooth gear, and a triple-thread worm as a three-tooth gear. This is stated in the formula:

$$\frac{\omega_T}{\omega_G} = \frac{N_G}{N_W}$$

where: ω_G = Speed of gear in rpm
        ω_T = Speed of worm in rpm
        N_G = Number of teeth on gear
        N_W = Number of threads on worm

Study Examples.

1. A triple-threaded worm meshes with a 39-tooth gear. How many rpm's of the worm will be required to turn the gear at 60 rpm?

   Step 1. Use formula:

   $$\frac{\omega_T}{\omega_G} = \frac{N_G}{N_W}$$

   Step 2. Substitute known values:

   $$\frac{\omega_T}{60} = \frac{39}{3}$$

   Step 3. Cross-multiply and solve:

   $$3\omega = 60 \cdot 39$$

   $$\omega_W = \frac{2340}{3}$$

   $$\omega_W = 780 \text{ rpm}$$

2. A worm gear is driven by a single-thread worm. If a speed of 2400 rpm of the worm turns the gear 60 rpm, how many teeth are on the gear?

   Step 1. Use formula:

   $$\frac{\omega_W}{\omega_G} = \frac{N_G}{N_W}$$

   Step 2. Substitute known values:

   $$\frac{2400}{60} = \frac{N_G}{1}$$

   Step 3. Cross-multiply and solve:

   $$60\,N_G = 2400$$
   $$N_G = 40 \text{ teeth on gear}$$

Study Problems.

1. A double-threaded worm drives a 30-tooth gear. When the worm is turning at 1750 rpm, how fast is the gear turning?

2. A triple reduction gear box is shown in Fig. 7-23. It consists of an input shaft with a triple-threaded worm meshing with a 33-tooth gear. This gear, in turn, is on a shaft with a double-threaded worm turning a 24-tooth gear. The 24-tooth gear is on the same shaft with a single-thread worm meshing with a 40-tooth gear. If the input speed is 1725 rpm, what is the output speed (the rpm of the 40-tooth gear)?

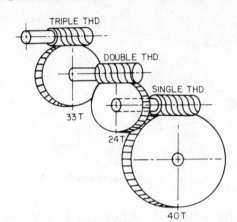

Fig. 7-23. Study problem 2.

3. The combination gear reduction set shown in Fig. 7-24 consists of a double-threaded worm meshing with a 26-tooth gear. The 26-tooth gear is on a shaft with a 22-tooth spur gear which meshes with a 28-tooth spur gear. What speed input is necessary to turn the 28-tooth gear at 250 rpm?

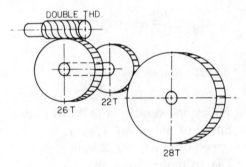

Fig. 7-24. Study problem 3.

Belt Drive Ratios

Many machine tools are driven by belt drives. While older machines may have flat belts, most modern machinery uses V-belts in straight drives and in speed-changing devices. Similar to gear ratios, the ratio of the speeds of the driving pulley to driven pulley equals the ratio of the diameter of the driven pulley to the diameter of the driving pulley in an *inverse* proportion. Figure 7-25 shows a simple belt drive. The pulley diameter shown refers to the underside surface of

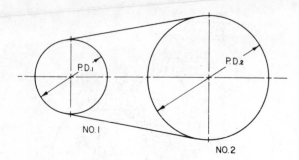

Fig. 7-25. A simple belt drive.

the flat belt or the pulley circumference. In a V-belt system, speed ratios are generally approximations from charts listing the various available V-belt sheaves. Normally the effective outside diameter of the V-sheave is used. This is not necessarily accurate, as true speed ratios of V-belt systems must use the effective pitch diameter of the sheave. (This can be calculated by subtracting twice the distance X between the pitch circle and the O.D. of the sheave.)*

The formula for a simple belt drive ratio is:

$$M_B = \frac{\omega_2}{\omega_1} \quad \text{or:} \quad M_B = \frac{D_1}{D_2}$$

$$\text{or:} \quad \omega_2 = \omega_1 \cdot \frac{D_1}{D_2}$$

where: M_B = Pulley speed ratio or belt drive ratio
ω_1 = Speed of pulley no. 1, rpm
ω_2 = Speed of pulley no. 2, rpm
D_1 = Dia. of pulley no. 1, in.
D_2 = Dia. of pulley no. 2, in.

Study Example.

1. A V-belt drive, Fig. 7-25, has a driving pulley with a pitch diameter (P.D. = O.D. − 2X) of 5 in. and a driven pulley of 8 in. P.D. If the driving pulley rotates at 1144 rpm, what is the speed of the driven pulley?

*See *Machinery's Handbook, 19th Ed.*, Table 9, page 1057.

Step 1. Use formula:

$$\omega_2 = \omega_1 \cdot \frac{D_1}{D_2}$$

Step 2. Substitute known values:

$$\omega_2 = 1144 \cdot \frac{5}{8}$$

$$\omega_2 = \frac{5720}{8} = 715 \text{ rpm}$$

Compound Belt Drive

A compound belt drive is shown in Fig. 7-26. The formula for solving these speed ratios is a proportion consisting of the ratio of the speed of the ultimate driven pulley to the speed of the initial driving pulley equated to the ratio of the product of the pitch diameters of the driving pulleys to the product of the pitch diameters of the driven pulleys:

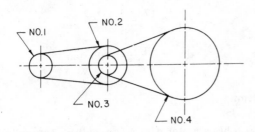

Fig. 7-26. A compound belt drive showing four pulleys.

$$M_B = \frac{\omega_4}{\omega_1} = \frac{D_1 D_3}{D_2 D_4}$$

where: M_B = Pulley speed ratio or belt drive ratio
ω_1 = Speed of pulley no. 1, rpm
ω_4 = Speed of pulley no. 4, rpm
D_1 = Diameter of pulley no. 1, in.
D_2 = Diameter of pulley no. 2, in.
D_3 = Diameter of pulley no. 3, in.
D_4 = Diameter of pulley no. 4, in.

NOTE: Pulleys no. 1 and no. 2 and pulleys no. 3 and no. 4, when operated as pairs independently, perform as simple belt drives.

Study Example.

1. Determine the speed of the final driven pulley in a compound belt drive, where pulley no. 1 is 4 in. in diameter and rotating at 1722 rpm. Pulley no. 2 is 12 in. in diameter, pulley no. 3 is 5 in. in diameter and pulley no. 4 is 14 in. in diameter.

Step 1.   Use formula:

$$M_B = \frac{\omega_4}{\omega_1} = \frac{D_1 D_3}{D_2 D_4}$$

Step 2.   Multiply both sides of formula by ω_1:

$$\omega_4 = \frac{\omega_1 D_1 D_3}{D_2 D_4}$$

Step 3.   Substitute known values:

$$\omega_4 = \frac{1722 \cdot 4 \cdot 5}{12 \cdot 14}$$

$$\omega_4 = \frac{8610}{42} = 205 \text{ rpm}$$

Study Problems.

1. A drill press has a motor with a 5-in. pulley. The motor is belted to the drill spindle through a 3-in. pulley. The motor runs at 1750 rpm. How fast does the spindle turn?

2. A grinder spindle is to be run from an overhead main shaft turning at 250 rpm as shown in Fig. 7-27. There is a countershaft placed between the main shaft and the spindle pulley. The pulley on the main shaft has a 15-in. dia. and is belted to a pulley on the countershaft which has a 6-in. dia. A second pulley on the countershaft is 10 in. in dia. and connects with the spindle pulley which is 6 in. in dia. How fast is the grinder spindle revolving?

3. A 6-in. dia. pulley on a motor drives a 9-in. dia. pulley on a countershaft. A 7-in. dia. pulley on the countershaft drives a 10-in. dia. pulley on a grinder spindle. The grinding wheel is 12 in. in dia. and its surface speed must not exceed 5500 ft/min. What should be the speed of the driving motor?

Hint:

$$N = \frac{12V}{\pi D}$$

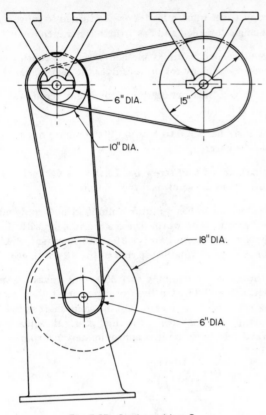

Fig. 7-27. Study problem 2.

where: N = Speed of grinding wheel, rpm
V = Surface speed of grinding wheel, fpm
D = Dia. of grinding wheel, in.

4. A machine to run at 600 rpm is driven by a belt from a 40-in. dia pulley on an overhead line shaft running at 150 rpm. What size pulley should be used on the machine?

5. In a machine shop, the main shaft is run at 330 rpm and a pulley $13\frac{1}{2}$ in. in dia. drives a 10-in. pulley on a countershaft. A second pulley on the countershaft drives a 12-in. dia. pulley on the machine. What is the size of the second pulley on the countershaft if the desired machine speed is 430 rpm?

Review Problems.

1. A machine designer has two shafts at right angles. One shaft has a 55-tooth bevel gear and turns at 10 rpm. How many teeth are needed on the matching bevel gear on the other shaft to rotate it at $27\frac{1}{2}$ rpm?

2. How many times must a lead screw be turned to move a slide 5 in. if the screw is single-threaded and has 7 threads per inch?

3. A double-threaded worm is meshed with a 37-tooth gear and is turned 123 times. How many revolutions does the gear make?

4. A 45-tooth gear turns at 62 rpm. How many teeth must the pinion have to turn at 186 rpm?

5. A set of gears has a 3.5 to 1 ratio. If the pinion has 14 teeth, how many teeth has the gear?

6. A double-threaded lead screw has 13 threads per inch. How far does the slide move when the screw is turned 6 times?

7. A crane mechanism has a motor running at 500 rpm and drives a single-threaded worm. The worm drives a worm gear with 30 teeth. On the same shaft as the worm gear is a 36-tooth spur gear which meshes with a 42-tooth spur gear. What is the rpm of the 42-tooth gear?

8. A two-thread worm, running at 1,200 rpm, drives a gear with 54 teeth as shown in Fig. 7-28. On the worm gear shaft is a pinion with 20 teeth engaging a 50-tooth spur gear. On the same shaft as the 50-tooth gear is a 27-tooth pinion that drives the final gear. How many teeth should be on the final gear to rotate it at approximately 1.5 rpm?

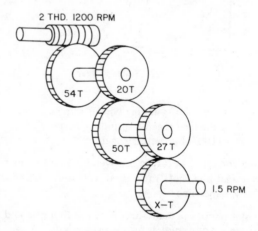

Fig. 7-28. Study problem 8.

9. In Fig. 7-29, a speed reduction system is shown. If the input speed to the worm is 1350 rpm, what is the speed of the 45-tooth bevel gear?

10. A drive for a milling machine table is shown in Fig. 7-30 with all of the thread, gear, and pulley sizes given.

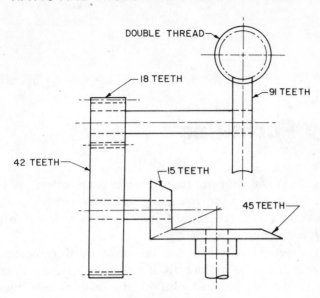

Fig. 7-29. Study problem 9.

a. How fast will the table travel, in terms of inches per minute?

b. If the table must travel 4 in. per minute, what should the speed of the electric motor be?

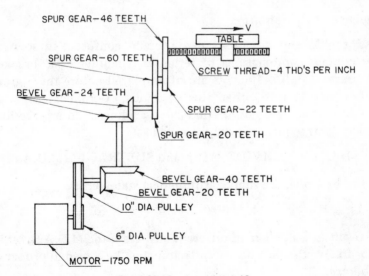

Fig. 7-30. Study problem 10.

CHAPTER 8

Shop Formulas

Many shop and drafting room problems are solved by the use of formulas. Handbooks provide many formulas which are used to calculate gearing, power, centers of gravity, areas, volumes, strength of materials, and other quantities.

Many practical problems can be solved by using an appropriate formula which can be looked up in a reference book. Formulas are algebraic equations, and the solution of equations uses many procedures that must be understood and properly applied. This chapter will present some of the basic procedures along with a large number of shop formulas. Some are solved as study examples. The other formulas are given as study problems to give you practice in learning to solve shop formulas.

Sequence of Operations

Formulas and equations can become confusing to solve when they contain combinations of terms in parentheses (), brackets [] and braces { }. Coefficients are often placed before these grouping symbols and can be either plus or minus. Complicated equations must be solved by doing the indicated arithmetic in a prescribed sequence. This sequence is simply stated as:

MULTIPLY, DIVIDE, ADD, and SUBTRACT   (M.D.A.S.)

For example, the simple arithmetical statement:

$$4 + 3 \times 2 - 6 \div 3 + 6$$

can result in a number of different answers if the M.D.A.S. sequence is not followed. The above expression must be solved in proper order as follows:

1. Multiplying: $4 + 3 \times 2 - 6 \div 3 + 6$
   giving: $4 + \quad 6 \quad - 6 \div 3 + 6$

2. Dividing: $4 +$ 6 $- 6 \div 3 + 6$
   giving: $4 +$ 6 $- 2$ $+ 6$

3. Adding: $4 + 6 + 6 -$ 2
   giving: 16 $- 2$

4. Subtracting: $16 - 2$
   giving: 14

When a complicated algebraic expression appears in an equation such as:

$$Z = 6\{a[2(a + b)(b)]\}$$

The M.D.A.S. sequence likewise applies. As the grouping symbols represent multiplication, they are worked on first. The suggested procedure to use is to remove the innermost symbol first. In the above equation, the parentheses $(a + b)(b)$ with the coefficient 2 are multiplied first, giving:

$$Z = 6\{a[2ab + 2b^2]\}$$

Then the terms within the bracket are multiplied by a, giving

$$Z = 6\{2a^2b + 2ab^2\}$$

Finally the terms in the braces are multiplied by the coefficient 6, to give:

$$Z = 12a^2b + 12ab^2$$

Many algebraic expressions are within grouping symbols that have a plus $(+)$ or minus $(-)$ coefficient. If no numerical coefficient appears outside of the grouping symbol, it means that a $+ 1$ or -1 is the coefficient. The following rules apply in the removal of parentheses, brackets and braces:

Rules for Removal of Grouping Symbols

1. PLUS coefficient of (1):
   When symbol is preceded by a plus $(+)$ sign and no other coefficient, simply drop the grouping symbol. For example:

   $3R + (2b - d)$ equals: $3R + 2b - d$

   This is equivalent to multiplying each term within the grouping symbol by $+1$.

2. MINUS coefficient of one (-1)

When symbol is preceded by a minus ($-$) sign and no other coefficient, multiply each term within the symbol by (-1) and then drop the symbol. For example:

$$3R - (2b - d) \text{ equals: } 3R - 2b + d$$

3. PLUS numerical or literal coefficient:

When the grouping symbol is preceded by a plus numerical or literal coefficient, multiply each term within the symbol: (), [], {}; by the coefficient and drop the symbol. For example:

$$3R + 2a(2b - d) \text{ equals: } 3R + 4ab - 2ad$$

4. MINUS numerical or literal coefficient:

When the grouping symbol is preceded by a minus numerical and/or literal coefficient, the grouping symbol may be removed in one of two ways:

a. Multiply each term within the symbol by the coefficient *and* its sign and then drop the symbol. For example:

$$3R - 6a(2b - d) \text{ equals: } 3R - 12ab + 6ad$$

b. Multiply each term within the symbol by *only* the coefficient (not its sign), retaining the symbol as the first step. For example:

$$3R - 6a(2b - d) \text{ equals: } 3R - (12ab - 6ad)$$

Then apply rule No. 2:

$$3R - (12ab - 6ad) \text{ equals: } 3R - 12ab + 6ad$$

Study Examples.

Simplify by removing parentheses:

1. $6x - (2y + 5) + (4z - 2)$
   $6x - 2y - 5 + 4z - 2$

2. $5a + 2(3b - 3) - 3(4c - 5d)$
   $5a + 6b - 6 - 12c + 15d$

Study Problems.

Simplify by removing grouping symbols and combining like terms.

1. $(6 + 3) \times (4 - 2 \div 4) =$

2. $(9 \times 6) \div (3 \times 2) =$

3. $4a - (2b + c) + 3a =$

4. $x(2 + 4) + 3(x - 3) =$

5. $(2x + 3y) - (x - 2y) =$

6. $3a + 4b - (2a - b) =$

7. $(4a - 2b + c) - a(2 + 3) =$

8. $41 + 16 - 2[30 - (8 - 6)(2 + 7) + 12 \div 4] =$

9. $6 \times 2 + 7(6 + 7) - 8(3 - 5)(4 + 3) - 2 =$

10. $5(6 + 5) - (9 + 3)(7 - 5) + 3(12 \div 3)(4 \times 2 - 6) =$

Combining Like Terms

In many expressions, the removal of grouping symbols results in a number of "like" terms. These are terms whose literal (letter) parts are the same, such as: $2a + 4a$, $6x^2 - 2x^2$. However, $3b + 2b^2$ are not like terms, as the powers (exponents) of b are different. Like terms may differ *only* in their signs and numerical coefficients. Expressions and equations having like terms may be simplified by combining the like terms, unknowns, and constants. The first step in combining like terms is the removal of the grouping symbols, then the adding of the like terms.

Study Examples.

Simplify and combine terms.

1. $6(5 - x) + 15 - 5(2x - 7)$
   $30 - 6x + 15 - 10x + 35$ Remove parentheses.
   $-6x - 10x + 30 + 15 + 35$ Group like terms.
   $-16x + 80$ Add like terms.

2. $3(2x - 3y) - 2(2x + 5y) + 10y$
   $6x - 9y - 4x - 10y + 10y$ Remove parentheses.
   $6x - 4x - 9y - 10y + 10y$ Group like terms.
   $2x - 9y$ Add like terms.

Study Problems.

Simplify and combine terms.

1. $(2x - 1) - (4x + 3)$

2. $(a - 5b) - (2a - 3b)$

3. $(4x - y) - (2x + 5y) + 9x - 5y$

4. $(4x^2 - 2x + 1) + (x^2 - 3x + 1) - (5x^2 - 2x - 3)$

5. $\{[(6a + 5b) - 3b] - (2a - 3b)\} - b$

6. $(2a^2 - 3b^2) - (4a^2 - 5b^2) + (a^2 - 2b^2)$

7. $3a(a + b) - 2a(a - b) - a^2$

8. $3 - (2 - x) + 2 - 3x$

9. $[(3x - 5y) - 4(2x - 2y)] + 2x$

10. $k(m - n) - 3k(n - m) + kmn$

Solving Simple Equations

    Equations are solved for the value of the unknown by performing a series of sequential operations. These are in order:

1. *Eliminate fractions.*

2. *Remove grouping symbols.*

3. *Simplify and combine.*

4. *Transpose terms, unknowns to left, knowns to the right side of the equal sign.*

5. *Divide both sides by coefficient (and sign) of the unknown.*

Study Examples.

1. Solve for x.

$$4(2x - 5) - 3(x - 5) = 0$$
$$8x - 20 - 3x + 15 = 0 \quad \text{Remove parentheses.}$$
$$8x - 3x = 20 - 15 \quad \text{Transpose unlike terms.}$$
$$5x = 5 \quad \text{Combine like terms.}$$
$$x = 1 \quad \text{Divide both sides by coefficient of}$$

                                                        unknown. (5)

2. Solve for W.

$$\frac{3(2W - 4)}{2} - \frac{6(3W + 5)}{3} = 0$$

$$\frac{6W - 12}{2} - \frac{18W + 30}{3} = 0 \quad \text{Remove parentheses.}$$

$$3W - 6 - 6W - 10 = 0 \quad \text{Eliminate the fractions by dividing}$$
                                         each term by its denominator.

$$3W - 6W - 6 - 10 = 0 \quad \text{Group like terms.}$$

$$-3W - 16 = 0$$
$$-3W = 16$$
$$W = -\frac{16}{3} = -5\frac{1}{3}$$

Add like terms.
Transpose (see pages 138 to 140, Chapter 6).
Divide both sides by coefficient of the unknown (including sign).

Study Problems.

Solve for x.

1. $(2x - 7) - (x - 5) = 0$

2. $5 + (x - 11) = 5x - 14$

3. $1 - (6 - x) = 5 - 3x$

4. $2(3x - 1) = 9(x - 2) - 7$

5. $3(6x - 5) - 7(3x + 10) = 0$

6. $5(x + 1) - 3(x - 2) + 2(x - 6) = 0$

7. $x - 2(x + 1) + 3(x + 2) = x - 4$

8. $7x - 3(x - 6) = 2$

9. $\dfrac{3(6 - x)}{2} = \dfrac{4(x - 2)}{2}$

10. $4(2x - 5) - 3(x - 5) = 0$

11. $\dfrac{9(x - 2)}{3} + 3x(4 + 3) = 0$

12. $16x - 3x(x + 4) + \dfrac{x(6x - 2)}{2} = 0$

Transforming and Solving Shop Formulas

Formulas used in the shops and engineering offices represent general statements. The letters and symbols used in the formulas must have definite values assigned to them in order to solve the formula. Known values are substituted for related literal terms in the solution of formulas. In many cases the unknown quantity in the formula is not in the left side of the formula. When the unknown is not in the left side of the formula, the terms must be transposed by the use of any of the six axioms explained earlier in Chapter 6.

Formulas can be solved when the values of the constants and the literal (letter) terms are known. For example, in the formula for the area of a circle:

$$\text{Area} = \frac{\pi D^2}{4}$$

the following values are known:

$\pi = 3.1416$ and the diameter, D, is 4.55 in.

Solution:

Step 1. Substitute values of π and D:

$$\text{Area} = \frac{3.1416 \cdot (4.55)^2}{4}$$

$$\text{Area} = \frac{3.1416 \cdot 20.7025}{4}$$

$$\text{Area} = \frac{65.039}{4}$$

$$\text{Area} = 16.26 \text{ sq in.}$$

When a shop formula is evaluated, the formula becomes an arithmetic problem; and the rules of the sequence of operations must be followed. The answer to the problem refers to specific values and units of measurement. Each substitution must be in the proper unit of measure called for in the formula. Feet must be used where "feet" are called for and pounds where "pounds" are called for.

The answers to solved formulas must always have the proper unit of measure attached; as, "pounds per square inch (psi)" or "miles per hour (mph)." Answers given to problems *without* the correct unit of measure are *wrong* answers even though the numerical value is correct.

Study Examples.

1. Solve for the area of a $3\frac{1}{4}$-in. dia. bar.

Step 1. Use formula:

$$\text{Area} = \frac{\pi D^2}{4}$$

Step 2. Substitute values: $pi(\pi) = 3.1416$, $D = 3\frac{1}{4}$

$$\text{Area} = \frac{3.1416 \cdot (3.25)^2}{4}$$

$$\text{Area} = \frac{3.1416 \cdot 10.5625}{4} = 8.30 \text{ sq in.}$$

2. Determine the quantity of water required per brake horsepower-hour in the evaporative cooling system of an auto engine which is given by the following formula:

$$W = \frac{xQ}{100(t_1 - t_2)}$$

where: W = Weight of water per horsepower-hour, lbs

x = Percent heat loss to jacket. (Note: this percentage should be expressed as a decimal; e.g., 20% = 0.20.)

t_1 = Temperature of cooling water out, degrees Fahrenheit

t_2 = Temperature of cooling water in, degrees Fahrenheit

Q = Heat liberation in Btu per hp-hr.

Given: Q = 12,500 Btu's per hp-hr.

$t_1 - t_2 = 30F$

x = 30% (Use 0.30 in calculation)

Solve for W:

Step 1. Substitute data in formula:

$$W = \frac{xQ}{100(t_1 - t_2)}$$

$$W = \frac{0.30 \times 12,500}{100 \times 30}$$

$$W = 1.25 \text{ lbs per hp-hr.}$$

Study Problems.

1. Solve for the voltage E in the formula:

$$E = I(R + r)$$

when: I = 32 amps, R = 8 ohms and r = 0.35 ohms.

2. The full depth of the thread on an external American National Standard Unified Thread is given in the formula:

$$D = \frac{0.61343}{N}$$

where: D = Full depth of thread, in.

N = Number of threads per in.

Calculate the full depth of the thread on a bolt having 11 threads per inch.

3. The outside diameter of a spur gear can be calculated from the pitch diameter by adding twice the addendum to the pitch diameter. It can also be calculated by using the formula:

$$D_o = \frac{N}{P} + (2 \times \frac{1}{P})$$

Calculate the outside diameter of a gear when the Diametral Pitch (P) is 14 and the number of teeth (N) is 42.

4. The diameters of blanks for drawing plain cylindrical shells can be obtained by using the following formula:

$$D = \sqrt{d^2 + 4dh}$$

Find the diameter (D) when the diameter of the finished shell (d) is 1.5 in. and the height (h) is 2 in.

5. The load capacity (P) in pounds of a round wire, helical compression spring is expressed in the formula:

$$P = \frac{\pi d^3 r}{8K_a D}$$

where: d = Dia. of wire, in.
         r = Shear stress, lb per sq in.
         K_a = Wahl stress correction factor
         D = Mean coil dia., in.

Solve for P when d = 0.192 in., r = 50,000 lbs per sq in., K_a = 1.19 and D = 2.0 in.

6. The length of a roll of belting is given in the formula:

$$L = \pi \left(\frac{D + d}{2} \right) n$$

where: D = Outside dia.
         d = Inside dia.
and n = Number of turns of belting.

Solve for L when $D = 37\frac{1}{4}$ in., $d = 7\frac{5}{8}$ in. and $n = 79\frac{1}{3}$ turns of belting.

7. Two pulleys, 85 in. center to center apart, are connected with a belt. The formula for length of belt is:

$$L = 2C + \frac{\pi}{2}(D + d) + \frac{(D - d)^2}{4C}$$

where: C = Center distance of 85 in.
         D = Dia. of larger pulley at 18 in.
         d = Dia. of smaller pulley at 12 in.

What is the length of the belting? See Fig. 8-1.

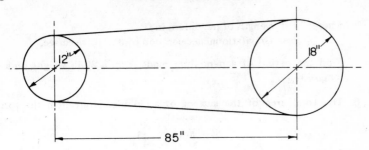

Fig. 8-1. Study problem 7.

8. The volume of a circular cylindrical tank is given as:

$$V = 0.7854D^2h$$

where: D = Dia. of the tank
          h = Height

Find the volume of a tank, 30 ft tall and 20 ft in dia.

General Formulas:

1. The distance attained by a falling object is given in the formula:

$$S = \frac{1}{2}gt^2$$

where: g = Acceleration due to the force of gravity (32.16 ft/sec/sec)
        t = Time of fall in seconds.

How far will a body fall in 22 seconds?

2. How long will it take for a body to fall 58,088 ft?

3. The centrifugal force in pounds exerted when a body is revolved about an axis is calculated by using the formula:

$$F = \frac{Wv^2}{gr}$$

where: W = Weight of body
        v = Linear velocity of the body
        g = Acceleration due to gravity (32.16 ft/sec/sec)
and r = Radius to the center of gravity.

Find the centrifugal force when a 10-lb weight is revolved at a linear velocity of 10 ft/sec and at a radius to the center of gravity of 10 ft.

4. The period of a pendulum, T, in seconds is given in the formula:

$$T = 2\pi\sqrt{\frac{l}{g}}$$

where: l = Length of pendulum in ft
g = Gravitational acceleration of 32.16 ft/sec/sec.

Find the period of a pendulum made from a 20-lb weight on a 40 ft long rope.

5. The total area of the surface of a cylinder is given in the formula:

$$S = \pi d(h + \frac{d}{2})$$

where: π = 3.1416
d = Diameter of the circular face
h = Height of the cylinder

Solve for S when $d = 8\frac{1}{2}$ in. and h = 6 in.

6. Solve for the height, h, when S = 147 sq in. and d = 6 in., using formula in problem no. 5.

7. The linear speed of a V-belt in thousands of feet per minute is given in the formula:

$$S = \frac{\pi \cdot (\text{pitch dia.})(\text{in.}) \cdot \text{rpm}}{12 \cdot 1000}$$

Compute the belt speed when pitch diameter of the sheave is 3 in. and it rotates at 1750 rpm.

8. The volume of a torus (a doughnut-shaped solid) is given in the formula:

$$V = 2(\pi)^2 \cdot R(r)^2$$

where: R = Radius of center of circular section
r = Radius of circular section

Solve for volume when R = 6 in. and r = 2 in.

9. The differential pulley is the chain-type pulley used in repair shops to lift heavy objects manually. The chain engages sprockets on two different diameter pulleys. The formula for the pounds of pull P on the chain to lift a weight is:

$$P = \frac{W(R - r)}{2R}$$

where: W = Weight in lbs
R and r = Radii of the sprockets

How much pull in lbs is necessary to lift 950 lbs when R = 8 in. and r = 7 in.?

10. The power lost by friction in a journal bearing is obtained by using the formula:

$$P_f = \frac{2\pi n}{12} fWr$$

where: f = Coefficient of friction
W = Load in pounds
n = Revolutions per second, rps
r = Radius of shaft, in.
P_f = Horsepower loss caused by friction.

Calculate the power lost in a journal bearing when:

$f = 0.05$
$W = 10,000$ lbs
$n = 10$ rps
$r = 3\frac{1}{2}$ in.

11. The strength required in a taper pin (safe unit stress in lbs per sq in.) is given in the formula:

$$S = \frac{1.27PR}{Dd^2}$$

where: S = Safe unit stress, lbs per sq in.
P = Weight in a turning moment, lbs
R = Length of lever arm of moment, in.
D = Dia. of shaft, in.
d = Dia. of taper pin at centerline of shaft, in.

What is the strength required of a taper pin to secure a lever to a 2 in. round shaft when $d = \frac{3}{8}$ in., $P = 50$ lbs, and $R = 30$ in.?

12. The coefficient of rolling friction c of a wheel with a load of W pounds and with a radius of r inches, moved at a uniform speed by a force of F pounds applied at its center is calculated by the formula:

$$c = \frac{Fr}{W} \text{ in.}$$

Calculate c when $F = 250$ lbs; $r = 15$ in.; and $W = 500$ lbs.

Gear Formulas

Gearing is a major field of mechanics, and many formulas are used in the design and the machining of the many different gears used in industry. Table 8-1 shows basic formulas for the dimensioning of standard spur gears. The notations at the top of the table are industry standards used by all designers, machinists, and mechanics in the building and maintenance of machinery. The formulas in the lower section of the table use these notations.

Table 8-1. Formulas for Dimensions of Standard Spur Gears

Notation		
ϕ = Pressure Angle	F = Face Width	
a = Addendum	h_k = Working Depth of Tooth	
b = Dedendum	h_t = Whole Depth of Tooth	
c = Clearance	N = Number of Teeth	
C = Center Distance	If both gear and pinion are referred to:	
D = Pitch Diameter	N_G = Number of Teeth in Gear	
D_b = Base Circle Diameter	N_P = Number of Teeth in Pinion	
D_O = Outside Diameter	p = Circular Pitch	
D_R = Root Diameter	P = Diametral Pitch	

No.	To Find	Formula	No.	To Find	Formula
		General Formulas			
1	Base Circle Diameter	$D_b = D \cos\phi$	6a	Outside Diameter (Full-depth Teeth)	$D_O = \dfrac{N+2}{P}$
2a	Circular Pitch	$p = \dfrac{3.1416D}{N}$	6b	Outside Diameter (Full-depth Teeth)	$D_O = \dfrac{(N+2)p}{3.1416}$
2b	Circular Pitch	$p = \dfrac{3.1416}{P}$	7a	Outside Diameter (Amer. Stnd. Stub Teeth)	$D_O = \dfrac{N+1.6}{P}$
3a	Center Distance	$C = \dfrac{N_G + N_P}{2P}$	7b	Outside Diameter (Amer. Stnd. Stub Teeth)	$D_O = \dfrac{(N+1.6)p}{3.1416}$
3b	Center Distance	$C = \dfrac{(N_G + N_P)p}{6.2832}$	8	Outside Diameter	$D_O = D + 2a$
4a	Diametral Pitch	$P = \dfrac{3.1416}{p}$	9a	Pitch Diameter	$D = \dfrac{N}{P}$
4b	Diametral Pitch	$P = \dfrac{N}{D}$	9b	Pitch Diameter	$D = \dfrac{Np}{3.1416}$
5a	Number of Teeth	$N = P \times D$			
5b	Number of Teeth	$N = \dfrac{3.1416D}{p}$	10	Root Diameter*	$D_R = D - 2b$

The following study example is solved by the use of the axioms described at the beginning of the chapter. The study problems will require the use of one or more of these axioms.

Study Example.

1. The formula for finding the outside diameter of a full depth gear (6a) is (see Table 8-1):

$$D_o = \frac{N + 2}{P}$$

Solve for P when $N = 22$ and $D_o = 4$ in.

$$4 = \frac{22 + 2}{P} \quad \text{(Substitute values)}$$

$$4P = 24 \qquad \text{(Divide both sides by } P\text{)}$$

$$P = 6$$

Study Problems.

1. In formula (3b)(Table 8-1), solve for circular pitch, p, when center distance $C = 4.76$ in., $N_G = 22, N_P = 16$.

2. In formula (2b), $p = \frac{3.1416}{P}$, solve for P when $p = 0.62852$ in.

3. A pinion with 23 teeth has an outside diameter of 3.125 in. What is the diametral pitch (6a)?

4. A 72-tooth gear is meshed with a 31-tooth pinion. The diametral pitch is 7. What is the pitch diameter of the pinion (9a)?

5. In the gears in problem 4, what is the outside diameter of the pinion if it is a full-depth tooth gear (6b)?

6. In the gears in problem 4, what is the pitch diameter of the gear (9b)?

7. In gear formula, 3b, solve for circular pitch when $C = 4.1666$ in., $N_G = 31$ teeth, and $N_P = 19$ teeth.

8. In gear formula, 3b, solve for center distance when the circular pitch $p = 0.6283, N_G = 92$ teeth, and $N_P = 45$ teeth.

9. In gear formula, 3b, solve for number of teeth on the gear when the center distance $C = 5.333$ in., circular pitch p is 0.5236 and number of teeth on pinion $= 19$.

10. Calculate the number of teeth on a gear with a pitch diameter of 10.286 in. and a circular pitch of 0.4488 in.

Horsepower Formulas

The term "horsepower," used for electrical and mechanical motors, was first defined as the rate of work done by a draft horse maintaining a pull of 150 lbs while walking at a rate of $2\frac{1}{2}$ miles per hour.

James Watts, the inventor, used the term "horsepower" less as a scientific unit than as a sales argument to sell his new steam engine in terms of the horses it would replace. Today a horsepower is defined as the work done at the rate of 550 ft-lbs per second or 33,000 ft-lbs per minute.

There are several formulas for calculating horsepower using a number of constants (known values). Some of these are:

$$\text{ihp} = \frac{PLAN_s}{33,000} \qquad \text{(indicated horsepower)} \qquad \text{(8-1)}$$

$$\text{bhp} = \frac{2\pi NFR}{33,000} \qquad \text{(brake horsepower)} \qquad \text{(8-2)}$$

$$\text{hp} = \frac{2\pi TN}{33,000} \qquad\qquad\qquad\qquad\quad \text{(8-3)}$$

$$\text{hp} = \frac{2\pi tN}{396,000} \qquad\qquad\qquad\qquad\quad \text{(8-4)}$$

$$\text{hp} = \frac{FV}{33,000} \qquad\qquad\qquad\qquad\quad \text{(8-5)}$$

$$\text{hp} = \frac{kw}{1.34} \qquad\qquad\qquad\qquad\quad \text{(8-6)}$$

In the above formulas, the constants used are described as:

P = mean effective pressure (mep) in lbs per sq in.
T = torque in ft lbs t = torque in in. lbs
F = force in lbs L = stroke in ft
V = velocity, ft per min. A = piston area, sq in.
kw = kilowatts (electric power) N_s = strokes per min.
N = shaft speed, rpm   = 2 × rpm
R = distance from center of engine shaft to knife edge, ft.

Study Example.

1. Determine the horsepower of a steam engine running at 83 rpm with a mean effective steam pressure of 110 lbs per sq in., an 8-in. dia. piston and a stroke of 16 in. Use formula 8-1.

     Step 1.   Solve for area of piston:

$$\text{Area} = \frac{\pi D^2}{4}$$

$$D = 8 \text{ in.}, \pi = 3.1416$$

$$\text{Area} = \frac{3.1416(8)^2}{4}$$

$$\text{Area} = \frac{3.1416 \cdot 64}{4} = 50.3 \text{ in.}$$

Step 2. Substitute values in formula 8-1:

$$\text{ihp} = \frac{PLAN_s}{33,000}$$

$P = 110$ lbs per sq in. (mep), $L = \frac{16}{12} \text{ ft} = 1.33 \text{ ft}$

$A = 50.3$ sq in., $N_s = 2(83) \text{ rpm} = 166$ strokes per min.

$$\text{ihp} = \frac{110 \cdot 1.33 \cdot 50.3 \cdot 166}{33,000}$$

$$\text{ihp} = 37.02$$

Study Problems.

1. In formula 8-1, determine the horsepower of a steam engine when the mep P is 78 psi, stroke L is 1.5 ft, piston area A 113 sq in. and the engine is running at 170 strokes per minute (85 rpm).

2. A motor is placed on a test stand to determine its bhp. Its speed N is 250 rpm, length of brake arm R is 2.5 ft and carries a weight F of 130 lbs. Using formula 8-2, determine the brake horsepower.

3. The torque T measurement of an engine on a test stand is 500 ft lbs while it runs at $N = 360$ rpm. Solve for the horsepower using formula 8-3.

4. A miniature gasoline engine used in a model plane has a torque t' of 110 in. oz at 12,000 rpm N. What is its horsepower?

$$\text{hp} = \frac{2\pi t' N}{6,336,000}$$

5. A 0.40 cid model gas engine running at 11,000 rpm generates 0.59 hp. What is its torque rating in in. oz?

6. An elevator motor must lift 15,750 lbs a distance of 90 ft in 5 min. What is the horsepower of the electric motor? Hint: Use formula 8-5.

7. Find the mean effective pressure (lbs per sq in.) of an engine with a 4-in. piston, 8-in. stroke running at 200 rpm and developing 30 ihp.

8. Find the torque (ft lbs) of a 500 ihp steam engine running at 100 rpm.

9. A pumped-water hydroelectric storage system has six pumps capable of lifting 750,000 cu ft of water per hour to a height of 250 ft. The weight of water is 62.4 lbs per cu ft. What is horsepower of each motor?

10. Find the horsepower necessary to lift a 300-lb casting 10 ft in 12 seconds. Hint: $\text{hp} = \dfrac{FS}{550t}$, where: F = weight in lbs, S = height in ft, t = time in seconds.

11. In the horsepower formula:

$$\text{hp} = \frac{2\pi TN}{33,000}$$

solve for torque, T, in ft lbs when hp = 100 and N is 1650 rpm.

12. An electric drive motor in a steel mill uses 670 kw at full load. What is the horsepower of this motor?

13. A drum on the shaft of a motor rotates at 750 rpm and has a diameter of 10 in. A nylon line is passed around the drum to lift a 325-lb load. Calculate the horsepower of the motor. Hint: Use formula 8-2.

Electrical Formulas

Ohm's law is the basic relationship from which electrical formulas in electronics and electrical engineering are formed. Ohm's law is stated in the formula:

$$E = IR$$

where: E = Voltage
           I = Current (amps)
           R = Resistance in ohms

Electric current flowing through a resistance produces heat which represents work done. The formula for work is:

$$W = I^2 Rt$$

where: W = Work (watt-seconds)
           I = Current (amps)
           R = Resistance (ohms)
           t = Time (seconds)

Power is the rate of doing work and is expressed in the formula:

$$P = I^2 R \text{ (watts)}$$

$$\text{or: } P = \frac{I^2 R}{1000} \text{ (kilowatts)}$$

where: P = Power (watts)
           I = Current (amps)
           R = Resistance (ohms)

Study Examples.

1. When 110 volts are applied to an incandescent lamp, a current of 0.5 amp flows. What is the resistance (R) of the lamp?

$$\text{Use formula: } R = \frac{E}{I} \text{ (from } E = IR)$$

Substitute: $E = 110,\ \ I = 0.5$

$$R = \frac{110}{0.5}$$

$$R = 220 \text{ ohms}$$

2. A current of 10 amperes flows through a resistance of 8 ohms. How much heat (in joules) is developed in 1 minute (60 sec) if a joule equals 1 watt-second?

Formula: $W = I^2Rt$

$W = 10 \times 10 \times 8 \times 60$

$W = 48,000 \text{ watt-seconds} = 48,000 \text{ joules}$

3. How many watts (power) are expended in a soldering iron operating on 110 volts and drawing 12 amps?

Use formula: $P = I^2R$ and $R = \frac{E}{I}$

Substitute: $I = 12,\ \ E = 110$

$R = 110 \div 12 = 9.166 \text{ ohms}$

$P = I^2R$

$P = 12 \times 12 \times 9.166 = 1319.904$

                 say, 1320 watts

Study Problems.

1. An electric oven has a resistance of 9.6 ohms. A current of 12.5 amps is flowing. How much power is used (watts)?

2. The formula for the sum of the separate resistances in a parallel circuit is:

$$\frac{1}{R} = \frac{1}{R_1} + \frac{1}{R_2} + \frac{1}{R_3}$$

What is the total resistance in a parallel circuit when: $R_1 = 2$ ohms, $R_2 = 6$ ohms and $R_3 = 12$ ohms?

3. The impedance, Z, of an electric circuit is given in the formula:

$$Z = \sqrt{R^2 + X^2}$$

where R is the resistance in ohms and X is reactance in ohms. Calculate the impedance of a circuit when $R = 342.00$ ohms and $X = 5.69$ ohms.

4. The voltage at the load end of a transmission line is given in the formula:

$$V_L = V_G - I_l R_l$$

where V_G is voltage of the generator, I_l is the current in amperes, and R_l the total resistance of the line in ohms. Solve for V_L when $V_G = 44,000$ volts, $I_l = 150.0$ amperes and $R_l = 49.5$ ohms.

Machine Shop Problems

The following formulas should be solved by first rearranging terms, if necessary, placing the unknown quantity on the left side. Then the known values and constants may be substituted.

Study Problems.

1. To find the best wire size W when measuring American National Standard Unified threads with three wires and a micrometer, the following formulas are used:

$$W = 0.57735P \quad \text{or:} \quad W = \frac{0.57735}{N}$$

where: $P =$ thread pitch in in., and $N =$ threads per in. Find the best wire size when $P = 0.1250$ in.; when $N = 8$.

2. The following formula is used to find the measurement M over three wires in the gaging of American National Standard (Unified) threads:

$$M = E - (0.866603P) + 3W$$

where: $E =$ pitch diameter of thread in in., $W =$ wire diameter in in., and $P =$ pitch of the thread in in.

Find M when $E = 0.4500$ in., $P = 0.0769$ in. and $W = 0.0500$ in.

3. The horsepower required at the cutter for turning a metal workpiece on a lathe is given as:

$$\text{hp}_c = \text{uhp} \cdot C \cdot 12 \cdot V \cdot f \cdot d$$

where: uhp = Unit horsepower: horsepower required to cut metal at the rate of 1 cu in. per minute
  $C =$ Correction factor for the feed
  $V =$ Cutting speed in feet per minute
  $f =$ Feed in inches per revolution
  $d =$ Depth of cut in inches

Find the horsepower necessary at the cutter when the unit horsepower (uhp) for mild machine steel (AISI 1020) equals 0.58, the correction factor C for a feed f of 0.015 ipr is 0.96, $V = 100$ fpm, $f = 0.015$ ipr and $d = 0.150$ in.

4. In problem no. 3, solve for the horsepower at the cutter when the uhp for gray cast-iron is 0.33, the correction factor C for a feed f of 0.018 ipr is 0.92, $V = 250$ fpm, $f = 0.018$ ipr and $d = 0.125$ in.

5. The total horsepower required to drive a lathe is determined by the formula:

$$hp_m = \frac{hp_c}{e_m}$$

where: hp_c = Horsepower at the cutting tool
          hp_m = Horsepower at the driving motor
          e_m = Mechanical efficiency factor

Solve for the horsepower of the driving motor of a lathe where $hp_c = 1.36$ and $e_m = 0.90$.

6. Solve for the horsepower at the motor of a lathe when the uhp for alloy steel (AISI 3120) is 0.50, the correction factor C for a feed of 0.030 ipr is 0.83, $V = 175$ fpm, $f = 0.030$ ipr, $d = 0.250$ in., and $e_m = 0.70$.

7. A milling machine must be set up so that the milling cutter will operate at the proper cutting speed. The formula used is:

$$N = \frac{12V}{\pi D}$$

where: N = Required spindle speed in rpm
          V = Required cutting speed in feet per minute (fpm)
          D = Diameter of milling cutter in inches
          π = 3.14

Calculate the spindle speed for a 0.500-in. diameter cutter when $V = 100$ fpm.

8. The correct spindle speed of a lathe in rpm is obtained from the following formula:

$$N = \frac{12V}{\pi D}$$

where: N = Spindle speed in revolutions per minute
          V = Cutting speed, feet per minute (fpm)
          D = Diameter of the workpiece, inches
          π = 3.14

Solve for the spindle speed when $V = 100$ fpm and $D = 0.7500$ in.

9. In the formula in problem no. 8 solve for the cutting speed when the spindle speed is 850 rpm and the work is 0.750 in. in diameter.

10. Solve for the diameter of the workpiece in a lathe when the spindle speed is 925 rpm and the cutting speed is 100 fpm.

11. The tap drill size for an American National Standard Unified Coarse thread is found by using the formula:

$$D = T - \frac{0.812}{N}$$

where: D = Size of tap drill, in.
T = Nominal diameter of the tap or thread
N = Number of threads per inch.

Find the size of the tap drill for a $\frac{5}{8}$-inch American National Standard Unified Coarse thread with 11 threads per inch.

12. Determine the size of the tap drill for a 1-in. American National Standard Unified Coarse thread with 8 threads per inch. Use formula in problem no. 11.

13. The cutting speed V for all metal cutting operations is given in feet per minute (fpm). The formula used to correct for vibrations in the feed and depth of cut is:

$$V = V_o F_f F_d$$

where: V_o = The original cutting speed from tables relating to material used, its condition, hardness and type of cutting tool used (in feet per minute, fpm)
F_f = Feed correction factor from tables based on test results.
F_d = Depth of cut correction from test tables.

Solve for V when V_o = 100 fpm, F_f = 0.74 and F_d = 0.87

14. The milling machine spindle speed N in revolutions per minute (rpm) is calculated by the formula:

$$N = \frac{12V}{\pi D}$$

where: V = Cutting speed in feet per minute
D = Dia. in inches, of the milling cutter.

Solve for N when V = 64.4 fpm and D = 3 in.

15. The table feed rate f_m for a milling operation is given in inches per minute (ipm), and the formula is:

$$f_m = f_t n_t N$$

where: f_t = feed rate, inches per tooth (ipt)
n_t = number of cutting teeth on cutter
N = spindle speed (speed of rotation of cutter)(rpm)

Solve for the table feed rate when f_t = 0.010 ipt, n_t = 10, and N = 350 rpm.

16. A machine shop foreman must recommend the speed at which a lathe should be turning when machining a 3.500-in. dia. part made from AISI 1040 steel with a cemented carbide tool. The feed is to be 0.020 in. per revolution and the depth of cut 0.200 in. From a table in the *Machinery's Handbook, 19th Ed.*, the cutting speed V_o is found to be 300 ft per minute, the feed correction factor F_f is 0.74 (for 0.020 ipr) and the depth of cut correction factor F_d is 0.91 (for 0.200 in.). Using the symbols and formulas in problem 13, calculate the lathe-spindle speed that should be recommended.

17. Calculate the indexing movement required to index a 60-tooth gear on a standard dividing head, using the 15-hole circle on the index plate. The formula for plain indexing is given as:

$$T = \frac{40}{N}$$

where: T = Number of complete and fractional parts of a turn of the dividing head crank

            N = Number of divisions required on the workpiece; in this case, $N = 60$

To solve this problem, the fraction $\frac{40}{N}$, must be reduced so that the final denominator is 15.

18. Threads are measured by a three-wire system. A $1\frac{1}{2}$-12 UNF-3A thread is to be measured using wires that are 0.050 in. in dia. The formula to use is:

$$M \text{ (distance over wires)} = D - (1.5155P) + 3W$$

where: D = nominal O.D. of bolt = $1\frac{1}{2}$ in., P = pitch of the threads which equal $1 \div$ threads per inch = $\frac{1}{12}$ = 0.0833 in. and W = diameter of the wires = 0.050 in.

Solve for M.

19. The drill spindle speed N in rpm is given in the formula:

$$N = \frac{12V}{\pi D}$$

where: V = The cutting speed in feet per minute, fpm, at which the drill can cut the workpiece material.

            π = 3.1416

            D = Diameter of the drill, in.

Calculate the spindle speed required to drill a $\frac{1}{8}$-in. hole into material requiring a cutting speed of 50 fpm.

20. When using a Cincinnati Universal Dividing Head, one turn of the crank rotates the spindle $\frac{1}{40}$ revolution. As there are $360°$ in one revolution,

one turn of the crank moves the spindle $\frac{360°}{40}$ or $9°$. When a definite angle is to be indexed, these formulas are used:

For angle in degrees: $T = \dfrac{D°}{9}$

For angle in minutes: $T = \dfrac{D'}{540}$

For angle in seconds: $T = \dfrac{D''}{32,400}$

where: $T =$ turns of crank and $D =$ angle to be indexed.

How many crank turns are required to index $14° \ 20'$?

Hydraulic Formulas

Study Problems.

The following basic formulas are widely used in hydraulic computations.

1. The basic pressure formula is:

$$P = \frac{F}{A}$$

where: $F =$ force in lbs, and $A =$ area in sq in.

In Fig. 8-2, calculate the pressure needed to raise a load F of 3 tons, if the piston diameter is 12 in.

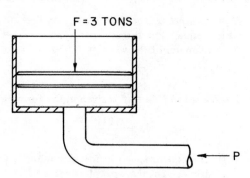

F = 3 TONS

Fig. 8-2. Study problem 1.

2. Pascal's law is given as: $\dfrac{F_1}{F_2} = \dfrac{A_1}{A_2}$

Figure 8-3 shows a simple hydraulic jack with a force F_1 of 150 lbs acting upon a piston of area A_1. Piston with area A_2 is 4 in. in diameter. Solve for F_2.

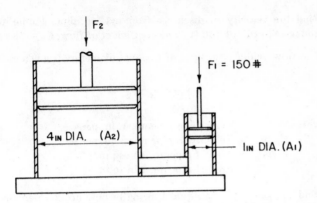

Fig. 8-3. Study problem 2.

3. The flow rate of a fluid is given in the formula:

$$q = Av$$

where: q = Flow rate in cu in. per sec.
          A = Inside area of the pipe
          v = Flow velocity in inches per second

Find the flow rate in a pipe that is 0.75 in. I.D. when the liquid is flowing at 12 in. per sec.

4. The metal thickness of a thin-wall cylinder subjected to internal pressure is given in the formula:

$$t = \frac{Dp}{2S}$$

where: t = Minimum wall thickness required, in.
          D = Inside diameter of cylinder, in.
          p = Internal pressure, lbs per sq in. (psi)
          S = Allowable tensile stress in metal, lbs per sq in. (psi)

Find the minimum wall thickness for a 3-ft dia. tank to retain 350 psi pressure and to be made from steel with an allowable tensile stress of 55,000 psi.

5. The flow of water in pipes is in feet per second V in the formula:

$$V = C \sqrt{\frac{hD}{L + 54D}}$$

where: C = Coefficient of flow (from tables)
          D = Diameter of pipe in feet
          h = Total "head" in feet
          L = Total length of pipeline in feet

Find the velocity of discharge from a 12-in. pipe, 1 mile long working under a "head" of 100 ft. The coefficient of flow, $C = 48$.

6. The flow velocity in pipes of unequal diameters is given in the formula:

$$v_1 = \frac{A_2}{A_1} v_2$$

where: v_1 = Velocity in pipe no. 1, in. per sec.
v_2 = Velocity in pipe no. 2, in. per sec.
A_1 = Area of pipe no. 1 in sq in.
A_2 = Area of pipe no. 2 in sq in.

Find the velocity of a liquid flowing in pipe no. 2 when the velocity in pipe no. 1 is 5 in. per sec. The area of pipe no. 1 is 2 sq in. and the area of pipe no. 2 is 1.25 sq in.

7. The flow through a pipe is given in the formula:

$$Q = 2.44 d^2 V$$

where: Q = Flow in gallons per minute
d = Inside diameter of pipe in inches
V = Flow velocity in feet per second

Find the flow from a $1\frac{1}{2}$ in. I.D. pipe when the flow velocity is 10 ft per second.

8. Barlow's Formula gives the wall thickness required for tubes and pipes. It is:

$$t = \frac{pD}{2S}$$

where: t = Wall thickness in inches
p = Hydraulic pressure in psi
D = Outside diameter of pipe in inches
S = Allowable stress in pipe wall in psi

Find the stress in the wall of a $1\frac{1}{2}$ in. O.D. pipe with an $\frac{1}{8}$ in. wall thickness when operating at 500 psi.

Horsepower of Pumps and Motors

Study Problems.

1. The horsepower required to drive a hydraulic pump is given in the formula:

$$hp = \frac{pq}{6600}$$

where:　　p = Pressure in psi
　　　　　　q = Fluid delivery in cu in. per sec.

Find the horsepower required to deliver 1500 gpm at 2000 psi.

2. An alternative formula for the one in problem no. 1, above, is:

$$hp = \frac{pq}{1715}$$

where:　　p = Pressure in psi
　　　　　　q = Fluid delivery in gallons per min

Using this alternative formula, find the horsepower required to deliver 3440 gallons per minute at 200 psi.

3. Another form of the formula in problem no. 1, above, is:

$$hp = 0.2618pq$$

where:　　p = Pressure in psi
　　　　　　q = Fluid delivery in cu ft per sec

Using this formula, find the horsepower required at a pumped water storage operation to pump 95 cu ft of water per sec at 5 psi.

Formulas Applying to Hydraulic Cylinders

Study Problems.

1. For single piston rod cylinders two formulas are used:

$$F = PA \quad \text{or:} \quad F = P\,\frac{\pi D^2}{4} \text{ at plain end}$$

$$F = P(A - a); \quad F = P\,\frac{\pi}{4}\,(D^2 - d^2)\text{ at rod end}$$

where:　　F = Force in lbs
　　　　　　p = Pressure in lbs per sq in. (psi)
　　　　　　A = Area of piston, sq in.
　　　　　　a = Area of piston rod, sq in.
　　　　　　D = Diameter of piston, in.
　　　　　　d = Diameter of rod, in.

Determine the force developed by a hydraulic cylinder 4 in. I.D. operated with line pressure of 150 psi at the plain end. See Figs. 8-2 through 8-4.

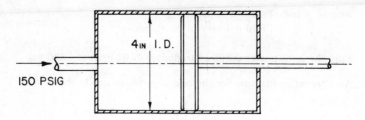

Fig. 8-4. Study problem 1.

2. Double-rod cylinders have the piston rod extending out of both ends of the cylinder and use the "rod-end" formula for both ends. Find the force developed by a cylinder $2\frac{1}{2}$ in. I.D. having a 0.875 in. dia. rod when working under 3000 psi line pressure.

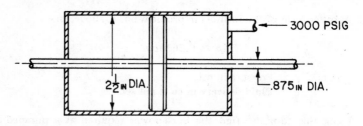

Fig. 8-5. Study problem 2.

3. The speed of operation of hydraulic cylinders is proportional to the flow rate of the hydraulic fluid. Formulas are given as:

$$q = \frac{\pi D^2 v}{4} \text{ for plain ends}$$

$$q = \frac{\pi (D^2 - d^2) v}{4} \text{ for rod ends}$$

where: q = Flow rate in cu in. per sec
D = Diameter of cylinder, in.
d = Diameter of piston rod, in.
v = Velocity of piston, in. per sec

Find the velocity of a 3-in. double rod end piston with a rod diameter of 0.75 in. and a flow rate of 230 cu in. per sec. Hint: transpose formula first, put v on the left.

4. For a velocity V given in ft per min, the formula for flow rates of fluid becomes:

$$q = \frac{\pi D^2 V}{20} \text{ for plain ends}$$

$$q = \frac{\pi(D^2 - d^2)V}{20} \text{ for rod ends}$$

where: q = Flow rate, cu in. per sec
         D = Diameter of cylinder, in.
         d = Diameter of rod, in.
         V = Velocity of piston, ft per min

Find the flow rate necessary to operate a single-rod piston at 20 ft per min if the I.D. = 6 in. and the piston rod is $1\frac{1}{2}$ in. dia. Give the plain end flow rate and the rod end flow rate.

5. When the flow rate of the hydraulic fluid is given in gallons per minute (gpm), the formulas for cylinders are:

$$Q = \frac{\pi D^2 V}{77} \text{ for plain ends}$$

$$Q = \frac{\pi(D^2 - d^2)V}{77} \text{ for rod ends}$$

where: Q = Flow rate, gpm
         D = Diameter of piston, in.
         d = Diameter of rod, in.
         V = Velocity of piston, ft per min

Calculate the flow rate necessary to operate a piston at 15 ft per min if the I.D. is 8 in. and the rod diameter is $1\frac{7}{8}$ in. Find the plain end flow rate and the rod end flow rate.

6. Hydraulic cylinders on earth moving equipment and other articulated machinery require flexible hosing to operate the moving cylinders.

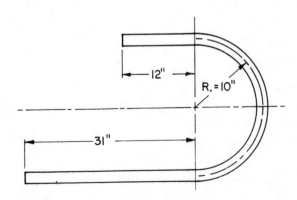

Fig. 8-6. Study problem 6.

The formula for the length of hose required is:

$$L = A + 3.1416R$$

where: L = length of hose, in.
$\quad\quad A$ = combined lengths beyond each end of bend, in.
$\quad\quad R$ = radius of bend, in.

Find required length of hose for Fig. 8.6.

CHAPTER 9

Tapers and Related Tooling Calculations

A taper is a gradual decrease of the width or the diameter of an object. Conical tapers have many applications in the design of machinery; i.e., taper shanks of cutting tools, machine tool spindle noses, taper pins, tapered bearings, etc. The conical taper provides an excellent method of alignment for external and internal surfaces. Flat tapers are really inclined planes that are used as adjusting wedges and keys. Tapered keys are used to hold two components together, while drift keys or drift pins, as shown in Fig. 9-1., are used to extract drills from sockets.

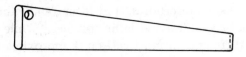

Fig. 9-1. Drift key.

Twist drills, end mills, arbors, lathe centers and other small tools and machine parts have tapered shanks which fit into spindles, sockets, or adaptors having a matching taper. This arrangement provides accurate alignment of the centers of the driving spindle and the tool. It also provides a convenient "friction" method for holding the tool in the spindle; however, the holding friction must not be depended on to drive the tool when cutting. The taper of the tool must be identical with the taper of the holder to provide the friction fit.

When a taper is in the range of 4 to 6 degrees included angle or less, the tool is so firmly seated that it is retained by the frictional resistance of the matching conical surfaces. The "tang" at the end of the taper shank tool, Fig. 9-2, is used to give positive rotational drive. The term "self-holding" is used to describe this type of taper.

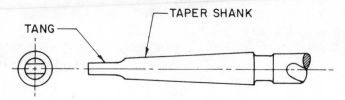

Fig. 9-2. Taper shank tool showing "tang."

Tapers with an included angle of about 18 to 20 degrees in steel are "self-releasing," depending on the condition of the tapered surfaces. These types of tapers provide accurate axial alignment but require a positive locking device to prevent their slipping out of the spindle.

The exact dimensions and specifications for standard tapers are tabulated in engineering handbooks such as *Machinery's Handbook*. There are five standard taper systems in common use in the United States. These are as follows:

Morse taper. The Morse taper is a self-locking taper used exclusively on twist drills and extensively on the shanks of reamers, end mills, countersinks, counterbores, spot facers, and lathe centers. It is used on drill-press spindles and lathe spindles. While each size of a Morse taper is different, in most cases it is approximately $\frac{5}{8}$ inch per foot. Figure 9-3 shows the critical dimensions of a Morse taper socket, shank, and tang. Note the method of ejecting the shank by means of a key or drift shown in the lower right of the figure.

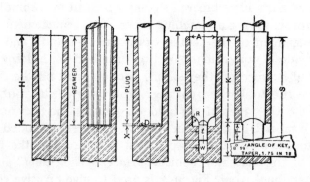

Fig. 9-3. Morse standard taper shanks.

Brown & Sharpe. This taper is used on the shanks of end mills and reamers, collets, milling machine spindles, and grinding machine spindles. The taper is approximately $\frac{1}{2}$ inch per foot. Figure 9-4 shows the main dimensions of the Brown & Sharpe taper shanks.

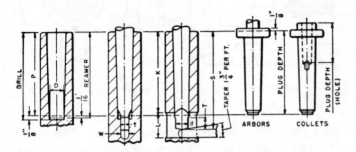

Fig. 9-4. Brown & Sharpe taper shanks.

Jarno. The Jarno taper has been used on some machine tools. The taper per foot of all Jarno Tapers is 0.600 inch. See Fig. 9-5 for details.

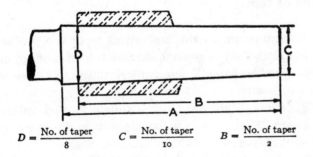

$$D = \frac{\text{No. of taper}}{8} \qquad C = \frac{\text{No. of taper}}{10} \qquad B = \frac{\text{No. of taper}}{2}$$

Fig. 9-5. The Jarno taper.

Milling-Machine Taper. This taper was adopted by the milling machine manufacturers of the National Machine Tool Builder's Association. It is an American standard and is used on almost all of the milling machine spindles built in the United States. The taper is $3\frac{1}{2}$ inches per foot in all sizes in order to assure easy release from the spindles of arbors, adaptors, and other tools. Figure 9-6 shows the basic dimensions of the standard milling machine arbor and the adaptors and draw-in bolts.

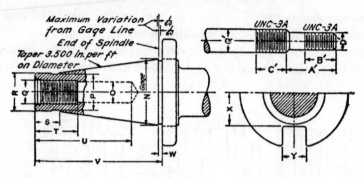

Fig. 9-6. Standard milling machine arbors, adaptors, and draw-in bolts.

American National Standard Machine Tapers. This standard includes a self-holding series and a steep taper (self-releasing) series. The self-holding tapers are, to a large extent, an adaption of the Morse and Brown & Sharpe tapers. The steep taper series corresponds closely to the Milling Machine Taper.

Calculation of Tapers

The determination of the amount of taper is a problem in proportion. For simplicity, a general formula is used that applies to both flat tapers (wedge-shaped) and conical tapers. Figure 9-7 shows both types of tapers.

In Fig. 9-7(a), a conical taper is shown as used on tapered drill shanks. Figure 9-7(b) shows a flat taper as in the shape of a "drift" used to remove tapered tools from spindles. Figure 9-7(c) is a drawing of a conical taper with the lines BD and FG drawn parallel to the centerline. The total taper is the distance $AB + FH$ for the length L. Figure 9-7(d) shows a flat taper with the line BC drawn parallel to the line DE. The total taper is the distance AC for the length CB.

As mentioned above, the calculation of tapers involves the use of a simple proportion involving the large and small dimensions at the ends of the tapers and the length. In Fig. 9-7(c), the taper per unit length is the sum of the distances $AB + FH$ over the length L. DG is the small diameter of the taper and is equal to the segment BF of the big end of the taper. Adding the segments of the large diameter, we get:

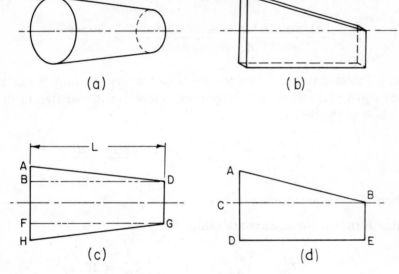

Fig. 9-7. Conical and wedge shape (flat) tapers.

$$BF + AB + FH = AH$$

$$\text{also } AH - AB - FH = BF$$

$$\text{and} \qquad AH - BF = AB + FH$$

The sum of $AB + FH$ is the difference between AH and DG, which are the large and small diameters of the taper. $AB + FH$ equals the total taper for the length, L.

Definitions for taper calculations.

T = Taper; the difference in size between the large and the small diameters.

T_{pi} = Taper per inch; the change in size of the taper in a 1-inch length.

T_{pf} = Taper per foot; the change in size of the taper in a 1-foot length.

D = Large diameter of the taper, inch.

d = Small diameter of the taper, inch.

L = Length of the taper in inches.

A proportion can be written for taper calculation;

$$\frac{\text{Large diameter} - \text{small diameter}}{\text{Length of taper in inches}} = \frac{\text{Taper}}{\text{Inch}}$$

Many tables give reference to tapers as "tapers per foot." As 1-foot equals 12 inches, the proportion above can be written in the symbols of the definition:

$$T_{pi} = \frac{D - d}{L} \quad \text{or} \quad T_{pf} = \frac{12(D - d)}{L}$$

Where D and d are given in inches.

Other formulas for taper relationships are:

$$T_{pi} = \frac{T_{pf}}{12} \qquad\qquad T_{pf} = 12T_{pi}$$

$$D = (LT_{pi}) + d \qquad d = D - (LT_{pi})$$

$$L = \frac{D - d}{T_{pi}}$$

Study Examples.

1. A taper is 4 in. long. The large diameter is 1 in. and the small diameter is $\frac{3}{4}$ in. Calculate the taper per inch. Solution:

$$T_{pi} = \frac{D - d}{L}$$

$$= \frac{1.00 - 0.75}{4}$$

$$= \frac{0.25}{4}$$

$$= 0.625 \text{ in. per in.}$$

2. A taper plug is measured on an optical comparator. $D = 0.9831$ in., $d = 0.8229$ in., $L = 6.0000$ in. Calculate the taper per inch. Solution:

$$T_{pi} = \frac{D - d}{L}$$

$$= \frac{0.9831 - 0.8229}{6.0000}$$

$$= \frac{0.1602}{6.0000}$$

$$= 0.0267 \text{ in. per in.}$$

3. Convert answer in example no. 2 to taper per foot. Solution:

$$T_{pf} = 12(T_{pi})$$

$$= 12(0.0267)$$

$$= 0.3204 \text{ in. per ft}$$

4. Determine the small diameter, d, of the taper shown in Fig. 9-8 when $D = 1.2206$ in., $L = 8.30$ in. and the taper per foot $= 1.1000$ in. Solution:

$$T_{pi} = \frac{T_{pf}}{12}$$

$$= \frac{1.1000}{12}$$

$$= 0.09167$$

Then: $d = D - (LT_{pi})$

$$= 1.2206 - (8.3 \times 0.09167)$$

$$= 1.2206 - 0.7609$$

$$= 0.4597 \text{ in.}$$

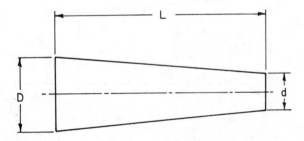

Fig. 9-8. Study example 4.

5. Calculate the diameter of the large end of a taper when the small diameter is 0.778 in., the taper per foot $= 0.60235$ in., the length of the taper is 3.250. Solution:

$$T_{pi} = \frac{T_{pf}}{12}$$

$$= \frac{0.60235}{12}$$

$$= 0.050196$$

Then: $D = (T_{pi} \times L) + d$

$$= (0.050196 \times 3.250) + 0.778$$

$$= 0.1631 + 0.778$$

$$= 0.941 \text{ in.}$$

Study Problems.

1. In the taper shown in Fig. 9-9 determine the taper per foot.

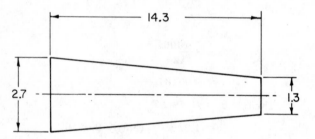

Fig. 9-9. Study problem 1.

2. Determine the taper per inch of the taper shown in Fig. 9-10.

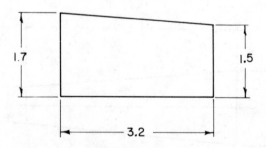

Fig. 9-10. Study problem 2.

3. The taper shown in Fig. 9-11 has a 0.4000 in. taper per foot. Solve for the length.

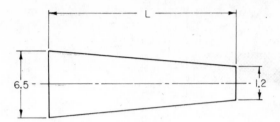

Fig. 9-11. Study problem 3.

4. In the taper gage shown in Fig. 9-12, solve for the length, L, when D = 1.2556 in., d = 1.0553 in. and the taper per foot is 1.205 in.

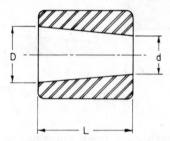

Fig. 9-12. Study problem 4.

5. Solve for the small diameter, d, in the taper ring gage shown in Fig. 9-12 when D = 1.3025, L = 1.2558 in. and the taper per foot is 1.925 in.

6. Solve for the taper per inch in the taper plug and ring gage set shown in Fig. 9-13 when dimension A = 1.125 in., B = 2.000 in., the length L = 5.000 in., dimension D = 1.250 in. and d = 0.7500 in.

7. In Fig. 9-13 and using the dimensions from problem 6 above, solve for C.

8. Solve for dimension c in Fig. 9-13 using the dimensions in problem 6 above.

9. In the taper plug and ring gage set shown in Fig. 9-14 the taper per foot is 0.3125 in. If d were machined 0.021 in. larger, how much farther would the taper plug advance into the ring gage, the taper per foot remaining constant?

10. In Fig. 9-14, the taper per foot is 0.875 in. How much must dimension d be increased to permit the taper plug to advance 0.011 in. farther, the taper per foot remaining constant?

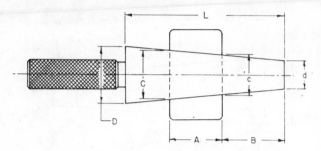

Fig. 9-13. Study problems 6, 7, and 8.

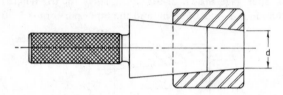

Fig. 9-14. Study problems 9 and 10.

Measuring Tapers

The method of determining the dimensions of tapers is by gaging and by measuring, using a checking method set up on a surface plate using sine bars. The sine bar method is treated in a later chapter, since it involves the use of trigonometry.

Taper plugs and taper ring gages are most often used in the tool-room and the machine shop as they are simple to use and dependable for accuracy. The plugs gage the taper angularity and the diameters of holes, while the rings check the taper angularity and diameters of tapered shafts.

In Fig. 9-15, a step-type plug and a ring "Go No-Go" gage is shown. The angularity of the hole or tapered shaft gaged is checked by the "bluing" process commonly used in shops. In the case of holes, the Prussian Blue is applied to the plug gage and the "rub-off" observed in the hole. Tapered shafts have the blue applied to the shafts and the "rub-off" observed on the interior of the ring gage.

Note the *steps* machined into the ends of both the plug and the ring. When the first step A is flush with the end of the shaft or the surface at the small diameter of the hole, the dimension stamped into the step is read. When the recessed step is flush with the surface, the stamped dimension on it is read. When taper gages are used in production, they are designed to be easily read by the operator without the need to calculate.

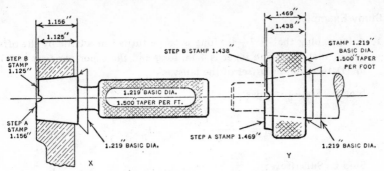

Fig. 9-15. Step plug and ring, go and not-go taper gages.

Taper Turning—Offset-Center Method

A common method of turning tapers is shown in Fig. 9-16. The workpiece is placed between the headstock and the tailstock centers. When both of the centers are in line, the movement of the cutting tool is parallel to the axis of the work. In this position a cylindrical piece would be cut. If, however, the tailstock center is moved out of axial alignment, as shown in Fig. 9-16 the work will then be tapered as the tool cuts from a to b because the axis of the workpiece, $x - x$, is at an angle with the movement of the tool. The total amount of taper, $D - d$, will depend on how much the tailstock center is offset.

The formula for taper turning by the offset-center method is:

$$\text{Offset} = \frac{\text{Taper per inch} \cdot \text{Length of workpiece}}{2}$$

or: $\text{Offset} = \dfrac{T_{pi} \cdot L_w}{2}$

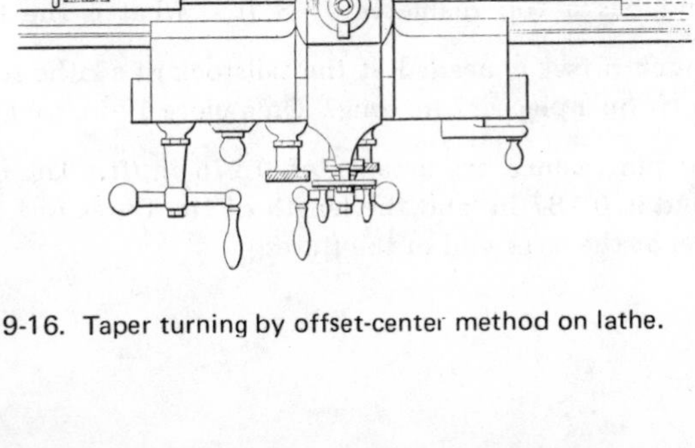

Fig. 9-16. Taper turning by offset-center method on lathe.

Study Example.

1. A taper plug that is $7\frac{1}{2}$ in. long is to be turned in a lathe by the offset-center method. The taper is 4 in. long and the taper per foot is 0.750 in. Determine the offset of the tailstock.

Step 1. State formulas:

$$\text{Offset} = \frac{T_{pi} \cdot L_w}{2} \qquad T_{pi} = \frac{T_{pf}}{12}$$

Step 2. Substitute:

$$T_{pi} = \frac{0.750}{12} = 0.0625 \text{ in.}$$

$$\text{Offset} = \frac{0.0625 \cdot 7.500}{2} = 0.234 \text{ in.}$$

Study Problems.

1. Calculate offset of the tailstock to cut a 0.625 T_{pf} taper on a workpiece that is 6.25 in. long.

2. What size taper would be cut on a 13-in. shaft if the tailstock were offset by 0.130 in. Give answer as taper per foot.

3. A shaft 3 ft long is to have a taper 5 in. long cut with a 0.625 taper per foot. Calculate the offset necessary at the tailstock.

4. A taper 7 in. long on a 16-in. shaft is cut when the tailstock is offset 0.256 in. What is the taper per foot of the taper?

Review Problems.

1. A taper may be turned in a lathe by shifting the tailstock a distance off center equal to half the taper in inches. A shaft 3 ft long (L) is to have a taper from a diameter of 4 in. on one end to a diameter of $3\frac{1}{2}$ in. on the other end. Find the amount of offset in inches. (Use formula: Offset = $\frac{D-d}{L}$.)

2. The shaft of a golf club is 38 in. long, $\frac{5}{8}$ in. in diameter at the large end, and $\frac{3}{8}$ in. at the small end. What is the T_{pi}?

3. A smokestack of a power plant is 200 ft high, 19 ft in diameter at the base and has a top diameter of 8 ft. What is the taper per foot?

4. How much offset is needed at the tailstock of a lathe to give a taper of 0.95 in./ft on a piece 27 in. long? On a piece 17 in. long?

5. A taper pin reamer has a taper of 0.275 in./ft. The diameter of the small end is 0.187 in. and the length of the flutes is 4.50 in. Find the diameter at the large end of the flutes.

6. How much is the offset of a tailstock to turn a taper on a piece 14 in. long, if the tapered portion is to be 6.5 in. long and the diameter of the shaft is 1.375 (*D*) in. and the small diameter is 0.915 in.?

7. The standard pipe thread taper is $\frac{3}{4}$ in. per foot. How much is this in inch per inch?

8. A taper plug is to be designed to check the taper of an exhaust fitting with a taper of 0.050 in. per in. The large diameter of the fitting must be held to 3.625±0.010 in. What will be the depth of the step at the large diameter of the plug?

9. A taper wedge used by millwrights in leveling machines measures 3 in. wide, 14 in. long and 1 in. thick at the butt (*D*) and tapers to 0 in. at the edge. How far must the wedge be inserted under a drill press to raise it 0.325 in.? (Assume vertical rise using two or more wedges.)

10. In Fig. 9-17, determine the distance *A*.

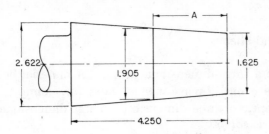

Fig. 9-17. Study problem 10.

CHAPTER 10

Geometry — Lines, Angles, and Plane Figures

Geometry deals with the properties and relationships of lines, surfaces, and areas and their measurement. Geometry has its origins in pre-history; ancient people were aware of basic geometric shapes and relations. Artifacts such as decorated prehistoric pottery from all over the world show the early use of circles, triangles, and parallel lines.

Plane and Solid Shapes

We live in a three-dimensional world. A material body occupies space and space has dimensions. A box, for example, has length, width, and depth — three dimensions. Three dimensional bodies are called *geometric solids.*

While our natural world is three-dimensional, the usual means of communication is on a flat or two-dimensional medium. The blackboard in the classroom, the drafting board, this page: all are examples of the two-dimensional world. The study of geometry begins in a flat, two-dimensional world called "planar" (plane) geometry. Our studies will focus on this area as it is the basis of communication between the designer and the shop man.

Shop drawings are made on a flat, two-dimensional sheet and concern a three-dimensional object to be made. Conventional methods of drawing and reading mechanical and architectural drawings involve plane geometry. Three-dimensional solids will be discussed in Chapter 14, "Surface Areas and Volumes of Solids."

Using Geometry

The study of geometry involves learning the definitions of geometric shapes and their relation to one another. It requires learning the basic, logical truths about these relationships which are called *axioms* and *postulates*. It requires learning drawing techniques to more easily construct geometric shapes.

When these "basics" of geometry have been acquired, the apprentice and journeyman will have new "tools" with which to solve a number of shop problems. Sheet metal workers, plumbers, pipefitters and electricians use geometry in laying-out and installing "runs" of ducts, pipe, and conduit. Machinists use geometry in making layouts and in setting up work on the various toolroom machines. Land surveyors, carpenters, and builders continually use geometric constructions in their work.

Geometry is also the basis for solving the more complicated trigonometric problems that arise in machining gears, cams, screws, and worm gears. The designer of mechanical products, the construction engineer, and the general contractor use basic geometry in connection with problems involving trig and compound angles. Problems of area, volumes, capacities, weights, and construction safety involve geometric principles.

The ability to use geometry effectively in your work will depend on the effort you make in learning the fundamentals and in working the study problems given in the text.

Definitions in Plane Geometry

To be able to study a subject, we must first define basic parts and constructions. The following are the basic geometric concepts, definitions, and relationships. These principles must be thoroughly understood and memorized so that geometry may be used as a practical tool in your daily work.

1. *The Point.* A point has only the characteristic of position. It has no dimensions, such as length, width, or thickness. A point is an origin or a beginning; for example, the starting place in constructing a circle. Two points are necessary to define a line and three points determine a flat surface or "plane" in space.

2. *The Line.* A line is one-dimensional and it has only length. A straight line is defined as the shortest distance between two points. A line, beginning at one point and ending at another, will have *finite* length, Fig. 10-1. If a line passes through one or both points, its length is undetermined and it is *indefinite* in length, Fig. 10-2. A line

having the same direction throughout its length is called a *straight* line. A line that continuously changes direction is a *curved* line, Fig. 10-3. A "broken" line is a series of connected straight lines extending in different directions, as shown in Fig. 10-4. Lines form boundaries and define geometric shapes and figures. In this text, unless otherwise stated, when the term *line* is used, a straight line is meant. A *horizontal* line describes a straight line which is level or even with the earth's horizon.

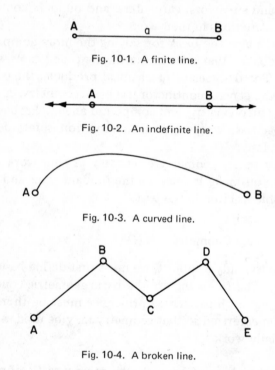

Fig. 10-1. A finite line.

Fig. 10-2. An indefinite line.

Fig. 10-3. A curved line.

Fig. 10-4. A broken line.

3. *The Circle.* A circle is a finite, curved line — all points on which are equidistant from a point called the *center* of the circle. The distance from the center of the circle to the curved line is called the radius. The diameter of a circle passes through the center, and its length is equal to twice the length of the radius. The diameter is the longest possible distance between any two points on the circle.

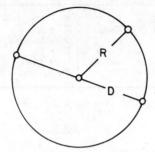

Fig. 10-5. The circle with radius *R* and diameter *D*.

4. *Angles.* Two straight lines meeting at a point (the vertex) form an *angle*. The two lines are the *sides* of the angle. An angle is conventionally identified by reading the letters placed at the ends of the sides, with the letter at the vertex placed between them. In Fig. 10-6, the angle is written; ∠*ABC*. The angle can also be referred to as *angle B* (∠*B*) using the letter of the apex. Angles many times are identified by using a Greek letter written within the vertex of the angle such as β in Fig. 10-6. Other Greek letters are used when more than one angle is to be identified.

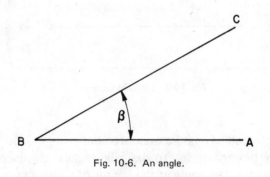

Fig. 10-6. An angle.

The Greek alphabet is used extensively in mathematics and in the sciences. It is reproduced in Fig. 10-7. Note that there are both capital and small letters in the Greek alphabet. It is recommended that the reader become familiar with these letters in order to properly identify them and pronounce them.

A	α	Alpha	II	η	Eta	N	ν	Nu	T	τ	Tau
B	β	Beta	Θ	ϑ θ	Theta	Ξ	ξ	Xi	Y	υ	Upsilon
Γ	γ	Gamma	I	ι	Iota	O	o	Omicron	Φ	φ	Phi
Δ	δ	Delta	K	κ	Kappa	Π	π	Pi	X	χ	Chi
E	ε	Epsilon	Λ	λ	Lambda	P	ρ	Rho	Ψ	ψ	Psi
Z	ζ	Zeta	M	μ	Mu	Σ	σ s	Sigma	Ω	ω	Omega

Fig. 10-7. The Greek alphabet.

The subject of angle measurement has been covered in detail in Chapter 5. The ability to add and subtract angles and to convert to and from decimal degrees is a necessary skill needed in the study and use of geometry and trigonometry. Angles are measured and constructed by the use of a *protractor* shown in Fig. 10-8.

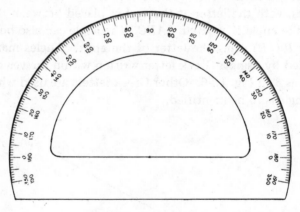

Fig. 10-8. The protractor.

An alternate unit of angular measurement is the *radian*. This unit is used primarily in engineering to describe angular displacement. The radian is defined as the angle at the center of a circle measured by an arc equal in length to the radius of the circle. This is shown in Fig. 10-9A.

In Proposition 36, discussed later in this chapter, the circumference of a circle will be shown to be equal to π times the diameter of the circle, or π times $2r$, where r is equal to the radius ($2r$ = diameter). Thus:

$$\text{Circumference} = 2\pi r$$

PERIMETER = 2πR

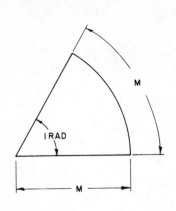

2π RADIANS
OR
360°

Fig. 10-9. A. Measurement of a radian.   B. The Circle. Circumference shown as the perimeter. The radian measure of 360° (one revolution).

Since the circumference is equal to 2π times r, it follows from the definition of a radian that the angle describing a complete revolution or a complete circle must be equal to 2π radians, as shown in Fig. 10-9B.

As this angle is also equal to 360 degrees, the relationship between radians and degrees is as given below:

$$2\pi \text{ radians } = 360°$$

$$1 \text{ radian } = \frac{360°}{2\pi}$$

$$\text{or: } 1 \text{ radian } = \frac{360°}{6.2832}$$

$$1 \text{ radian } = 57.2958°$$

$$1 \text{ radian } = 57° \ 17' \ 44.8''$$

5. *A Right Angle.* The angle described by one-quarter of a full revolution is $\frac{360°}{4} = 90°$. This is called a *right* angle or a 90° angle. The two sides of the 90° angle are referred to as being *perpendicular*. See Fig. 10-10.

6. *A Straight Angle.* The angle described by a half-revolution is $\frac{360°}{4} = 180°$. This is a straight line, and therefore the angle is called a *straight* angle. A straight angle = 180°. See Fig. 10-11.

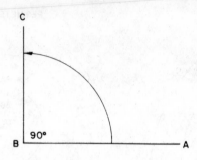

Fig. 10-10. The right angle.

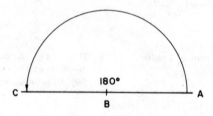

Fig. 10-11. A straight angle.

7. *An Acute Angle.* An *angle* less than 90° is called an *acute* angle. See Fig. 10-12.

8. *An Obtuse Angle.* An angle *greater* than 90° but less than 180° is called an *obtuse* angle. See Fig. 10-12.

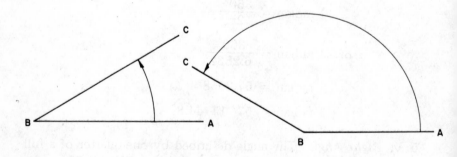

Fig. 10-12. (Left) An acute angle. (Right) An obtuse angle.

9. *Complementary Angles.* Two angles whose sum is one right angle (90°) are called *complementary* angles. In Fig. 10-13, angle *CBD* is complementary to angle *ABC* and their sum is 90°.

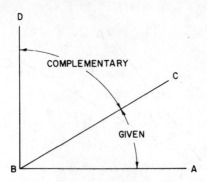

Fig. 10-13. A complementary angle.

10. *Supplementary Angles.* Two angles whose sum is one straight angle (180°) are called *supplementary*. In Fig. 10-14, angle *CBD* is supplementary to angle *ABC* and their sum is 180°.

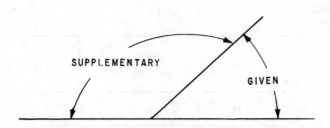

Fig. 10-14. A supplementary angle.

11. *Adjacent Angles.* When three or more lines meet at a point, adjacent angles are formed. In Fig. 10-15, ∠*ABC* is adjacent to ∠*CBD*. Line *BC* is said to be a side common to each of the pair of adjacent angles.

12. *Distance from a Point to a Line.* The distance from a point to a line is the measure of the perpendicular line drawn from the point to the line.

Lines and Their Relationships

While two points determine the length and direction of a straight line, three or more straight lines determine the size and shape of many plane geometric figures. These relationships between straight lines and the geometric figures they form are explained as follows:

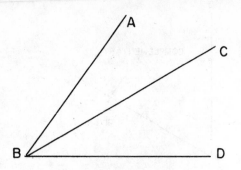

Fig. 10-15. Adjacent angles.

1. *Parallel lines.* Two or more straight lines in the same plane that are the same distance apart at any point are called *parallel* lines. Parallel lines do not intersect. Figure 10-16 shows three parallel lines. The convention of using two dimension lines, with the same letter in each, signifies that the two lines are parallel; i.e., equally distant from one another.

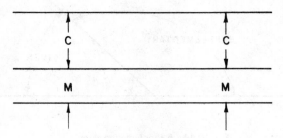

Fig. 10-16. Parallel lines.

2. *Perpendicular Lines.* A line that forms a right angle (90°) with one or more lines is said to be *perpendicular* to the line(s). Perpendicular lines erected to a horizontal line are said to be *vertical*. A "plumb-bob" on the end of a length of cord, Fig. 10-17, is a very ancient method of erecting perpendiculars during the process of building many kinds of structures. Figure 10-18 shows the relationship of perpendicular lines.

Plane Figures

Three or more straight lines, intersecting in a regular pattern, can form plane figures which are called *polygons* from the Greek words:

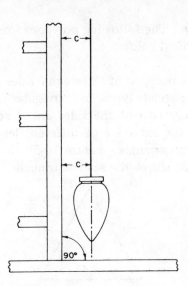

Fig. 10-17. A plumb bob and line as used in construction.

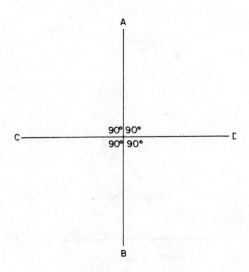

Fig. 10-18. Perpendicular lines.

poly meaning "many," and *gonia* meaning "angle." Many-angled figures have, of course, many sides.

Some of the more familiar polygons are called by special names referring to the number of sides that the figures contain. When *all* sides are of equal length and *all* internal angles are equal, the figure is a *regular polygon*. The following polygons frequently appearing in geometry are described below:

1. *Triangles*. A polygon of three equal sides is known as an *equilateral triangle*. Special types of "irregular" triangles are: the *isosceles*, where only two of the sides are of equal length, and the *scalene*, where all the sides are of different lengths. Figure 10-19 shows the equilateral triangle. Figure 10-20 shows the isosceles triangle, and Fig. 10-21 shows the scalene triangle.

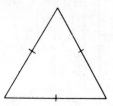

Fig. 10-19. An equilateral triangle.

Fig. 10-20. An isosceles triangle.

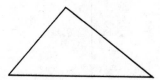

Fig. 10-21. A scalene triangle.

Triangles are also classified in regards to special angular conditions. Where one of the three internal angles is 90°, as shown in Fig. 10-22, the triangle is called a *right* triangle.

When one of the three internal angles of a triangle is *greater* than 90°, the triangle is referred to as an *obtuse* triangle. This is shown in Fig. 10-23.

When each of the interior angles of a triangle is *less* than 90°, then the triangle is called an *acute* triangle, as shown in Fig. 10-24. Both acute and obtuse triangles are called oblique triangles.

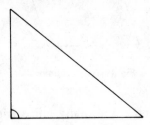

Fig. 10-22. A right triangle.

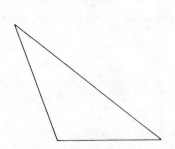

Fig. 10-23. An obtuse oblique triangle.

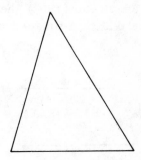

Fig. 10-24. An acute oblique triangle.

2. *Quadrilaterals.* A quadrilateral is a polygon of four sides. When all four sides are equal and all of the interior angles are equal (90°) the figure is called a *square*. See Fig. 10-25.

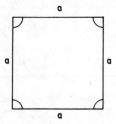

Fig. 10-25. A square.

When the four sides are equal but the interior angles diagonally across from each other are equal but not 90°, the figure is defined as a *rhombus*. See Fig. 10-26.

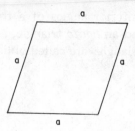

Fig. 10-26. A rhombus.

Irregular quadrilaterals are many and varied. A common one is the *parallelogram* shown in Fig. 10-27, where the opposite sides are equal and the diagonally opposite angles are equal. When the interior angles are all equal (90°), the figure is called a *rectangle* and is shown in Fig. 10-28. (A square is a rectangle with all of its sides of equal length.)

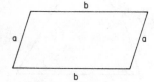

Fig. 10-27. A parallelogram.

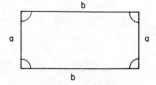

Fig. 10-28. A rectangle.

A *trapezoid*, Fig. 10-29, is an irregular quadrilateral with two sides parallel which are referred to as "bases" of the trapezoid.

Fig. 10-29. A trapezoid.

3. *Pentagons.* While not a common polygon, the five-sided figure is familiar to the general public as the name for the building in Washington, D.C. called The Pentagon. This building, which houses the United States Department of Defense, is a mammoth, five-sided structure. Figure 10-30 shows a pentagon.

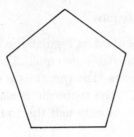

Fig. 10-30. A pentagon.

4. *Hexagons.* This is probably the most ancient of polygons, as this six-sided figure, Fig. 10-31, was in use *before* mankind arrived on the scene. The hexagon is the form used by honey bees to build their nests. The hexagon is unique in that it "nests" with other hexagons without waste space and forms a strong structure. Many aircraft have structural units made of a "honeycomb" (hexagonal) core covered with metal to form a strong, rigid, ply structure.

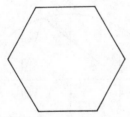

Fig. 10-31. A hexagon.

5. *Octagons.* A polygon of eight sides is called an *octagon*. This regular, eight-sided figure, Fig. 10-32, is used in architectural designs. Bar steel comes in octagonal sections shapes as well as squares, rectangles (flats), hexagons, and rounds.

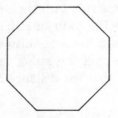

Fig. 10-32. An octagon.

Regular and Irregular Polygons

Polygons can be classified as regular if *all* the sides are of equal length and *all* the interior angles are equal. The square and equilateral triangle are regular polygons. The rhombus is irregular, as the interior angles are not all equal. The rectangle is also irregular because the pairs of sides are not equal although the four interior angles are all equal (90°).

6. *The Ellipse.* Of special interest to certain trades such as sheet metal work, the ellipse is a commonly occurring figure. An ellipse is a figure bounded by a curved line such that the sum of the distances of any point on the periphery from two fixed points called *foci* is constant.

In Fig. 10-33, F_1 and F_2 are the two fixed points or foci (singular, focus). Point O is the center of the ellipse. AC is the *minor axis* while BD is called the *major axis*. OA and OB are called *semi-axes*.

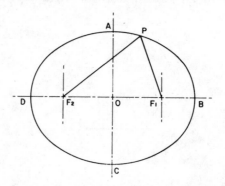

Fig. 10-33. An ellipse.

The ellipse is drawn by first making a continuous loop of string from a piece that is a little less than twice the length of the major axis DB. Two pins are inserted into the drawing at the foci F_1 and F_2 and are encircled by the loop of string. Place a pencil point P inside the loop and extend the loop outward until it is taut. Move the pencil clockwise, keeping the string taut, and the drawn curve will be an ellipse. Note that the distance $F_1P + F_2P$ remains constant; therefore:

$$F_1P + F_2P = \text{constant} = DB$$

The ellipse is a regular curve related to the circle. It is the shape of the resulting end section on a pipe that has been cut on a slant.

Mathematical Symbols Used in Geometry

In mathematics, particularly in geometry, many symbols are used rather than spelled-out words.

Some of the most common symbols are given below:

⌒	arc	△	triangle
⊥	perpendicular	≥	greater than or equal to
‖	parallel	≤	less than or equal to
∴	therefore	∞	infinity
×	multiply by (not used in algebra)	α	alpha
		β	beta
·	multiply by (preferred)	γ	gamma
		θ	theta
÷	divide by	ϕ	phi
—	divide by (as in a fraction)	π	pi = 3.1416
∷	proportional	Σ	sigma, sign of summation
⊙	circle	∠	angle
∟	right angle	▱	parallelogram
≅	congruent	∿	similar

Study Problems.

1. Using a protractor and a ruler, draw a triangle ABC with side $AB = 3$ in., angle A ($\angle A$) = $100°$ and angle B ($\angle B$) = $40°$.

2. Identify triangle ABC in problem no. 1 with both the descriptive names applying to it.

3. Calculate the complementary angles for the following angles: $\angle A = 42°$, $\angle B = 61°$, $\angle C = 22°$ and $\angle D = 19°$,

4. Calculate supplementary angles for the following angles: $\angle \alpha = 35°$, $\angle \beta = 110°$, $\angle \gamma = 89°$, $\angle \delta = 33°$.

Draw a sketch for each of the following geometrical situations:

5. ab and cd are two lines intersecting at P.

6. Point P is PE distant from line AB.

7. P is not a point on line AB, but point D is on that line.

8. Three lines intersect. What is the resulting plane figure?

9. A six-sided figure has all sides of equal length. What is its specific name?

10. Two points determine a line. How many points determine a plane?

11. Draw a triangle with all sides equal. What is the name of this triangle?

12. Draw a triangle with three different length sides. What is the name of this triangle?

13. Two four-sided figures are drawn. In one all the sides are of equal length and the interior angles are each $90°$. How many names can this figure be called? What are the names?

14. Draw an ellipse whose major axis is two times its minor axis.

15. Draw an ellipse whose major and minor axes are equal. What figure has been drawn?

CHAPTER 11

Geometry – Axioms, Postulates, and Propositions

In the study of geometry for use in solving practical shop problems, the formal proving of geometric statements (theorems) will be given in some cases. However, the practical use of the axioms, postulates, and propositions in solving problems and not the learning of the formal proofs is the aim of the text.

Axioms

Geometry is based on certain mathematical statements called *axioms*. These are statements of "truths" which have been observed by early mathematicians but not rigorously proved. The following eight axioms will form the basis of the geometric "tools" you will use in working geometry problems and later in solving trigonometry problems:

AXIOM I: Things equal to the same thing, or to equal things, are equal to each other.

AXIOM II: Any quantity may be substituted for its equal in any mathematical statement.

AXIOM III: If equals are added to equals, their sums are equal.

AXIOM IV: If equals are subtracted from equals, their remainders are equal.

AXIOM V: If equals are multiplied by equals, their products are equal.

AXIOM VI: If equals are divided by equals, their quotients are equal.

AXIOM VII: The whole is greater than any of its parts.

AXIOM VIII: The whole is equal to the sum of its parts.

Postulates

A *postulate* is an assumption applicable to a particular branch of mathematics, such as geometry. The following postulates should also be carefully studied and learned as a means to understand and solve problems:

POSTULATE 1: One and only one straight line can be drawn between any two points.

Fig. 11-1. The straight line.

POSTULATE 2: Two straight lines can intersect in one and only one point.

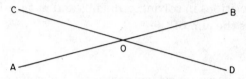

Fig. 11-2. The intersection of two lines.

POSTULATE 3: A straight line is the shortest line between two points.

Fig. 11-3. The shortest distance between two points.

POSTULATE 4: One and only one circle can be drawn with any given point as a center and a given line segment as a radius.

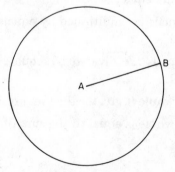

Fig. 11-4. One circle per given center and radius.

POSTULATE 5: Any geometric figure can be moved without change in size or shape.

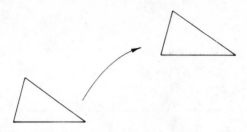

Fig. 11-5. Moving a geometric figure.

POSTULATE 6: A straight line segment has one and only one midpoint.

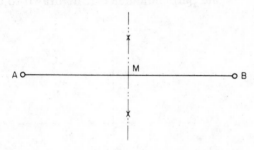

Fig. 11-6. The midpoint of a line.

POSTULATE 7: An angle has one and only one bisector.

(A bisector is a line that divides an angle into two equal angles. In Fig. 11-7, AF is the bisector of $\angle BAC$.)

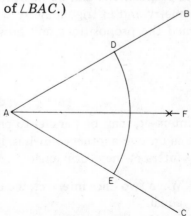

Fig. 11-7. An angle bisector.

POSTULATE 8: Through any point on a line, one and only one perpendicular can be drawn to the line.

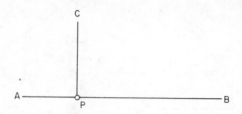

Fig. 11-8. A perpendicular erected to a point on a line.

POSTULATE 9: Through any point outside a line, one and only one perpendicular can be drawn to the given line.

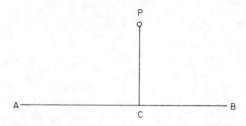

Fig. 11-9. A perpendicular to a line from a point.

Eight axioms and nine postulates have been stated above. In this text the statements that are derived from the axioms and postulates will be referred to as *propositions* and be identified as P-1, P-2, etc. Your success in geometry and in trigonometry will depend on how well you have learned the propositions and how they are used in problem solutions.

Propositions

Two lines may intersect, may be parallel to one another, or may be coincident (lying upon one another) which is, in reality, only one line. When two lines intersect they form angles.

PROPOSITION 1: When two lines intersect, the opposite angles are equal.

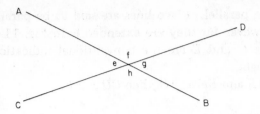

Fig. 11-10. Two lines intersecting; opposite angles are equal.

$\angle f$ is the supplement of $\angle e$; $180° - \angle e = \angle f$
$\angle h$ is the supplement of $\angle e$; $180° - \angle e = \angle h$
Therefore ($\therefore$) $\angle f = \angle h$ (Axiom 1)
Similarly, $\angle e = \angle g$

PROPOSITION 2: The supplements or the complements of the same or equal angles are equal.

This proposition is a corollary of P-1. A corollary is a statement, the truth of which is seen to be a direct consequence of the related proposition. See Figs. 10-13 and 10-14 defining complementary and supplementary angles.

PROPOSITION 3: When two lines are perpendicular to a third line, the two lines are parallel.

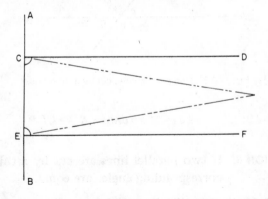

Fig. 11-11. Two lines perpendicular to a third line are parallel.

If CD and EF were extended and could meet, there would be two perpendiculars from a point to the same line. This is not possible according to Postulate 9. Therefore, CD and EF cannot meet and hence must be parallel. (Two lines are said to be parallel if they do not meet, however far they are extended.) In Fig. 11-11 the small arcs at points C and E are the conventional indication of a right (90-degree) angle.

If $CD \perp AB$ and $EF \perp AB$, then $CD \parallel EF$.

PROPOSITION 4: If a line is perpendicular to one of two parallel lines, it is perpendicular to the other also.

In Fig. 11-11, if AB is perpendicular to CD (one of a pair of parallel lines) it is perpendicular to EF also. This is evident if we consider the condition existing if EF was *not* at right angles with AB. This would mean that EF was not parallel to CD, contrary to Proposition 3.

If $CD \parallel EF$ and $AB \perp CD$, then $AB \perp EF$.

PROPOSITION 5: If two parallel lines are cut by a third line, the alternate interior angles are equal.

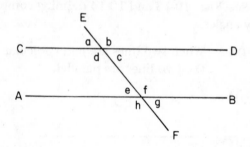

Fig. 11-12. Two parallel lines cut by a third line.

In Fig. 11-12: $\angle e = \angle c$ and; $\angle d = \angle f$

PROPOSITION 6: If two parallel lines are cut by a third line, the corresponding angles are equal.

In Fig. 11-12 the parallel lines AB and CD are cut by line EF.

$$\angle a = \angle e, \quad \angle b = \angle f$$
$$\angle c = \angle g, \quad \angle d = \angle h$$

PROPOSITION 7: The sum of the degrees of the interior angles of any triangle is equal to $180°$.

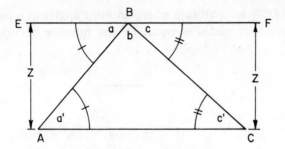

Fig. 11-13. The sum of the degrees of the interior angles of a triangle equals 180°.

In Fig. 11-13, line EF is drawn through B, parallel to AC.

$$\angle a + \angle b + \angle c = 180° \quad \text{(Straight Angle)}$$
$$\angle a = \angle a' \quad \text{(P-5)}$$
$$\angle c = \angle c' \quad \text{(P-5)}$$

Hence: $\angle a' + \angle b + \angle c' = 180°$ (Axiom II)

In Fig. 11-13, the equal angles a and a' are identified as being equal by placing a single mark on each angle-arc symbol. Angles c and c' are likewise identified as being equal to each other by a double mark on the symbol, etc.

PROPOSITION 8: The exterior angle formed by prolonging one side of a triangle is equal to the sum of the two opposite interior angles.

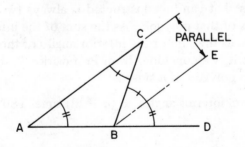

Fig. 11-14. An exterior angle of a triangle equals the sum of the two opposite interior angles.

In Fig. 11-14, BE is parallel to AC by construction. Then:

$$\angle BAC = \angle DBE \text{ and } \angle BCA = \angle EBC \text{ (P-6 and P-5)}$$
$$\angle BAC + \angle BCA = \angle DBE + \angle EBC \text{ (Axiom III)}$$
$$\angle DBC = \angle DBE + \angle EBC \text{ (Axiom VIII)}$$
$$\angle DBC = \angle BAC + \angle BCA \text{ (Axiom I)}$$

PROPOSITION 9:   The sum of the interior angles of a polygon of n sides is equal to the number of sides less two multiplied by $180°$.

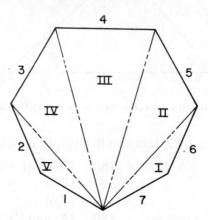

Fig. 11-15. The sum of the interior angles of a polygon.

A polygon of seven sides (heptagon) is shown in Fig. 11-15. If diagonals are drawn from any vertex to each of the other vertices, the polygon is divided into five triangles. Each triangle includes one side of the original polygon except the first and last triangles (I and V) which have two such sides. The seven sided polygon is thus divided into seven minus two, or five triangles and therefore, the number of triangles that can be constructed is always two less than the number of sides of that polygon. As the sum of the interior angles of the polygon equals the sum of the interior angles of the triangles constructed within it, we conclude, from Proposition 7, that the sum of the interior of a polygon of n sides is:

$$\text{Sum interior angles} = (n - 2) \text{ times } 180°$$

Solving Geometry Problems

The method used in solving the following geometry problems involves the use of one or more of the axioms, postulates, and propositions.

The first step in a solution is to carefully determine what answer is required. In the study example below, the problem asks the size of the angles x and y. Next, carefully determine the known conditions in the problem. The problem states that the line CD bisects (cuts in two equal parts) the angle C. Finally the drawing shows that $\angle A = 70°$ and $\angle B = 50°$.

Study Example.

1. Solve for angles x and y in Fig. 11-16 when line CD bisects angle C.

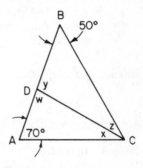

Fig. 11-16. Study example 1.

Step 1. In $\triangle ABC$:

$$\angle A + \angle B + \angle C = 180°$$
$$\angle A + \angle B = 70° + 50° = 120°$$
$$\angle C = 180° - (\angle A + \angle B) = 180° - 120°$$
$$\angle C = 60°$$

P-7

Step 2.
$$\angle x = \frac{1}{2}\angle C \quad \text{(Bisected)}$$

$$\angle x = \frac{1}{2}(60°) = 30°$$

Step 3. In $\triangle ACD$:

$$\angle A + \angle x + \angle w = 180°$$
$$\angle w = 180° - (\angle A + \angle x)$$
$$= 180° - (70° + 30°)$$
$$\angle w = 180° - 100° = 80°$$

Step 4. Line ADB is a straight angle and equals $180°$

$$\angle w + \angle y = \angle ADB = 180°$$
$$\angle y = 180° - \angle w$$
$$= 180° - 80°$$
$$\angle y = 100°$$

NOTE: This problem can also be solved by using $\angle z$ in step 3.

The above study example is often solved by "charting" the data. This is the classical approach and very useful in both geometry and trigonometry. Study example 1 is charted on the next page:

Solution:

Analysis	Proof or Reason
Step 1. $\angle A + \angle B + \angle C = 180°$ $\angle C = 180 - \angle A - \angle B$ $\angle C = 180 - 70 - 50$ $\angle C = 60°$	Proposition 7:
Step 2. $\angle x = \dfrac{1}{2}\angle C$ $\angle x = \dfrac{1}{2} \times 60°$ $\angle x = 30°$	In problem: DC bisects $\angle C$
Step 3. In $\triangle ADC$ $\angle A + \angle x + \angle w = 180°$ $\angle w = 180° - \angle A - \angle x$ $\angle w = 180° - 70° - 30°$ $\angle w = 80°$	Proposition 7:
Step 4. $\angle y = 180° - \angle w$ $\angle y = 180° - 80°$ $\angle y = 100°$	ADB is a straight line. A straight line $= 180°$.
Summary: $\angle x = 30°$, $\angle y = 100°$	

From the above example, it can be seen that the method of charting is an organized method of solving problems that can be easily checked. For this reason, this method is recommended.

Study Problems.

The following problems involve the use of the axioms, postulates, and propositions 1 through 9.

1. In the oblique triangle shown in Fig. 11-17, solve for $\angle B$.

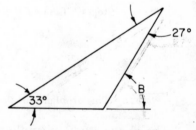

Fig. 11-17. Study problem 1.

2. In the polygon shown in Fig. 11-18, solve for angle A using the (n − 2) formula. Hint: Be sure to use *only* interior angles.

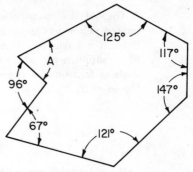

Fig. 11-18. Study problem 2.

3. Determine angle B in Fig. 11-19.

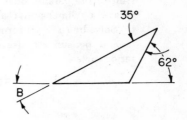

Fig. 11-19. Study problem 3.

4. Figure 11-20 may be resolved into a polygon. Solve for angle C.

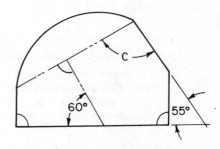

Fig. 11-20. Study problem 4.

5. In the right triangle ACB shown in Fig. 11-21, solve for angle ABC.

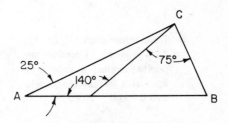

Fig. 11-21. Study problem 5.

6. A triangle is circumscribed around a circle in Fig. 11-22. Solve for angle β. Hint: Find supplement of $114°$ angle in four-sided polygon. Use $(n - 2)$.

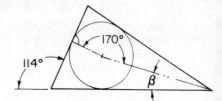

Fig. 11-22. Study problem 6.

7. In the polygon shown in Fig. 11-23, determine angle A. Hint: Note parallel sides indicated by dimensions C and C. Solve by $(n - 2)$ formula and be sure to use interior angles.

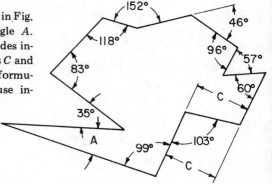

Fig. 11-23. Study problem 7.

8. Solve for angle B in the polygon shown in Fig. 11-24. Hint: Extend short sides of "ears" inward to form a four-sided polygon. Solve by $(n - 2)$.

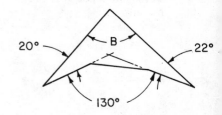

Fig. 11-24. Study problem 8.

9. Determine angle C in Fig. 11-25. Hint: Consider opposite angles, supplements and sum of interior angles in a triangle.

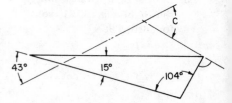

Fig. 11-25. Study problem 9.

10. The dimensions C, C in Fig. 11-26 indicate the lines are parallel. Solve for angle A. Hint: Supplements, opposites, and parallel lines cut by a third line.

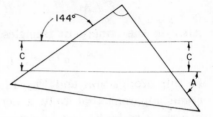

Fig. 11-26. Study problem 10.

Congruent Triangles

The word *congruent* is used to describe plane geometric figures which have the same size and shape. Congruent figures are "carbon-copies" of one another, much the same as mass-produced metal stampings. Many problems in geometry can be solved by the use of basic propositions relating to the condition of congruency. These propositions are:

PROPOSITION 10: Congruent triangles have their corresponding sides and angles equal.

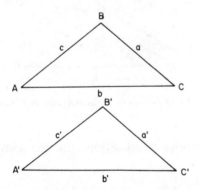

Fig. 11-27. Congruent triangles.

In Fig. 11-27, the triangles are defined as being the same size and shape, congruent. Size relates to length of lines, and the corresponding sides of the triangles are the same or are of equal length. Likewise, the shape of the triangles is related to the interior angles. Shape relates to the angles and therefore the corresponding angles are equal.

$$\angle A = \angle A'; \quad \angle B = \angle B'; \quad \angle C = \angle C'$$

Also, in the congruent triangles, the following sides must be equal:

$$a = a'; \quad b = b'; \quad c = c'$$

If this proposition and Fig. 11-27 are studied carefully, it will become apparent that only a certain number of sides or angles are necessary to be known to determine if two triangles are indeed congruent. Congruent triangles have all their corresponding parts (angles and sides) equal. Therefore:

PROPOSITION 11: **If two sides and the included angle of one triangle are equal to the corresponding two sides and the included angle of another triangle, the figures are congruent.**

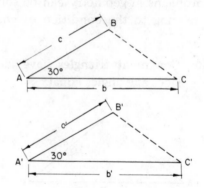

Fig. 11-28. Congruent triangles; side-angle-side.

According to the above proposition, the triangles in Fig. 11-28 are congruent when:

$$b = b'; \quad c = c'; \quad \text{and} \quad \angle A = \angle A'$$

PROPOSITION 12: **If two angles and the included side of one triangle are equal to the corresponding two angles and the included side of another triangle, the figures are congruent.**

According to the above proposition, the triangles in Fig. 11-29 are congruent when:

$$b = b'; \quad \angle A = \angle A'; \quad \text{and} \quad \angle C = \angle C$$

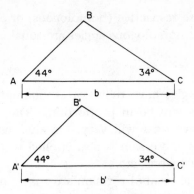

Fig. 11-29. Congruent triangles; angle-side-angle.

PROPOSITION 13:   If three sides of one triangle equal the corresponding three sides of another triangle, the
figures are congruent.

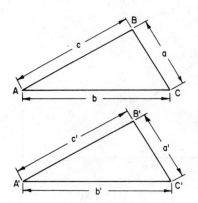

Fig. 11-30. Congruent triangles; side-side-side.

According to the above proposition, the triangles in Fig. 11-30 are
congruent when:

$$a = a'; \quad b = b'; \quad \text{and} \quad c = c'$$

Note that there cannot be a similar proposition relating to three
angles.   Knowing only the angles would permit the construction of
similar triangles of the same shape but of various sizes.

While the foregoing discussion of congruency dealt with triangles, the same propositions apply to other plane figures.   For example, two squares whose sides are equal will be congruent as will be

two hexagons whose respective (homologous) or corresponding sides are equal and whose homologous angles are equal.

Similar Triangles

When two triangles have their corresponding angles equal, their corresponding sides will be in proportion. The shapes will be the same but the relative sizes will vary. In this case the triangles are called *similar*. This expression is also applied to the other polygons such as hexagons, octagons, etc.

The conventional symbol for similarity is ~, and the expression; "$\triangle ABC \sim \triangle A'B'C'$" is read: "Triangle ABC is similar to triangle A prime, B prime, C prime.

In Fig. 11-31, the two similar triangles noted above are shown. Note that the capital letters are used conventionally to designate the angle vertices and that lower case (small) letters are used for the sides. The side opposite $\angle A$ is side a, the side opposite $\angle B$ is side b and the side opposite $\angle C$ is side c.

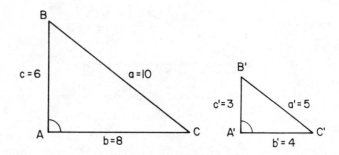

Fig. 11-31. Similar triangles.

When similar triangles are further examined, additional facts are discernible which can be stated as propositions:

PROPOSITION 14:   **The corresponding angles of similar triangles are equal.**

In Fig. 11-31, for example:

$$\angle A = \angle A', \quad \angle B = \angle B' \quad \text{and} \quad \angle C = \angle C'$$

PROPOSITION 15:   **The corresponding sides of similar triangles are in proportion.**

Again, in Fig. 11-31, it can be noted that lengths of corresponding sides form equal ratios:

$$\frac{a'}{a} = \frac{b'}{b} = \frac{c'}{c} \quad \text{or} \quad \frac{5}{10} = \frac{4}{8} = \frac{3}{6}$$

PROPOSITION 16: Two triangles are similar if two angles of one triangle are equal respectively to two angles of the other.

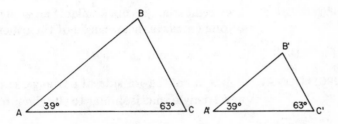

Fig. 11-32. Similar triangles; two angles of each respectively equal.

In Fig. 11-32:

$$\angle A = \angle A' \quad \text{and} \quad \angle C = \angle C'$$
$$\angle B = 180° - (\angle A + \angle C)$$
$$\angle B' = 180° - (\angle A' + \angle C')$$
$$\therefore \angle B = \angle B' \quad \text{(Axiom I)}$$

and: the triangles are similar.   (P-14)

PROPOSITION 17: Two triangles are similar if an angle of one is equal to a corresponding angle of the other and the corresponding sides including these angles are in proportion.

In. Fig. 11-31:

$$\angle B = \angle B'$$
$$\frac{AB}{A'B'} = \frac{BC}{B'C'}$$

Thus: $\triangle ABC \sim \triangle A'B'C'$

In this and the next three propositions an examination of triangles ABC and $A'B'C'$ in Fig. 11-31 and the drawing of the triangles as described in each proposition will demonstrate its truth.

PROPOSITION 18: Two triangles are similar if their three corresponding sides are in proportion.

In Fig. 11-31: $$\frac{AB}{A'B'} = \frac{BC}{B'C'} = \frac{AC}{A'C'}$$

PROPOSITION 19: Triangles similar to the same triangle are similar to each other.

Compare: Axiom I

PROPOSITION 20: Two right triangles are similar if an acute angle of one equals an acute angle of the other.

In Fig. 11-31: $\angle A = \angle A'$ and $\angle C = \angle C'$

PROPOSITION 21: A line parallel to a side of a triangle cuts off a new triangle that is similar to the given triangle.

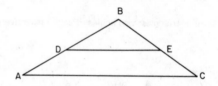

Fig. 11-33. A line parallel to a side of a triangle cutting off a similar triangle.

In Fig. 11-33: $\triangle DBE \sim \triangle ABC$

PROPOSITION 22: The altitude to the hypotenuse of a right triangle divides it into two triangles which are similar to the given triangle and to each other.

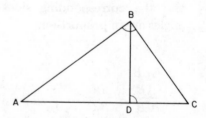

Fig. 11-34. An altitude to the hypotenuse creates two similar triangles.

In Fig. 11-34, BD is drawn from B, perpendicular to the hypotenuse AC. Line BD is known as an altitude, which is a line drawn from the apex of one angle of a triangle perpendicular to the side opposite.

$$\triangle BDC \sim \triangle ABC, \triangle ADB \sim \triangle ABC, \text{ and } \triangle ADB \sim \triangle BDC$$

PROPOSITION 23:   Triangles are similar if their sides are respectively parallel to each other.

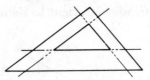

Fig. 11-35. Triangles are similar when their sides are respectively parallel.

In Fig. 11-35:   The corresponding angles are equal (P-6).
Hence the triangles are similar. (P-16).

PROPOSITION 24:   Triangles are similar if their sides are respectively perpendicular to each other.

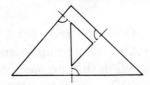

Fig. 11-36. Triangles are similar when their sides are respectively perpendicular.

In Fig. 11-36 if one triangle is rotated 90° with respect to the other, the respective sides will be parallel and Proposition 23 will apply to prove similarity.

PROPOSITION 25:   Two angles are equal when their sides are parallel, right to right and left to left.

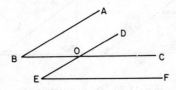

Fig. 11-37. Two angles are equal when their sides are parallel, right to right and left to left.

In Fig. 11-37, the side AB is parallel to side ED and side BC is parallel to side EF.

$$\angle DEF = \angle DOC \quad \text{and} \quad \angle ABC = \angle DOC \quad \text{(P-6)}$$
Hence: $\angle ABC = \angle DEF$

PROPOSITION 26: Two angles are equal when their sides are perpendicular, right to right and left to left.

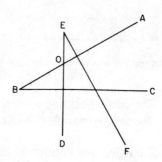

Fig. 11-38. Two angles are equal when their sides are perpendicular, right to right and left to left.

In Fig. 11-27: $AB \perp EF$ and $BC \perp ED$
$\angle ABC$ is the complement of $\angle BOD$
$\angle DEF$ is the complement of $\angle EOA$
but $\angle BOD = \angle EOA$ Proposition 1
Hence: $\angle ABC = \angle DEF$ (Axiom I)

Study Examples.

1. In the congruent triangles ABC and $A'B'C'$ shown in Fig. 11-39, determine angle C' and the length of side $A'C'$.

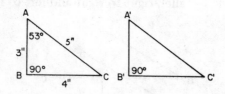

Fig. 11-39. Study example 1.

Solution:

Analysis	Proof or Reason
Step 1. $\angle A + \angle B + \angle C = 180°$ $\angle C = 180° - (\angle A + \angle B)$ $\angle C = 180° - (53° + 90°)$ $\angle C = 180° - 143° = 37°$	Proposition 7
Step 2. $\angle C = \angle C' = 37°$	Proposition 10
Step 3. $A'C' = AC = 5$	Proposition 10

2. In Fig. 11-40 a right triangle ABC is cut by a line DE which is perpendicular to the base BC at E. If $AB = 6$, $BC = 30$, and $EC = 15$, what is the length of DE?

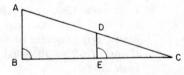

Fig. 11-40. Study example 2.

Solution:

Analysis	Proof or Reason
Step 1. $\triangle ABC \sim \triangle DEC$	(P-21)
Step 2. $\dfrac{AB}{DE} = \dfrac{BC}{EC}$	(P-15)
Step 3. $\dfrac{6}{DE} = \dfrac{30}{15}$	Substitute values.
Step 4. $30 \cdot DE = 6 \cdot 15$ $30DE = 90$ $DE = \dfrac{90}{30} = 3$	Cross multiply.

3. In Fig. 11-41 two similar triangles are shown with $AC \perp$ EF and $AB \perp ED$. Solve for $\angle FDE$ when $\angle ABC = 70°$.

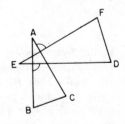

Fig. 11-41. Study example 3.

Solution:

Analysis	Proof or Reason
Step 1. $\triangle ABC \sim \triangle EFD$	Given
Step 2. $\angle ABC = \angle FDE$	(P-15)
Step 3. $\angle FDE = 70°$	
	Axiom I

Study Problems.

1. In triangle ABC, angle ABC is a right angle, and BD is drawn perpendicular to AC. If angle $DBC = 30°$ what is angle BAD?

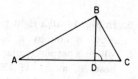

Fig. 11-42. Study problem 1.

2. In triangle ABC, $AC = 25''$, BD is perpendicular to AC and equals $10''$. Line EG is parallel to AC and is $7''$ above AC. Find length of EG.

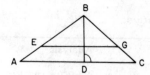

Fig. 11-43. Study problem 2.

3. In triangle ABC, line AD bisects angle BAC and is perpendicular to line BC. If side BC is $6''$ long, what is the length of BD? of DC? Are triangles ABD and ACD congruent? Why?

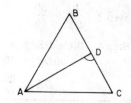

Fig. 11-44. Study problem 3.

4. In the figure shown the interior angles of triangle ABC are equal. Line BE bisects angle ABC and is perpendicular to line AC. Line CE bisects angle ACD. Solve for angle BEC.

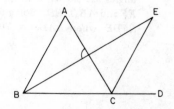

Fig. 11-45. Study problem 4.

5. In the figure shown, $AB =$ AD and $BC = CD$. Solve for angles BCD and BDC.

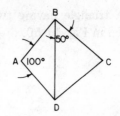

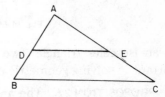

Fig. 11-46. Study problem 5.

6. In triangle ABC, DE is parallel to BC, $DA = 4''$, $AB =$ $8''$, $DE = 10''$. Solve for BC.

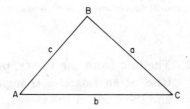

Fig. 11-47. Study problem 6.

Isosceles and Equilateral Triangles

Triangles can be classified according to the equality of their sides or to the type of angles they have. An irregular shaped triangle having no sides or angles that are equal is called a *scalene* triangle. See Fig. 11-48.

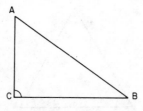

Fig. 11-48. A scalene triangle.

A scalene triangle having one angle equal to $90°$ is called a *right* triangle, as shown in Fig. 11-49.

Fig. 11-49. A right triangle.

A triangle having two equal sides is called an *isosceles* triangle, as seen in Fig. 11-50.

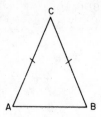

Fig. 11-50. An isosceles triangle.

In an isosceles triangle, two principles are developed which are important in solving geometric problems. These are:

PROPOSITION 27:   The angles opposite the equal sides of an isosceles triangle are equal.

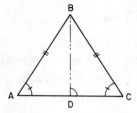

Fig. 11-51. Angles opposite equal sides of an isosceles triangle and the bisector of the base.

In Fig. 11-51: $\angle A = \angle C$

PROPOSITION 28:   The line from the vertex, perpendicular to the base of an isosceles triangle, bisects the base and the angle at the vertex.

In Fig. 11-51: $\angle ABD = \angle CBD$;   and $AD = DC$

Definition: A triangle having three equal sides, as in Fig. 11-52, is called an *equilateral* triangle. An equilateral triangle is also equiangular with the three interior angles each equal to 60°.

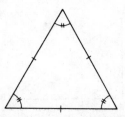

Fig. 11-52. Equal interior angles in an equilateral triangle.

PROPOSITION 29: If a line is drawn through two sides of a triangle and is parallel to the base, it divides those
sides proportionally.

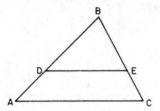

Fig. 11-53. A line parallel to the base of a triangle dividing the sides proportionally.

In Fig. 11-53: $\dfrac{AD}{DB} = \dfrac{EC}{BE}$

Study Examples.

1. In the parallelogram $ABDC$ shown below, AB is parallel to CD and AC is
   parallel to BD. Line AD bisects line BC. Are triangles AEB and CED
   congruent?

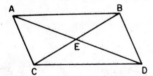

Fig. 11-54. Study example 1.

Solution:

Analysis	Proof or Reason
Step 1. $CE = EB$	Bisected.
Step 2. $\angle AEB = \angle CED$	Proposition 1.
Step 3. $\angle ABE = \angle ECD$	Proposition 5.
Step 4. $\triangle AEB = \triangle CED$	Proposition 12.

2. In the triangle ABC shown
   in Fig. 11-55, $AB = AC$.
   Solve for angle ABC and
   angle ACB.

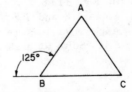

Fig. 11-55. Study example 2.

Solution:

Analysis	Proof or Reason
Step 1. $\angle ACB = \angle ABC$	Proposition 27.
Step 2. $\angle ABC + 125° = 180°$	Straight angle $= 180°$
Step 3. $\angle ABC = 180° - 125° = 55°$	
Step 4. $\angle ACB = \angle ABC = 55°$	

Study Problems.

1. In Fig. 11-56, angle CAE equals angle DBE, $AE = EB$, and angle AEC equals angle DEB. Side CE is given as: $2x - 5$ and side BD as: $3y + 2$. Using geometric principles and algebra, solve for x and y.

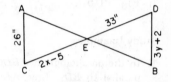

Fig. 11-56. Study problem 1.

2. In Fig. 11-57, angle ADC equals angle BEC, side CD equals side CE, and angle ACD equals angle ECB. Side AC is given as $3y - 5$, side CB as $2y + 7$, side AD as $x + 8$ and side EB as $3x$. Solve for x and y using geometric principles and algebra.

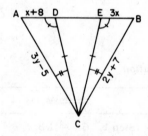

Fig. 11-57. Study problem 2.

3. In Fig. 11-58, $DE = BC$, $EF = AC$, and $DF = AB$. Angle ABC is twice angle BAC. Solve for angle DFE.

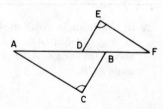

Fig. 11-58. Study problem 3.

4. In triangle *ABC*, shown in Fig. 11-59, the sizes of the acute angles are given as *CAB* = 3*x* and *CBA* = 2*x*. Solve for the three angles in degrees.

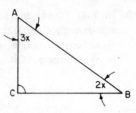

Fig. 11-59. Study problem 4.

5. In Fig. 11-60, side *AB* equals side *AC*. Angle *BAC* = 80°, *BD* bisects angle *ABC*, and *DC* bisects angle *ACB*. Solve for angle *ABC* and angle *BDC*.

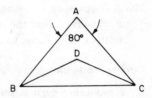

Fig. 11-60. Study problem 5.

6. In Fig. 11-61, *AB* = *BC* = *AC*, *BD* = *AD*. Angle *BDC* = 45° Solve for angle *ABD* and angle *BCD*.

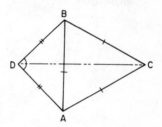

Fig. 11-61. Study problem 6.

7. Triangle *ABC* in Fig. 11-62 is equilateral. Angle *EAD* = 40° and angle *ADE* = 90°. Solve for angle *AEB* and angle *CED*.

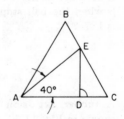

Fig. 11-62. Study problem 7.

8. In right triangle *BAC*, in Fig. 11-63, lines *DE* and *FG* are parallel to the base *BC* and cut the side *AB* in three equal parts and the side *AC* in equal parts. *FG* = 22 in. Solve for the length of lines *DE* and *BC*.

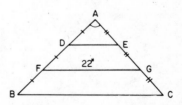

Fig. 11-63. Study problem 8.

9. In Fig. 11-64, angle *ABC* equals angle *CBD* and angle *BAC* equals angle *CAE*. Solve for angle *ACB*.

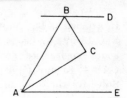

Fig. 11-64. Study problem 9.

10. A flagpole casts a shadow of 15′ at the same time that a 6′ post casts a 2′ shadow. Find the height of the flagpole if the pole and the post are both at right angles to the ground.

Review Problems.

1. A moulding gage is shown in Fig. 11-65. Solve for angle α.

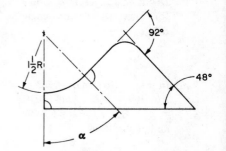

Fig. 11-65. Review problem 1.

2. A lift-stop on a machine is shown in Fig. 11-66. What is the angle of the lift wedge β when the ball stop angle is 117°?

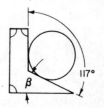

Fig. 11-66. Review problem 2.

3. A rubber extrusion die is shown in Fig. 11-67. Solve for the angle θ.

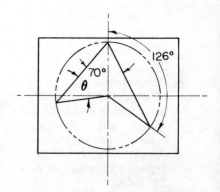

Fig. 11-67. Review problem 3.

4. A truss is constructed as shown in Fig. 11-68. What is the length X of the right side member?

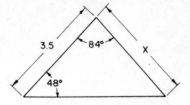

Fig. 11-68. Review problem 4.

5. An A-frame truss is shown in Fig. 11-69. From the data given, solve for angles α and β.

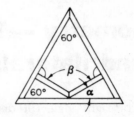

Fig. 11-69. Review problem 5.

CHAPTER 12

Geometry – Triangles and Quadrilaterals

The right-angle (90-degree) triangle is one of the most common elements in modern technology. Its use in mechanics and structures makes it necessary to be able to solve right-angle problems. A typical right triangle in standard position is shown in Fig. 12-1. The conventional terminology used to identify its parts is also shown. Note the small arc in the corner of the right angle. This is used to identify a right angle or a perpendicular position where a line is positioned 90° in relation to a second line.

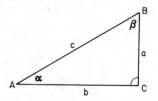

Fig. 12-1. Right triangle with right angle symbol shown.

The right angle of a right triangle is generally labeled "C" which relates to the small "c" on the side opposite angle C. This side is called the *hypotenuse*. The sides opposite the other two angles likewise take their identification as a small "a" or "b." You will note that the hypotenuse, being opposite the largest angle (90°), is the longest side of the triangle. The other two sides of the triangle are identified as being *adjacent* or next to the right angle. Adjacent sides are also opposite the two acute angles of a right triangle.

Pythagorean Theorem

The Pythagorean Theorem or Law is one of the most important and useful theorems in mathematics. It was proposed by the Greek philosopher, Pythagoras, in the Fifth Century B.C. This theorem should be committed to memory.

This law applies specifically to right triangles. It was observed that if the three sides of a right triangle were in the proportion of 3, 4, and 5, with the construction of a square on each of the sides as shown in Fig. 12-2, and with the subdivision of each of the three squares into smaller unit squares, the number of unit squares in the square erected on the hypotenuse would equal the sum of the unit squares in the other two larger squares erected on the other two sides. While this condition is apparent in a 3, 4, 5 right triangle, the Pythagorean Theorem can be applied to *any* right triangle.

PROPOSITION 30: In a right triangle, the square of the hypotenuse is equal to the sum of the squares of the other two sides. (Pythagorean Theorem.)

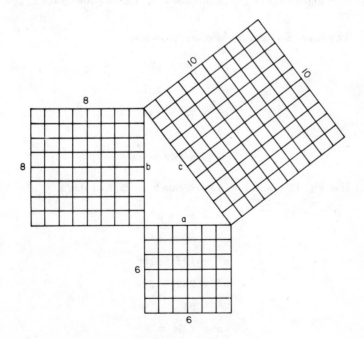

Fig. 12-2. The Pythagorean Law.

This relationship is shown in the formula:

$$c^2 = a^2 + b^2$$

If the lengths of the sides of the triangle in Fig. 12-2 are substituted into this formula, we get:

$$10^2 = 6^2 + 8^2$$

$$100 = 36 + 64$$

$$100 = 100$$

The use of this very important proposition is demonstrated in the following examples:

Study Examples.

1. If in Fig. 12-1, side $a = 5$ and side $b = 12$, how long is the hypotenuse c?

   The figure is a right triangle, and therefore:

   $$c^2 = a^2 + b^2$$
   $$c^2 = 5^2 + 12^2$$
   $$c^2 = 25 + 144$$
   $$c^2 = 169$$
   $$c = \sqrt{169} = 13$$

2. If in Fig. 12-1, side $c = 17$ and side $b = 15$, find side a.

   $$c^2 = a^2 + b^2$$
   $$a^2 = c^2 - b^2$$
   $$a^2 = 17^2 - 15^2$$
   $$a^2 = 289 - 225$$
   $$a^2 = 64$$
   $$a = \sqrt{64} = 8$$

3. Solve for length of side AE in Fig. 12-3:

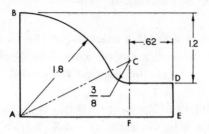

Fig. 12-3. Study example 3.

Step 1. Draw line AC connecting centers of both arcs.

Step 2. Draw line CF from center $C \perp$ to line AE.

Step 3. Line AC = radius 1.8 + radius $\frac{3}{8}$

$$AC = 1.80 + 0.375 = 2.175$$

Step 4. In $\triangle AFC$, $\overline{AC}^2 = \overline{AF}^2 + \overline{CF}^2$ (Pythagorean Law)

Step 5. $CF = AB - 1.2 + 0.375$

$$AB = \text{radius} = 1.8$$

$$CF = 1.8 - 1.2 + 0.375 = 0.975$$

Step 6. $AF = \sqrt{\overline{AC}^2 - \overline{CF}^2}$

$$AF = \sqrt{(2.175 \times 2.175) - (0.975 \times 0.975)}$$

$$AF = \sqrt{4.731 - 0.951}$$

$$AF = \sqrt{3.780}$$

$$AF = 1.94$$

$$AE = AF + FE = 1.94 + 0.62 = 2.56$$

There are two useful corollary principles related to the Pythagorean Theorem. These are:

PROPOSITION 30A: **In a right triangle with acute interior angles of 30° and 60°, the hypotenuse is equal to two times the length of the side opposite the 30° angle.**

In Fig. 12-1, if $\angle\alpha = 30°$, $\angle\beta = 60°$ and side $a = 5$ inches; then side c (hypotenuse) will equal 10 inches.

PROPOSITION 30B: **If a triangle has the length of its sides in the ratio of 3 to 4 to 5, the angle opposite the longest side will be 90°.**

In Fig. 12-2, the three sides of the triangle are 6, 8 and 10 units in length. This is in the ratio of 3-4-5. Hence, the triangle is a right triangle and the angle opposite side c is 90°.

Study Problems.

Refer to Fig. 12-1 for problems 1 through 7:

1. Find c when $a = 12$ and $b = 16$.

2. Solve for a when $b = 6$ and $c = 8$.

3. Find b when $a = 2$ and $c = 4$.

4. Solve for b when $a = 12$ and $c = 20$.

5. If $\angle B = 30°$, what does $\angle A$ equal?

6. Find a and b if $c = 15$ and a and b are in the ratio of 3 to 4.

7. If a and b are both equal to 7, how much is c?

8. In Fig. 12-4 solve for X.

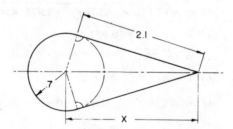

Fig. 12-4. Study problem 8.

9. Determine the distance A in Fig. 12-5.

10. In Fig. 12-6, find the depth X to which a milling cutter must be sunk to cut a keyway with a width of 0.500 in. and a depth A of 0.375 in. The shaft is 3.625 in. in diameter.

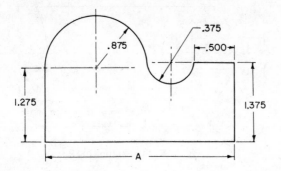

Fig. 12-5. Study problem 9.

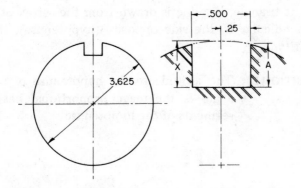

Fig. 12-6. Study problem 10.

11. Solve for radius *r* in the following template (Fig. 12-7).

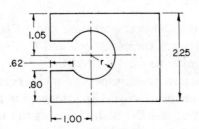

Fig. 12-7. Study problem 11.

12. Solve for dimension *X* in Fig. 12-8.

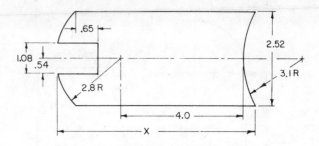

Fig. 12-8. Study problem 12.

Mean Proportionals

In a right triangle, if a line is drawn from the vertex of the right angle, perpendicular to the side opposite (hypotenuse), the line is called an *altitude*.

PROPOSITION 31: The altitude to the hypotenuse of a right triangle is the mean proportional between the segments of the hypotenuse.

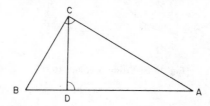

Fig. 12-9. An altitude from the right-angle vertex.

In Fig. 12-9, line CD is the altitude and $\angle CDA$ is a right angle. Two similar triangles are formed; ADC and CDB (P-22). These two triangles are also similar to $\triangle ACB$ (P-22).

Similar triangles have their respective angles equal (P-14) and their respective sides in proportion (P-15). Hence, the ratio of side BD of $\triangle BDC$ to side CD of $\triangle ADC$ is the same as the ratio of side CD of $\triangle BDC$ to side DA of $\triangle ADC$. This condition of proportionality is stated:

$$\frac{BD}{CD} = \frac{CD}{DA}$$

Note that the term CD is in both of the ratios of the proportion. This position of the CD terms is referred to as the *means* of the proportion while the terms BD and DA are referred to as the *extremes* of the proportion.

When the proportion above is cross-multiplied, the means are squared and we get:

$$\overline{CD}^2 = BD \cdot DA$$

This proposition provides a method of solving for the length of an altitude erected to the hypotenuse of a right triangle.

PROPOSITION 32: Either side adjacent to the right angle is a mean-proportional between the hypotenuse and the projection of that side on the hypotenuse.

In Fig. 12-9, the segment BD is the projection of side BC upon the hypotenuse BA. Likewise, DA is the projection of side CA upon the hypotenuse BA. The respective proportions for each of the sides are:

$$\frac{AB}{BC} = \frac{BC}{BD} \text{ and } \frac{AB}{AC} = \frac{AC}{AD}$$

PROPOSITION 33: If a perpendicular is dropped from the vertex of the right angle to the hypotenuse, the angle opposite the perpendicular in one triangle is equal to the angle adjacent to this perpendicular in the other triangle.

In Fig. 12-9:

$$\angle CBD = \angle DCA \qquad \text{(P-14)}$$

$$\angle BCD = \angle CAD \qquad \text{(P-14)}$$

Study Examples.

1. Solve for CD in Fig. 12-10.

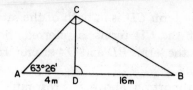

Fig. 12-10. Study example 1.

Step 1. $\dfrac{AD}{CD} = \dfrac{CD}{DB}$ (Prop. 31)

Step 2. Substitute values for AD and DB:

$$\frac{4}{CD} = \frac{CD}{16}$$

Step 3. Cross-multiply:

$$CD^2 = 64$$
$$CD = \sqrt{64}$$
$$CD = \pm 8\text{m}$$

2. In triangle ABC of study example 1, above:

If angle $CAD = 63°26'$, solve for angles ACD, DCB and CBD.

Step 1. In rt. $\triangle ACD$, the sum of the interior angles $= 180°$.   (Prop. 7)

Step 2. $\angle ADC + \angle CAD + \angle ACD = 180°$
$90° + 63°26' + \angle ACD = 180°$
$\angle ACD = 180° - (90° + 63°26')$
$\angle ACD = 180° - 153°26'$
$\angle ACD = 26°34'$

Step 3. $\angle DCB = \angle CAD = 63°26'$ (Prop. 33)

Step 4. $\angle CBD = \angle ACD = 26°34'$ (Prop. 33)

Study Problems.

1. In Fig. 12-11 when $p = 2$ and $q = 6$, solve for a and h.

2. When $p = 4$ and $a = 6$, solve for c and h.

3. When $p = 16$ and $h = 8$, solve for q and b.

4. When $b = 12$ and $q = 6$, solve for p and h.

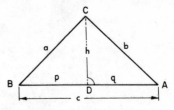

Fig. 12-11. Study problem 1.

5. In right triangle ABC, Fig. 12-12, solve for CD.

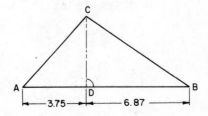

Fig. 12-12. Study problem 5.

6. Solve for angle β in Fig. 12-13.

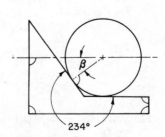

Fig. 12-13. Study problem 6.

7. Solve for angle β in Fig. 12-14.

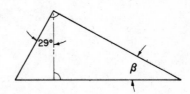

Fig. 12-14. Study problem 7.

8. Solve for angle α in Fig. 12-15.

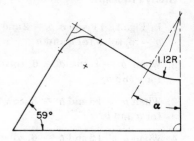

Fig. 12-15. Study problem 8.

9. In the trapezoid *ABCD*, *AD* = 5, *CB* = 14, *DB* = 13, solve for *DC* (Fig. 12-16).

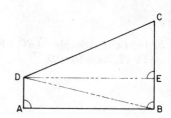

Fig. 12-16. Study problem 9.

10. In an isosceles trapezoid, *ABCD*, find *AD* when *AB* = 32, *DC* = 20 and *DE* = 8. See Fig. 12-17.

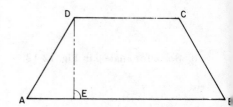

Fig. 12-17. Study problem 10.

11. Solve for angle *ACO* in Fig. 12-18.

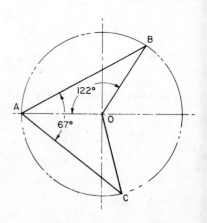

Fig. 12-18. Study problem 10.

12. Solve for angle α in Fig. 12-19.

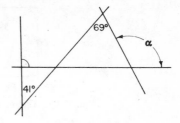

Fig. 12-19. Study problem 12.

Parallelograms and Trapezoids

Any geometric figure having four sides is called a *quadrilateral*. This includes the parallelogram, which has opposite sides equal and parallel; the trapezoid, which has only two of its sides parallel; and all other four-sided figures whose sides are neither equal nor parallel. Parallelograms include the square, the rectangle, and the rhombus.

The Square. A square is an equilateral, equiangular parallelogram. (Fig. 12-20). All four sides are the same length, and all four angles are right (90°). Diagonals drawn from opposite vertices are of equal length, intersect at 90°, and bisect the vertex angles and each other. The diagonals form four right, isosceles triangles which are congruent.

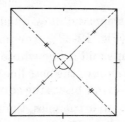

Fig. 12-20. The square; equal sides and equal angles.

The Rectangle. A rectangle is an equiangular parallelogram (Fig. 12-21). All four interior angles are right (90°), and the opposite sides are equal in pairs and parallel to one another. The rectangle is an elongated square, and the diagonals drawn from opposite right angles are of equal length, bisect each other, and form pairs of congruent, isosceles triangles.

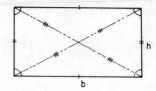

Fig. 12-21. The rectangle; equiangular parallelogram.

The Rhombus. A rhombus is an equilateral parallelogram (Fig. 12-22) with four sides of equal length, but with none of the interior angles equal to 90°. The opposite angles, in pairs, are equal to each other. Diagonals bisect each other, but are of unequal length. The segments of one diagonal do not equal the segments of the other. The diagonals form four congruent triangles and bisect the four interior angles of the rhombus.

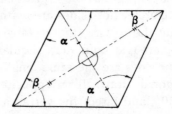

Fig. 12-22. The rhombus; equal sides and opposite angles equal.

The Trapezoid. A trapezoid is a quadrilateral having two and only two sides parallel (Fig. 12-23). The non-parallel sides are called the *legs*. An isosceles trapezoid is one whose legs are equal and whose base angles (angles at the ends of the base line) are equal. The diagonals of an isosceles trapezoid do *not* bisect the corner angles nor each other. They do form pairs of similar triangles, the triangles formed with the parallel sides being isosceles. The triangles formed with the legs are scalene. In a non-isosceles trapezoid (legs unequal) the diagonals generate unrelated triangles.

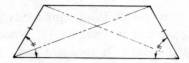

Fig. 12-23. The trapezoid; quadrilateral with only two sides parallel.

Areas of Polygons

A plane geometric figure of three sides or more bounds a surface whose area can be calculated. The system of land measure—whether in square feet, square meters, square rods, acres, hectares, or square miles—is based on a geometric method of calculation. The SI (metric) unit for area is the square meter, although for land areas the hectare may be used. (See Chapter 5.) As the measure of area is in "square" units, only a rectangle or a square can be readily measured for area.

In the case of a square, the area is the product of the lengths of the two equal sides (Fig. 12-20):

$$\text{Area} = a \cdot a = a^2$$

In the case of a rectangle, the area is the product of the short side times the long side (Fig. 12-21):

$$\text{Area} = h \cdot b$$

The calculation of other polygonal areas is made by reducing the polygon to a series of squares, rectangles, and/or triangles as shown in Fig. 12-24.

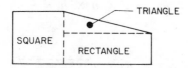

Fig. 12-24. Area calculation of a polygon.

Area of a Parallelogram. The area of a parallelogram equals the product of one side and the altitude to that side. Expressed in a formula it is: $A = bh$. In Fig. 12-25, the parallelogram $ABCD$ has line AD extended to F. By dropping perpendiculars BE and CF to line AF, a rectangle, $EFCB$, is formed having the same base and altitude as the parallelogram. Triangles AEB and DFC are congruent. If AEB is subtracted from $ABCD$ and its equal DFC is added to the remainder, $BCDE$, the rectangle $EFCB$ will equal parallelogram $ABCD$ in area. The area of $EFCB$ equals the altitude h times the base b. This equals the area of $ABCD$ also (Fig. 12-25). Hence:

$$\text{Area} = b \cdot h$$

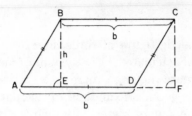

Fig. 12-25. Area calculation of a parallelogram.

Area of a Triangle. The area of a triangle is equal to one-half the product of one side and the altitude to that side. An altitude is a perpendicular drawn from the apex of one angle of the triangle to the opposite side. In Fig. 12-10, the triangle ABC has a base b and an altitude h. By drawing BE through B and parallel to AC, and EC through C parallel to AB, a parallelogram is formed. Parallelogram $ABEC$ has the same base AC (b) and altitude h as the triangle ABC. Line BC, being a diagonal, cuts the parallelogram in half. Thus the area of triangle ABC equals the area of triangle BCE, both of which equal one-half of the area of parallelogram $ABEC$. Hence the formula for the area of a triangle is (Fig. 12-26):

$$\text{Area} = \frac{1}{2} \times b \times h = \frac{1}{2}bh$$

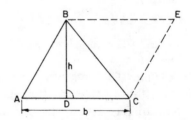

Fig. 12-26. Area calculation of a triangle.

Area of a Trapezoid. The area of a trapezoid is equal to one-half the product of the altitude and the *sum* of its bases. In Fig. 12-27, the trapezoid $ABCD$ has two bases, b and b', and an altitude of h. Note that the altitude from D is drawn perpendicular to the base BC which has been extended to point F. A diagonal BD is drawn, dividing the trapezoid into two triangles having a common altitude h.

The area of triangle $ABD = \frac{1}{2}bh$ and the area of triangle BCD $= \frac{1}{2}b'h$. The total area of the trapezoid $ABCD$ is the sum of the areas of the two triangles, Fig. 12-27:

$$\text{Area} = \frac{1}{2}b \times h + \frac{1}{2}b' \times h = \frac{1}{2}h\,(b + b')$$

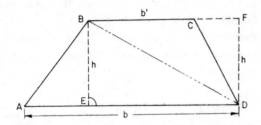

Fig. 12-27. Area calculation of a trapezoid.

Area of Regular Polygons. Regular polygons are geometric figures of many sides whose sides are all equal in length and whose interior angles are all equal to each other. The *hexagon* is a typical regular polygon. In Fig. 12-28, the hexagon $ABCDEF$ is shown with its altitude h or *apothem*, which is drawn from the center of the hexagon and at right angles to one side. Note that the polygon is divided into equal triangles. Each triangle's area equals $\frac{1}{2}bh$. In a hexagon there are six triangles. Therefore:

$$\text{Area} = 6\left(\frac{1}{2}bh\right) = 3bh$$

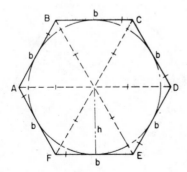

Fig. 12-28. Area calculation of a regular hexagon.

Notice, however, that the six b sides form the perimeter p of the hexagon. Therefore, $p = 6b$ and the formula will also read:

$$\text{Area} = \frac{1}{2}ph$$

As regular hexagons are common in such things as bar stock, etc., several special formulas have been derived to aid in calculating areas and related dimensions of a hexagon. These formulas, which are given below for reference and use in your calculations, were derived from basic geometric principles.

When: A = Area of a regular hexagon

R = Radius of the circumscribed circle (distance across points = $2R$)

r = Radius of inscribed circle

F = Distance across-the-flats ($2r$)

S = Length of a side (Fig. 12-29):

$$A = 2.598\,S^2 = 2.598\,R^2 = 3.464\,r^2 = 3.464\left(\frac{F}{2}\right)^2$$

$$R = S = 1.155\,r = 1.155\,\frac{F}{2}$$

$$r = 0.866\,S = 0.866\,R$$

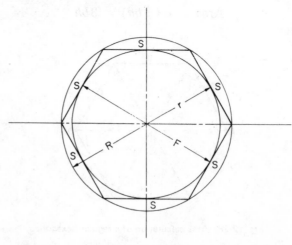

Fig. 12-29. A hexagon

Area of Irregular Polygons. Irregular polygons can be subdivided into component parts—which are triangles, squares, or rectangles. The area of a given irregular polygon equals the sum of the areas of the parts of the subdivided polygon. Shown below are some irregular polygons and a method for solving for their areas.

Study Examples.

1. The irregular polygon in Fig. 12-30 consists of a square, 20 cm on a side plus an isosceles triangle with a height of 20 cm.

    Solution:

    $$\text{Area}_p = \text{Area}_{sq} + \text{Area}_\triangle$$

    $$\text{Area}_{sq} = 20 \times 20 = 400 \text{ cm}^2$$

    $$\text{Area}_\triangle = \frac{1}{2}(20 \times 20) = 200 \text{ cm}^2$$

    $$\text{Area of polygon} = 600 \text{ cm}^2$$

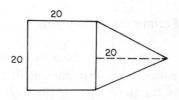

Fig. 12-30. Study example 1.

2. Figure 12-31 is made up of a series of rectangles. The total area of the figure equals the sum of the areas of the component figures. All dimensions are in meters.

    Solution:

    1. $11.8 \times 26 = 306.8 \text{ m}^2$

    2. $7.2 \times 12 = 86.4 \text{ m}^2$

    3. $4.0 \times 12 = 48.0 \text{ m}^2$

    Total area $= 441.2 \text{ m}^2$

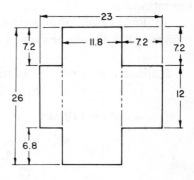

Fig. 12-31. Study example 2.

3. The irregular polygon in Fig. 12-32 can be divided into two triangles.

The total area equals:

1. $\dfrac{1}{2}eh + \dfrac{1}{2}eh'$

2. $A = \dfrac{1}{2}e(h + h')$

When $e = 20$ mm, $h = 5$ mm
and $h' = 9$ mm

3. $A = \dfrac{20}{2}(5 + 9)$

4. $A = 10 \times 14 = 140$ mm^2

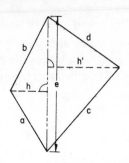

Fig. 12-32. Study example 3.

Perimeters of Polygons

One of the basic geometric concepts used in land survey is the *perimeter*. The amount of fencing a rancher needs to enclose a field is the perimeter or the distance around the field. In irregular polygons, the perimeters are the sums of the many sides involved; hence it is necessary to measure or solve for the length of each side.

The perimeter of a metal stamping is a very important dimension in calculating the force required to blank the part from the sheet or coil of sheet metal from which it is made. This force is important in estimating the size of the press required to perform the blanking operation.

In Fig. 12-33, the perimeter is the sum of the sides, nine in number. All dimensions are in millimeters. Therefore:

Perimeter = 7 mm + 14 mm + 13 mm + 26 mm + 10 mm
+ 11 mm + 9 mm + 14 mm + 15 mm

$P = 119$ mm

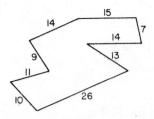

Fig. 12-33. Perimeter calculation of a polygon.

The perimeters of many of the regular polygons can be determined by the use of the following formulas:

1. Fig. 12-34, triangle with sides a, b, and c:

$$p = a + b + c$$

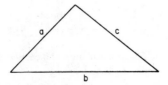

Fig. 12-34. Perimeter calculation of a triangle.

2. Fig. 12-35, equilateral triangle, 3 equal sides s:

$$p = 3s$$

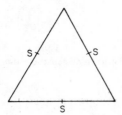

Fig. 12-35. Perimeter calculation of an equilateral triangle.

3. Fig. 12-36, quadrilateral with sides a, b, c, and d:

$$p = a + b + c + d$$

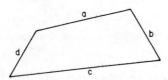

Fig. 12-36. Perimeter calculation of a quadrilateral.

4. Fig. 12-37, parallelogram with sides a and b:

$$p = 2a + 2b$$

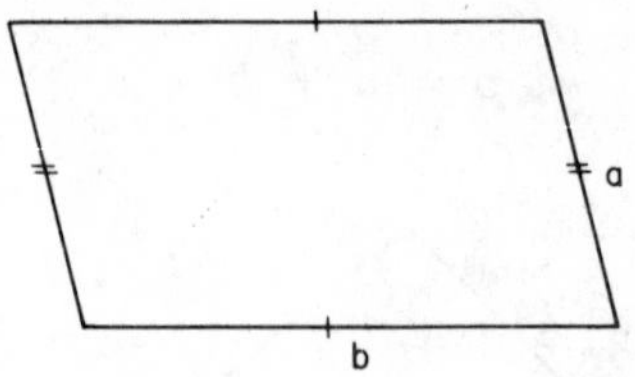

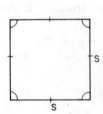

Fig. 12-37. Perimeter calculation of a parallelogram.

5. Fig. 12-38, square of side s:

$$p = 4s$$

Fig. 12-38. Perimeter calculation of a square.

6. Fig. 12-39, regular hexagon, side s:

$$p = 6s$$

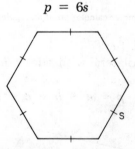

Fig. 12-39. Perimeter calculation of a regular hexagon.

7. Any given regular polygon of n sides of s:

$$p = ns$$

Study Example.

1. A painter charges 25 cents per square foot for exterior house painting. A house has one side, as shown in Fig. 12-40. Allowing for three $4' \times 5'$ windows, how much will the painter charge to paint this side of the house?

     Step 1. Area of lower half: $10 \times 42 = 420$ sq ft

     Step 2. Area of upper half: $\frac{1}{2} \times 8 \times 42 = 168$ sq ft

     Step 3. Gross area of side: $= 588$ sq ft

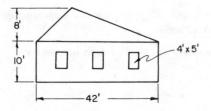

Fig. 12-40. Study example 1.

     Step 4. Area of 3 windows: $3 \times 4 \times 5 = 60$ sq ft

     Step 5. Net area of side: $588 - 60 = 528$ sq ft

     Step 6. Painting cost $=$ net area $\times$ \$.25

                    $= 528 \times \$.25 = \$132.00$

Study Problems.

1. A swimming pool is $40'$ long and $20'$ wide with a depth from $3'$ at one end to $10'$ at the other.

    a. What is the perimeter at ground level?

    b. What is the area of one $40'$ wall?

    c. What is the total area of all four walls?

    d. What is the perimeter of one $40'$ wall?

    e. What is the perimeter of the bottom?

2. In problem 1, how much will it cost to paint the total interior surface with a waterproof paint at \$.875 per square foot?

3. In Fig. 12-41 (a rectangle with a triangular part removed):

    a. Calculate the perimeter.

    b. Calculate the area.

    (All dimensions are in millimeters.)

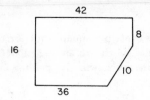

Fig. 12-41. Study problem 3.

4. Find the surface area of a rectangular solid 32 cm × 22 cm × 8.3 cm.

5. Find the surface area of a cube 26 cm on an edge.

6. Calculate the length of perimeter for a rectangular press blank 30 in. long by 14.25 in. wide.

7. What is the perimeter of the blank shown below?

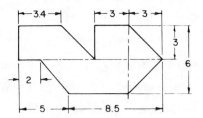

Fig. 12-42. Study problem 7.

8. A screen has 96 openings with dimensions as shown in Fig. 12-43. Find the total area of the openings in the screen.

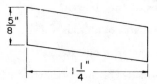

Fig. 12-43. Study problem 8.

9. A steam engine cylinder is to have 9 port openings as shown in Fig. 12-44. If the total area of the port openings is to be 10.125 sq in. find the width x of each port. Answer to the nearest 0.001 in.

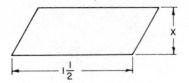

Fig. 12-44. Study problem 9.

10. Calculate the cross-sectional area of a storm drain shown in **Fig. 12-45.**
Give answer in square feet and also in square meters.

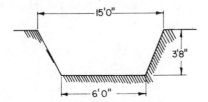

Fig. 12-45. Study problem 10.

11. Calculate the cross-sectional area and the perimeter of the rail **section**
shown in Fig. 12-46. Give answer in square millimeters.

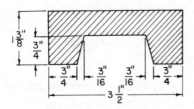

Fig. 12-46. Study problem 11.

Oblique Triangles

Triangles in which no interior angle is equal to 90° are termed
oblique or *non-right-angle* triangles. These triangles are of two shapes:

1. *Obtuse*, which are oblique triangles having one angle greater
than 90°. In Fig. 12-47, △ *ABC* is obtuse with ∠*B* greater
than 90°.

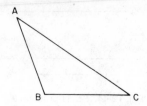

Fig. 12-47. An obtuse oblique triangle.

2. *Acute*, which are oblique triangles having *all* interior angles less than 90°. Fig. 12-48 shows an acute triangle.

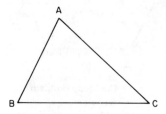

Fig. 12-48. An acute oblique triangle.

As in all triangles, the sum of the interior angles of an obtuse or an acute triangle equals 180°.

Special Lines in Triangles

In the solution of problems involving triangles, certain auxiliary lines may be drawn to help in the solution. These are:

Angle Bisector. In Fig. 12-49, $\angle B$ in $\triangle ABC$ is bisected (cut in half) by line *BD*, the angle bisector.

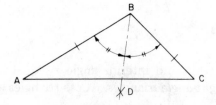

Fig. 12-49. The angle bisector.

In the case of an equilateral triangle, Fig. 12-50, the angle bisector also bisects the side opposite.

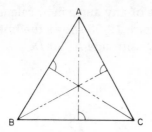

Fig. 12-50. Angle bisector in an equilateral triangle.

The three bisectors of an equilateral triangle meet in a common point which is the center of a circumscribed circle, shown in Fig. 12-51.

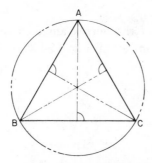

Fig. 12-51. Intersection of angle bisectors in an equilateral triangle is the center of a circumscribed circle.

This center is also the center of the inscribed circle shown in Fig. 12-52. In cutting tools, the inscribed circle is a very important element in metal cutting tool geometry.

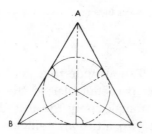

Fig. 12-52. Intersection of angle bisectors in an equilateral triangle in the center of an inscribed circle.

The angle bisectors of any angle in a scalene triangle do *not* bisect the opposite sides. Figure 12-53 shows the bisector of $\angle B$, BD, which does not bisect the opposite side AC at D.

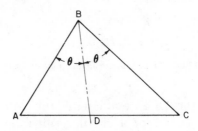

Fig. 12-53. Angle bisector in a scalene triangle.

The angle bisector of the vertex angle of the two equal sides of an isosceles triangle as shown in Fig. 12-54, bisects the opposite side, BC.

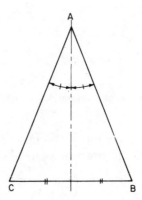

Fig. 12-54. Angle bisector in an isosceles triangle.

Altitude to a side. The altitude to a side of any triangle is a line drawn from any vertex perpendicular to the opposite side. In the special case of a right triangle, only the altitude from the 90° vertex is useful, as the other two altitudes coincide with the sides of the right triangle. See Fig. 12-55.

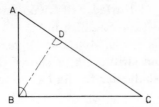

Fig. 12-55. An altitude to an hypotenuse.

The altitudes in a typical scalene triangle are shown in Fig. 12-56. Altitude *BD* is perpendicular to line *AC* and is *within* the boundaries of the triangle. This is always true of an altitude from the obtuse angle. The altitudes from the acute angles, *A* and *C*, fall *outside* the boundaries of the triangle and their opposite sides must be extended: *BC* to *E* for altitude *AE*, and *AB* to *F* for altitude *CF*.

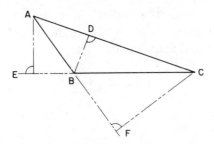

Fig. 12-56. Altitudes in a scalene triangle.

The Projection Formula.

The oblique triangle does not appear as often in practical shop problems as does the right triangle. However, for oblique triangles with three sides given, the altitudes may be calculated by using the *projection* formula; and with an altitude known, the area of the triangle may be determined.

Oblique triangles may be defined in three ways:

1. When the three sides are known.

2. When two sides and the included angle are known.

3. When two angles and the included side are known.

Cases 2 and 3 require a knowledge of trigonometry to solve, but case 1 may be solved by the projection formula.

Projection Formula for an Acute Triangle. In the acute triangle shown in Fig. 12-57, an altitude, h, is shown drawn from the vertex of angle B. h is perpendicular to the base b and cuts the base into two segments p and p_1.

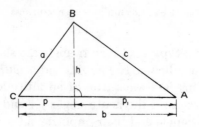

Fig. 12-57. Projection formula in an acute oblique triangle.

The distance p from the vertex of angle C to the foot of the altitude h is called the *projection* of side BC on the base b. The projection formula for this acute triangle is:

$$p = \frac{a^2 + b^2 - c^2}{2b}$$

and may be stated:

The length of the projection of a side of an acute triangle equals the sum of the squares of the base and the side adjacent to the projection minus the square of the side opposite the angle made by the projecting side and its projection, all divided by two times the base.

Study Example.

1. In Fig. 12-58, an acute scalene triangle, solve for p and the altitude h; when $a = 9$ in., $b = p_1 + p = 16$ in. and $c = 13$ in. Also, find the area of the triangle.

Step 1. Using the formula for an acute scalene triangle:

$$p = \frac{a^2 + b^2 - c^2}{2b}$$

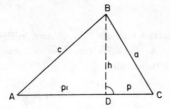

Fig. 12-58. Study example 1.

Substitute values given in problem:

$$p = \frac{9 \times 9 + 16 \times 16 - 13 \times 13}{2 \times 16}$$

$$p = \frac{81 + 256 - 169}{32} = \frac{168}{32} = 5.25 \text{ in.}$$

Step 2. Using the Pythagorean Law:

$$h = \sqrt{a^2 - p^2} = \sqrt{81 - 27.56}$$

$$h = \sqrt{53.44} = 7.31 \text{ in.}$$

Step 3. Using formula: Area $\triangle = \frac{1}{2}bh$

$$A = \frac{16 \times 7.31}{2} = \frac{116.96}{2} = 58.48 \text{ sq in.}$$

Projection Formula for an Obtuse Triangle. In the obtuse triangle shown in Fig. 12-59, an altitude, h, is shown drawn from the vertex of angle B. h is perpendicular to the base AC extended to D. $AC = b$ and $AB = c$. The projection of side c on base CD is p_1. The projection formula for this obtuse triangle is:

$$p_1 = \frac{a^2 - b^2 - c^2}{2b}$$

NOTE:. This formula varies from the acute triangle formula only by the sign of b^2.

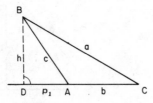

Fig. 12-59. Projection formula in an obtuse oblique triangle.

Study Example.

1. In the obtuse scalene triangle ABC shown in Fig. 12-59, solve for p_1 and altitude h, when $a = 13.75$ cm, $b = 8.5$ cm and $c = 7.5$ cm. Also find the area of the triangle.

Step 1. Using the formula for an obtuse scalene triangle:

$$p_1 = \frac{a^2 - b^2 - c^2}{2b}$$

Substitute values given in problem:

$$p_1 = \frac{(13.75)^2 - (8.5)^2 - (7.5)^2}{2 \times 8.5}$$

$$p_1 = \frac{189.06 - 72.25 - 56.25}{17} = \frac{60.56}{17}$$

$$p_1 = 3.56 \text{ cm}$$

Step 2. Using the Pythagorean Law:

$$h = \sqrt{c^2 - p_1^2} = \sqrt{(7.5)^2 - (3.56)^2}$$
$$h = \sqrt{56.25 - 12.67} = \sqrt{43.58}$$
$$h = 6.60 \text{ cm}$$

Step 3. Using formula:

$$\text{Area} \triangle ABC = \frac{1}{2}bh$$

$$A = \frac{(8.5)(6.6)}{2}$$

$$A = 28.05 \text{ sq cm}$$

Study Problems.

1. In the acute triangle shown in Fig. 12-60 solve for the altitude and the area.

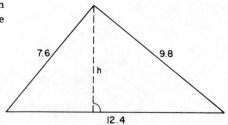

Fig. 12-60. Study problem 1.

2. In the obtuse triangle shown in Fig. 12-61 solve for X and the area of the triangle.

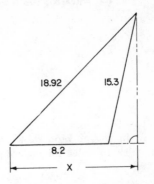

Fig. 12-61. Study problem 2.

3. In Fig. 12-62 a numerical control drilling machine will be used to drill the three holes shown. In order to position the table to drill the upper hole, the distances X and Y must be known. Calculate these distances to 0.001 in. accuracy.

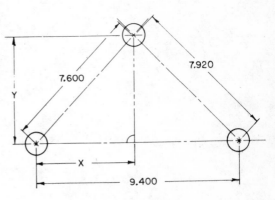

Fig. 12-62. Study problem 3.

4. In the obtuse triangle shown in Fig. 12-63, solve for h and the area.

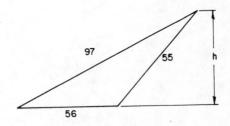

Fig. 12-63. Study problem 4.

5. In the polygon shown in Fig. 12-64, solve for the total area.

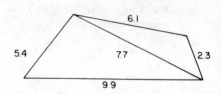

Fig. 12-64. Study problem 5.

6. In the polygon shown in Fig. 12-65, solve for the total area.

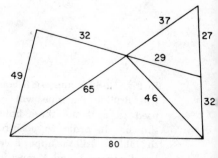

Fig. 12-65. Study problem 6.

CHAPTER 13

Geometry – The Circle

Often called the "perfect" geometric form, the circle occurs frequently in mechanical design, architecture, and applied mathematics. The circle forms the base for systems of navigation on the sea and in the air. The study of geometry involves not only the realtionships of the principal parts of the circle but also the relation of the circle to lines, angles, and polygons.

The circle has been defined earlier in the text as: "a finite, curved line that is equidistant around a point in all directions. This point is called the *center* of the circle. Figure 13-1 depicts the principal parts of a circle, which are defined as follows:

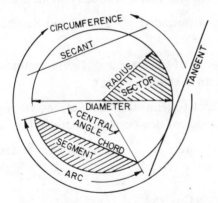

Fig. 13-1. Principal parts of a circle.

Circumference. This is the perimeter of a circle, the distance around it. The circumference is measured both in linear terms such as inches and millimeters, or in degrees. The complete circumference of a circle is measured as 360 degrees, which is equal to one revolution of the radius in describing the circle.

Radius. A radius (plural - radii) of a circle is a line joining the center of a circle with a point on the circumference.

Diameter. The diameter of a circle is a line through the center of the circle joining two opposite points on the circumference. The diameter is a specialized chord that passes through the center, and is the longest chord that can be drawn in a circle. The diameter is equal to twice the length of the radius.

Chord. A chord of a circle is a line joining any two points on the circumference. A chord is said to *subtend* an arc.

Secant. A secant is any line that intersects a circle at two points.

Arc. An arc is a portion of the circumference measured around the circumference between two points on that circumference. A straight line connecting the points at each end of an arc is a chord. An arc may be measured in linear units or in degrees of the angle formed by radii to each of the points. This angle, at the center of the circle, is described as a *central* angle.

Central Angle. An angle formed by two radii is called a central angle. A central angle defines both a chord and an arc.

Segment. The area bounded by an arc and its chord is called a segment.

Sector. The "wedge-of-pie" shaped figure bounded by an arc and its two related radii is called a sector.

Tangent. A line that touches the circle at one and only one point is called a tangent. A tangent does not cut through the circumference.

Semicircle. A semicircle is an arc equal to one-half the circumference of a circle, a half circle.

Constructions Involving Circles

Inscribed Polygon. A polygon whose sides are all chords of a circle is said to be *inscribed*, as shown in Fig. 13-2.

Circumscribed Polygon. A polygon whose sides are all tangent to an internal circle is said to be *circumscribed*, as shown in Fig. 13-3.

Inscribed Circle. A circle to which all sides of a polygon are tangent is said to be "inscribed," as shown in Fig. 13-3.

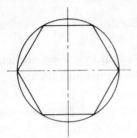

Fig. 13-2. Inscribed polygon and circumscribed circle.

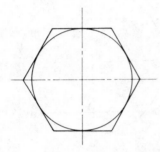

Fig. 13-3. Circumscribed polygon and inscribed circle.

Circumscribed Circle. A circle passing through each vertex of a polygon is said to be circumscribed, as shown in Fig. 13-2.

Concentric Circles. Circles are said to be "concentric" when they have a common center as shown in Fig. 13-4.

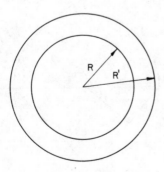

Fig. 13-4. Concentric circles.

Propositions Involving Circles

PROPOSITION 34: **A diameter perpendicular to a chord bisects the chord and its arc.**

In Fig. 13-5, diameter *CD* is perpendicular to chord *AB* at point *E*. Chord *AB* is bisected by *CD* and line *AE* equals line *EB*. Arc *ADB* is also bisected by *CD* and arc *AD* equals arc *DB*.

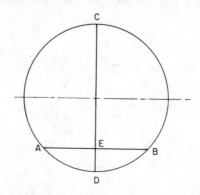

Fig. 13-5. Diameter perpendicular to a chord.

PROPOSITION 35: **In the same or equal circles, equal chords are equally distant from the center and conversely.**

In Fig. 13-6, chords *AB* and *CD* are of equal length. Lines *OE* and *OF* are segments of radii drawn perpendicular to the chords and are also of equal length.

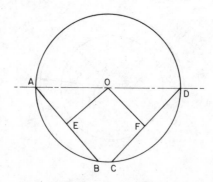

Fig. 13-6. Equal chords equally distant from center of circle.

PROPOSITION 36: The circumference of a circle is equal to the product of the diameter and the constant pi(π).

The Greek letter *pi* (π) is used to describe the ratio of the circumference of any circle to its diameter:

$$\pi = \frac{\text{Circumference}}{\text{Diameter}}$$

or: Circumference $= \pi \cdot D$

The approximate value of π, normally used in working shop math problems is 3.1416. This is approximately $\frac{22}{7}$.

A circle can be regarded as a regular polygon, of an infinite number of sides. In Fig. 13-7, a square $ABCD$ is inscribed in the base circle of 1-inch radius. A second square, $EFGH$, is circumscribed on the same base circle.

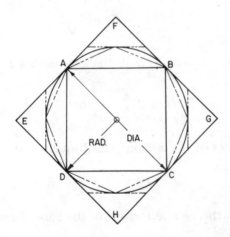

Fig. 13-7. Development of numerical value for *pi* (π).

The perimeter of the circumscribed square, $EFGH$, is $4s_c$, where s_c is the length of one side. Length s_c equals the diameter D of the base circle and, hence, the perimeter of the circumscribed square is $4D$.

The perimeter of the inscribed square, $ABCD$, is also equal to $4 \cdot s_i$ where s_i is the length of one side. In $\triangle ABC$, AC is the diameter D. Hence, by Pythagorean law:

$$\overline{AC}^2 = \overline{AB}^2 + \overline{BC}^2$$

But side $AB =$ side $BC = s_i$. Therefore:

$$\overline{AC}^2 = s_i^2 + s_i^2 = 2s_i^2$$

or: $\overline{AC}^2 = 2s_i^2;$ or: $2s_i^2 = D^2$

then: $s_i^2 = \dfrac{D^2}{2}$

and: $s_i = \sqrt{\dfrac{D^2}{2}}$ or: $s_i = \dfrac{D}{\sqrt{2}}$

The perimeter of the inscribed square equals $4 \cdot s_i$, or:

$$P_i = 4\sqrt{\frac{D^2}{2}}$$

In Fig. 13-7, if the base circle diameter equals 2,

$$P_i = 4\sqrt{\frac{4}{2}}   \text{or:}   P_i = 4\sqrt{2}$$

The perimeter of the circumscribed square is $P_c = 4D$, or:

$$P_c = 4 \cdot 2 = 8$$

The ratio of the two perimeters to the base diameter is:

$$\frac{8}{2} = 4.00000 \text{ for the larger square.}$$

$$\frac{4\sqrt{2}}{2} = \frac{5.656}{2} = 2.828 \text{ for the smaller.}$$

If the squares were made into polygons of ever increasing numbers of sides, the two polygons would approach an infinite number of sides. They would coincide as a circle and the ratio of their perimeters to the base diameter would reach:

$$3.1415926535847932384626\ldots\ldots$$

or 3.1416 in more practical terms. This ratio is pi (π) and is an extremely important mathematical constant that should be memorized. The table below illustrates the ratio changes as the number of sides increases:

No. of Sides	Perimeter Circumscribed Polygon ÷ D	Perimeter of Inscribed Polygon ÷ D	Tabular Difference
4	4.00000	2.82843	1.17157
8	3.31368	3.08144	0.23222
16	3.18256	3.12144	0.06112
32	3.15168	3.13664	0.01504
36	3.14874	3.13776	0.01098
64	3.14412	3.14033	0.00379
128	3.14398	3.14170	0.00228
256	3.14324	3.14141	0.00183
360	3.14203	3.14148	0.00055
512	3.14163	3.14157	0.00006
720	3.14161	3.14159	0.00002

PROPOSITION 37: The area of a circle is equal to the product of π and the radius squared: $A = \pi r^2$; or the product of π and one-quarter of the square of the diameter: $A = \pi \dfrac{D^2}{4}$.

These are the well-known formulas for finding the area of a circle. The proof is illustrated in Figs. 13-8 and 13-9. In Fig. 13-8, two semicircles are formed by the diameter AB. Radii are drawn, cutting the semicircles into connected segments. If each semicircle of segments were to be "flattened" along the semicircumference, the segments would separate like the segments in a slice of orange. The two, flattened, separated, segment semicircles can then be fitted together as in Fig. 13-9 to approximate a rectangle whose width is radius r and whose length is a half-circumference of $\frac{1}{2} C$. While this proof is approximate, the formulas that follow are exact.

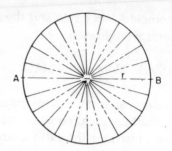

Fig. 13-8. Segmented circle.

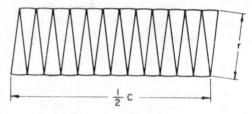

Fig. 13-9. Segmented semicircles fitted together.

The formula for the approximated rectangular area would be:

$$A = \frac{1}{2}C \cdot r$$

however: $C = \pi D$ or: $C = 2\pi r$ Prop. 36

$$A = \frac{1}{2}(2\pi r)(r) = \pi r^2$$

PROPOSITION 38: **A straight line, perpendicular to a radius at its extremity, is tangent to the circle. Conversely, the tangent at the extremity of the radius is perpendicular to the radius.**

A tangent to a circle is a line that touches the circumference at a point. In Fig. 13-10A, AB is perpendicular to radius OC at point C which is at the end of the radius. Therefore, AB is tangent to the circle.

COROLLARY: **If two or more tangents are drawn to a circle, the lines drawn perpendicular to each tangent at the point of tangency will intersect each other at the center of the circle unless they are colinear.**

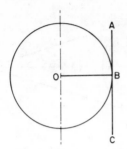

Fig. 13-10A: A tangent to a circle.

In Fig. 13-10B *EC* and *E'C'* are perpendicular to tangents *AB* and *A'B'*, respectively, and they pass through the respective points of tangency, *E* and *E'*. The center of the circle is located at 0, the point of intersection of *EC* and *E'C'*.

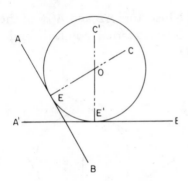

Fig. 13-10B. Two tangents to a circle.

PROPOSITION 39: If two tangents are drawn to a circle from an external point, they are equal in length and make equal angles with a line drawn from the center of the circle to the external point.

In Fig. 13-11A, lines *AB* and *AC* are drawn tangent to the circle from an external point *A*. Line *AB* equals line *AC* and ∠*BAO* equals ∠*CAO*. This is an important proposition that is used in machine shop practice.

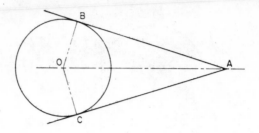

Fig. 13-11A. Equal length of tangents.

COROLLARY: If two or more pairs of tangents are drawn from external points around a circle, the bisectors of the angles formed by each pair of tangents will intersect at the center of the circle, except when the bisectors are colinear.

In Fig. 13-11B the bisectors AB and $A'B'$ of the angles formed by the tangents drawn from external points A and A' intersect at the center of the circle, O.

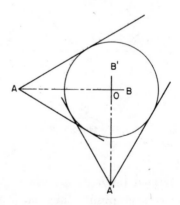

Fig. 13-11B. Intersection of bisectors at the center of the circle.

Proposition 39 is the basis of the tool called a *center head* shown in Fig. 13-12. This is used to find the center of the end faces of cylindrical bars or shafts.

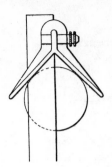

Fig. 13-12. Use of center head to find center of end face of shaft.

Study Examples.

1   1. In the accompanying drawing
in Fig. 13-13, lines AB and AC
are tangent to the circle. If
$\angle BAC = 30°$, find $\angle COD$.

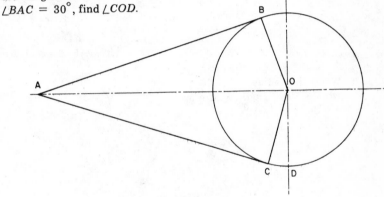

Fig. 13-13. Study example 1.

Step 1. From Proposition 39, AO bisects $\angle BAC$ and $\angle OAC = \frac{1}{2}\angle BAC$.
$\angle OAC = \frac{30°}{2} = 15°$.

Step 2. $\angle ACO$ is a right $\angle$ (Prop. 38) and equals $90°$. The sum of the
interior $\angle s$ in $AOC = 180°$ (Prop. 7). $\angle AOC = 180° -$
$(15° + 90°) = 180° - 105° = 75°$.

Step 3. $\angle AOD$ is a right $\angle$ and $\angle AOC + \angle COD = 90°$.

Therefore: $\angle COD = 90° - 75° = 15°$

NOTE: Step 3 could also be referred to Proposition 26.

Study Problems.

1. In Fig. 13-14 solve for angle α.

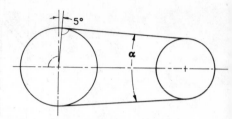

Fig. 13-14. Study problem 1.

2. The circle in Fig. 13-15 is inscribed in a scalene triangle.
Solve for angle β.

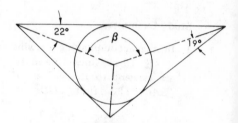

Fig. 13-15. Study problem 2.

3. When the radius $r = 1.25$ in.,
what is the length of the arc
X? See Fig. 13-16.

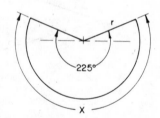

Fig. 13-16. Study problem 3.

4. Determine the size of angles A,
B, C, D and E in Fig. 13-17.

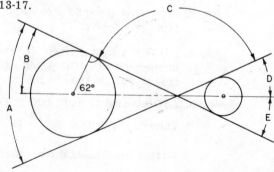

Fig. 13-17. Study problem 4.

5. Solve for the diameter in Fig. 13-18. The circumference is in meters.

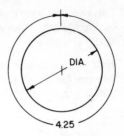

Fig. 13-18. Study problem 5.

6. *AP*, *BQ* and *AB* are tangents to the circle, $AP = BQ$, in Fig. 13-19. Solve for *y* when $AP = 6$ in.

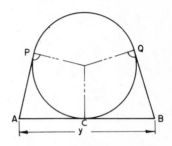

Fig. 13-19. Study problem 6.

7. In Fig. 13-20, solve for $\angle PAO$ where $\angle POQ = 140°$.

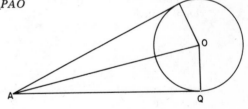

Fig. 13-20. Study problem 7.

8. In Fig. 13-21 solve for *AB* when $BC = 11$ cm, $CD = 5.5$ cm, $CN = 2$ cm, and $AD = 9.5$ cm.

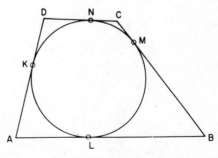

Fig. 13-21. Study problem 8.

9. Solve for the length of arc AB in Fig. 13-22. The radius is 3.12 in.

Hint: Equate ratios of circumference for $360°$ and AB for $135°$.

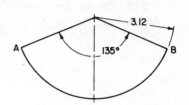

Fig. 13-22. Study problem 9.

10. Solve for angle α in Fig. 13-23.

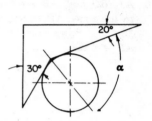

Fig. 13-23. Study problem 10.

11. Determine the distance X in Fig. 13-24. The given dimension is in inches.

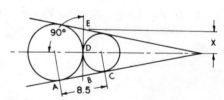

Fig. 13-24. Study problem 11.

12. Determine the angle α in Fig. 13-25.

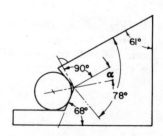

Fig. 13-25. Study problem 12.

PROPOSITION 40: In the same or equal circles, equal central angles subtend equal chords.

In Fig. 13-26 triangles *AOD* and *BOD* are congruent (two sides and included angles of one equal two sides and included angle of the other). Hence, *AD* = *DB*.

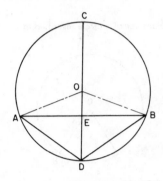

Fig. 13-26. Diameter bisecting chord and arc.

PROPOSITION 41: In the same or equal circles, equal central angles subtend equal arcs.

If in Fig. 13-26 angles *AOD* and *BOD* are equal, arcs *AD* and *BD* are equal, since a central angle has the same number of degrees as the arc it intercepts (Proposition 43).

PROPOSITION 42: The diameter of a circle inscribed in a right triangle is equal to the difference of the sum of the legs of the triangle (sides adjacent to the right angle) and the hypotenuse.

In Fig. 13-27, *AB* + *BC* − *AC* = Diameter of inscribed circle.

PROPOSITION 43: A central angle has the same measurement as its intercepted arc (in degrees).

Angle measurement is known as *angle degrees* but arc measurement is in *arc degrees* or *degrees of arc*. In Fig. 13-28, the angle degrees of ∠*AOB* equals the arc degrees in *AB*. This is true as there

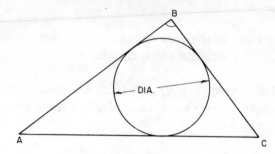

Fig. 13-27. Diameter of a circle inscribed in a right triangle.

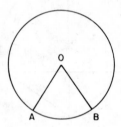

Fig. 13-28. Measurement of a central angle.

are four right angles or 360° at the center of a circle, and the circumference of a circle also contains 360° of arc. Therefore:

$$\text{Central angle degrees} = \text{Degrees of arc}$$

Be careful not to think of a degree of angle and a degree of arc as
the same thing. It is only when the vertex of an angle is at the center
of a circle that this principle applies.

PROPOSITION 44: An inscribed angle is measured by one-half of
its intercepted arc.

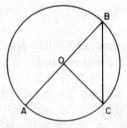

Fig. 13-29. Measurement of an inscribed angle.

Proof	
Analysis	Reason
Angle *ABC* is an inscribed angle.	
Step 1. $\angle AOC$ is measured by arc *AC*.	P-43
Step 2. $\angle AOC = \angle ABC + \angle BCO$	P-8
Step 3. $\angle ABC = \angle BCO$	P-27
Step 4. $\angle AOC = 2\angle ABC$	Axiom II
Step 5. Therefore: $\angle ABC = \frac{1}{2} \angle AOC$	

Proposition 44 is a very important proposition in machine shop practice. The proposition can also be stated as:

> An inscribed angle is equal to one-half of the central angle subtended by the intercepted arc of the inscribed angle.

In Fig. 13-29:

$$\angle ABC = \frac{1}{2} \angle AOC$$

also:

> A central angle is equal to two times the inscribed angle subtended by the intercepted arc of the central angle.

In Fig. 13-29:

$$\angle AOC = 2\angle ABC$$

There are also two corollaries (related truths) to P-44. These are:

> All inscribed angles subtending the same arc are equal.

In Fig. 13-30:

$$\angle ABE = \angle ACE = \angle ADE$$

also:

> An inscribed angle in a semicircle is a right angle.

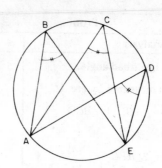

Fig. 13-30. Inscribed angles subtending same arc.

In Fig. 13-31: When AE = Diameter of circle; $\angle ABE$ = 90°; $\angle ACE$ = 90°; $\angle ADE$ = 90°.

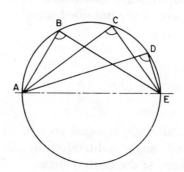

Fig. 13-31. Inscribed angle in a semicircle.

Study Example.

Find the diameter of an inscribed circle in a right triangle with sides 8.6 in. and 10.2 in. long, as seen in Fig. 13-32.

A. Solve for length of the hypotenuse.

$$c^2 = \overline{8.6}^2 + \overline{10.2}^2$$

$$c^2 = 73.96 + 104.04 = 178.00$$

$$c = 13.34 \text{ in.}$$

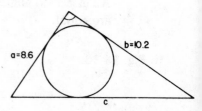

Fig. 13-32. Study example 1.

B. From P-42:

Dia. $= a + b - c$

Dia. $= 8.6 + 10.2 - 13.34$

Dia. $= 18.8 - 13.34$

Dia. $= 5.46$ in.

Study Problems.

1. Solve for $\angle ABC$, as shown in Fig. 13-33.

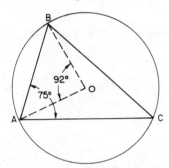

Fig. 13-33. Study problem 1.

2. In Fig. 13-34, solve for diameter, D.

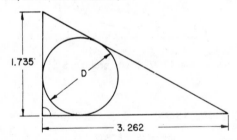

Fig. 13-34. Study problem 2.

3. Solve for dimension x in Fig. 13-35. The given dimension is in inches.

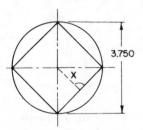

Fig. 13-35. Study problem 3.

4. In Fig. 13-36 solve for dimension A. The given dimension is in inches.

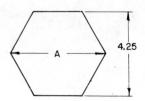

Fig. 13-36. Study problem 4.

5. Solve for dimensions X and Y in Fig. 13-37. The given dimension is in inches.

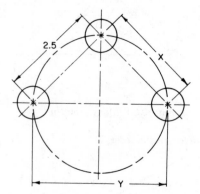

Fig. 13-37. Study problem 5.

6. Determine $\angle \beta$ in Fig. 13-38.

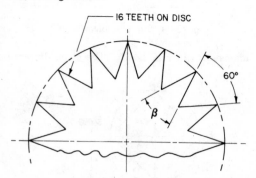

Fig. 13-38. Study problem 6.

7. Solve for the length of the other two sides of the isosceles triangle in Fig. 13-39. The given dimensions are in inches.

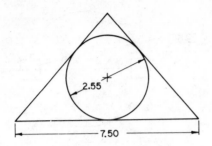

Fig. 13-39. Study problem 7.

8. Solve for $\angle ABC$ and $\angle BCA$ in Fig. 13-40.

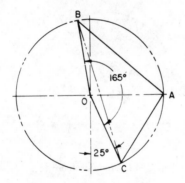

Fig. 13-40. Study problem 8.

9. Solve for angle α in Fig. 13-41.

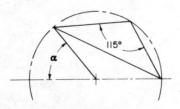

Fig. 13-41. Study problem 9.

10. Solve for dimensions X and Y in Fig. 13-42. All dimensions are in inches.

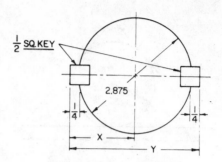

Fig. 13-42. Study problem 10.

11. Solve for $\angle\alpha$ and $\angle\beta$ in Fig. 13-43.

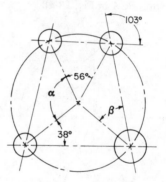

Fig. 13-43. Study problem 11.

12. Solve for angles α, β and ω in Fig. 13-44. Small circles are not equally spaced on hole circle nor are they symmetrically located (top half to bottom half).

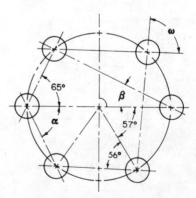

Fig. 13-44. Study problem 12.

PROPOSITION 45:   The line of centers of two circles that are internally or externally tangent passes through the point of tangency.

The two circles shown in Fig. 13-45 are tangent at point T. The line O' passes through the centers of both circles and through point T. Line O' is perpendicular to tangent AB.

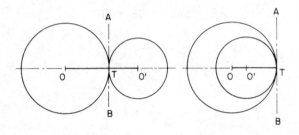

Fig. 13-45. Line of centers of two tangent circles.

PROPOSITION 46:   If two chords intersect within a circle, the product of the two segments of one chord equals the product of the two segments of the other chord.

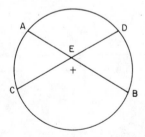

Fig. 13-46. Intersection of two chords.

In Fig. 13-46, AB and CD are chords which intersect at E. The product of the two segments of AB is $AE \cdot EB$. The product of the two segments of CD is $CE \cdot ED$. The proposition can be stated:

$$AE \cdot EB = CE \cdot ED$$

PROPOSITION 47: If a triangle is inscribed in a circle, the product of two of the sides is equal to the product of the diameter of the circle and the altitude erected to the third side.

The triangle ABC is inscribed in a circle whose diameter is BE as shown in Fig. 13-47. The altitude erected to side AC is BD. The principle stated as a formula is:

$$AB \cdot BC = BE \cdot BD$$

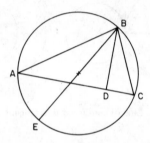

Fig. 13-47. Triangle inscribed in a circle; altitude and diameter.

Study Example.

1. A sector of a circle is shown in Fig. 13-48. The radius of the base circle is required.

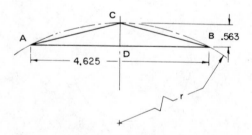

Fig. 13-48. Study example 1.

Step 1.   Lines AC and BC are constructed to form an isosceles triangle, ABC. The given dimensions are in inches.

Step 2. CD is the altitude of $\triangle ABC$ and equals 0.563 in.

Step 3. $AD = DB$; $DB = \dfrac{4.625}{2} = 2.3125$ in.

Step 4. Solve for BC by use of the Pythagorean Formula:

$$\overline{BC}^2 = \overline{BD}^2 + \overline{DC}^2$$

$$\overline{BC}^2 = (0.563)^2 + (2.3125)^2$$

$$\overline{BC}^2 = 0.31697 + 5.34766 = 5.66463$$

$$BC = 2.380 \text{ in.}$$

Step 5. In $\triangle ABC$, $AC = BC$ (Isosceles triangle).

Step 6. Apply Proposition 47:

$$AC \cdot BC = CD \cdot 2r$$

$$2.380 \cdot 2.380 = 0.563 \cdot 2 \cdot r$$

$$5.6644 = 1.126 \, r$$

$$r = 5.031 \text{ in.}$$

Study Problems.

1. Solve for X in Fig. 13-49. The given dimension is in inches.

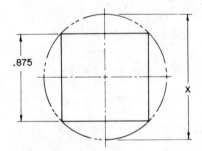

Fig. 13-49. Study problem 1.

2. Determine radius r in Fig. 13-50. All dimensions are in inches.

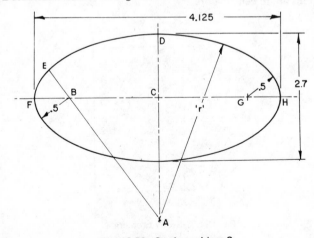

Fig. 13-50. Study problem 2.

Hint: See Proposition 46. Draw arc through *B* and *G* with its center at *A*.

3. In the approximate ellipse shown in Fig. 13-51, find the minor diameter, *d*. All dimensions are in inches.

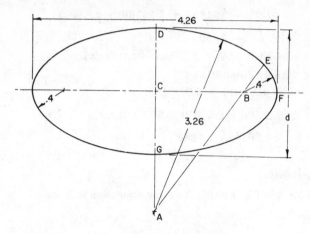

Fig. 13-51. Study problem 3.

Hint: In $\triangle ACB$: $\quad AB = 3.26 - 0.4 = 2.86$

$$CB = \frac{1}{2}(4.26) - 0.4 = 1.73$$

$$AC = 3.26 - \frac{1}{2}d$$

4. Determine the diameter of the base circle, *D*, in Fig. 13-52. All dimensions are in centimeters.

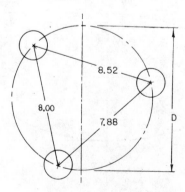

Fig. 13-52. Study problem 4.

5. Solve for the distance across the flats, D, in Fig. 13-53. The given dimension is in inches.

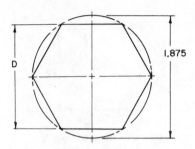

Fig. 13-53. Study problem 5.

6. In Fig. 13-54 solve for X when X is the radius passing through the centers of the three holes. All dimensions are in inches.

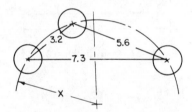

Fig. 13-54. Study problem 6.

7. Determine the radius r in Fig. 13-55. All dimensions are in millimeters.

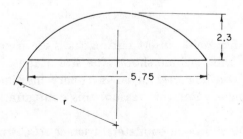

Fig. 13-55. Study problem 7.

8. Determine the diameter of the circumscribed circle in Fig. 13-56. The given dimension is in inches.

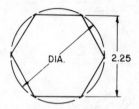

Fig. 13-56. Study problem 8.

9. In Fig. 13-57 determine the diameter D.   All dimensions are in millimeters.

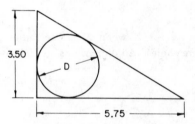

Fig. 13-57. Study problem 9.

10. Determine $\angle \alpha$ in Fig. 13-58.

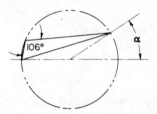

Fig. 13-58. Study problem 10.

Triangular disposable insert cutting tools are frequently used in cutting tools in the machine shop. The size of these cutting tools is determined by the diameter of an imaginary inscribed circle in an equilateral triangle. For this reason, this configuration is of some importance.

Figure 13-59 shows an equilateral triangle ABC with an inscribed circle:

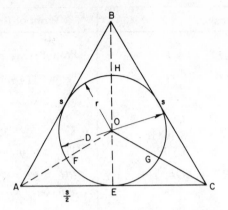

Fig. 13-59. Equilateral triangle with inscribed circle.

In this configuration the following conditions prevail:

Analysis	Proof or Reason
1. $\angle BAC = \angle ABC = \angle ACB = 60°$	Equilateral $\triangle$
2. $\angle OAC = \frac{1}{2}\angle BAC = 30°$	
3. $BE \perp AC$	
4. $AE = EC = \frac{1}{2}s$	s = length of each side of $\triangle$.
5. In $\triangle AEB$; $AE = \frac{1}{2}AB$	
6. In $\triangle AEO$; $EO = \frac{1}{2}AO$	Proposition 30A
$AO = 2EO$	
7. Also $EO = GO = HO = FO = r$	r = radius of the inscribed circle.
8. Thus: $AO = 2EO$	From step 6.
$AO = 2r$	From step 7.
$AO = D$	$2r = D$ (This is a very important relationship)
9. $\quad\quad AO = AF + FO$	
$AF = AO - FO$	
$AF = \;\; 2r - r$	From steps 6 and 7.
$AF = r$	This is an important relationship.

Analysis	Proof or Reason
10. In △ AEO: $\quad \overline{AE}^2 = \overline{AO}^2 - \overline{OE}^2$	Prop. Pythagorean Law.
$(\frac{1}{2}s)^2 = (2r)^2 - r^2$	From steps 4, 8 and 9.
$\frac{1}{4}s^2 = 4r^2 - r^2$	
$s^2 = 12r^2$	
$s = 3.4641r$	
$s = 1.732D$	This is an important relationship.
11. $BE = BH + HO + EO = r + r + r$	$BH = AF = r$, by construction and step 7.
$BE = 3r$	
$BE = \frac{3}{2}D$	This is an important relationship.

In summary, the following important relationships have been shown:

1. A line (AO) extending from the apex of the triangle to the center is equal to the diameter of the inscribed circle.

2. A line (AO) extending from the apex to the center is divided into two equal parts (AF and FO) by the intersection of the inscribed circle, and each part (AF and FO) is equal to the radius of the inscribed circle. Of particular interest is the distance AF, which is equal to the radius of the inscribed circle.

3. The length (s) of each side of the equilateral triangle is equal to 1.732 times the diameter of the inscribed circle. It is of interest to note that 1.732 is equal to the square root of 3 ($\sqrt{3}$).

4. The length of the height (BE), or the perpendicular line from the base to the opposite apex, is equal to $\frac{3}{2}$ times the diameter of the inscribed circle.

Study Example.

1. A typical triangular disposable type carbide cutting tool tip is shown in Fig. 13-60. Solve for h when Dia. = 0.50 in. and $r = \frac{1}{32}$ in.

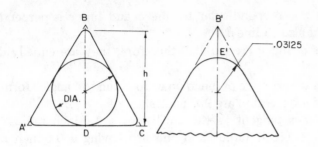

Fig. 13-60. Study example 1.

Step 1. The perpendicular distance $BD = \dfrac{3}{2}$ dia. $= 0.75$

Step 2. In enlarged view, $B'E' = r = \dfrac{1}{32}$

Step 3. Distance $h = BD - B'E'$
$$h = 0.75 - 0.03125$$
$$h = 0.71875, \text{ say } 0.719 \text{ in.}$$

Study Problems.

1. Solve the triangular tip problem in example when $D = \dfrac{3}{8}$ in. and $r = \dfrac{3}{64}$ in.

2. Solve the triangular tip problem in example when $D = \dfrac{5}{8}$ in. and $r = \dfrac{3}{64}$ in.

3. Solve the triangular tip problem in example when $D = \dfrac{9}{16}$ in. and $r = \dfrac{1}{8}$ in.

4. In the study example, what would be the full length of a side of the triangle?

5. Calculate the diameter of an inscribed circle in an equilateral triangle when $s = 1.25$ in.

Relations of Lines, Angles, and Arcs

Shop drawings are sometimes hard to understand if you do not know the conventional relationships of lines, angles, and arcs and the information that is not on the drawing but is understood because of design conventions.

The front view of a gage is shown in Fig. 13-61. Unless otherwise stated corners are assumed to be 90° angles. Line a is parallel to line c and line b is parallel to line d.

Line *a* is perpendicular to line *b* and line *c* is perpendicular to line *b* and also to line *d*.

Lines *a* and *c* are parallel; therefore, they are equally distant at all points.

Lines *a* and *c* are perpendicular to *b* and the angles formed at the points of intersection are 90° angles.

Line *a* is tangent to the arc having a 1.0-inch radius. The arc having a 1.0-inch radius and the arc having a 0.5-inch radius are tangent to each other.

Line *d* is tangent to the arc having a 0.500-inch radius. Note that the dimensions relate mainly to the lower right corner. This design could not be drawn nor the data from it be used if these fundamental facts and their relationships were not known.

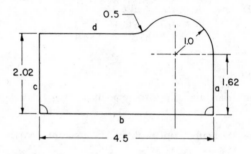

Fig. 13-61. Relations of lines, angles, and arcs.

CHAPTER 14

Geometry – Surface Areas and Volumes of Solids

The figures in the preceding chapters on geometry have been two-dimensional figures or *plane* figures, as all the points, lines, and angles lie in the same plane.

When a third dimension is added, we get *geometric solids*, which are the objects of our real world of three dimensions. The word *solid* used here has a different meaning than it does in everyday conversation. Solid does not refer to the material of the object, but to the shape formed by points in space. Solids are defined as: "enclosed portions of space bounded by plane (flat) and/or curved surfaces."

Six common solids used in commerce and industry are illustrated in Fig. 14-1. These are: the cube, the prism, the cylinder, the cone, the sphere, and the pyramid.

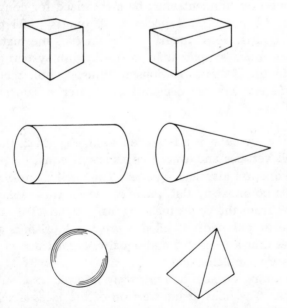

Fig. 14-1. The six common solids: cube, prism, cylinder, cone, sphere, and pyramid.

Solid Measurements

There are two very important measurements that are related to solids in general and these are: surface areas and volumes. Surface areas are two-dimensional, and the surface measurement is stated in *square* units. Volumes are the result of adding a third dimension (depth, thickness, or height), and volume measurement is stated in *cubic* units or in special volume description terms such as gallons or liters.

Surface measurements of solids are generally given as a square of the basic measurement unit such as: square inches, square feet, square yards, etc., in the English system. In the S.I. (Metric) system, areas of solids are usually given in square millimeters (mm^2), square centimeters (cm^2), or square meters (m^2).

Volume measurements are given in two ways: a cubic measure such as cubic inches, cubic feet, etc., in the English system or cubic millimeters (mm^3), etc., in the S.I. system; or in special dry or liquid measure such as pecks, bushels, gallons, barrels, etc., in the English system. The basic unit for volume measurements in the S.I. system is the cubic meter although the liter is recognized for practical use in liquid measure (1000 liters = 1 cubic meter).

The relationship between the special English units of dry and liquid volume measurement must be memorized for conversion purposes, thus: 1 bushel = 4 pecks = 32 quarts = 64 pints; 1 US gallon = 4 quarts, and 1 quart = 2 pints. In the metric system, conversion is much simpler than in the English system of measurement due to the decimal relationship, thus, 1 cubic meter = 1000 cubic decimeters; 1 cubic decimeter = 1 liter = 100 centiliters = 1000 milliliters.

The Cube. A cube is a flat-sided solid, Fig. 14-2, that has right angles at all vertices and whose edges are of equal length. The six faces of a cube are identical squares. The cube is but one of a number of solids bounded by flat, plane surfaces. These solids are called *polyhedrons* from the Greek meaning many-sided. The cube is also a special type of polyhedron called a *prism.* A prism is a solid with two surfaces that are parallel and equal polygons and whose remaining surfaces are parallelograms.

The bounding surfaces of a prism are referred to as *faces.* A cube has six faces. The lines of intersection of the faces are called *edges*, and a cube has twelve edges. The points of intersection of the edges of a prism are called the *vertices* (singular:vertex).

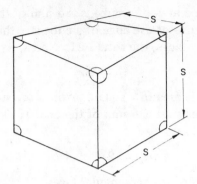

Fig. 14-2. The cube.

A variation of the cube is the *square prism* shown in Fig. 14-3. This is a box-like solid having four edges that are equal in length but are longer (or shorter) than the other eight edges

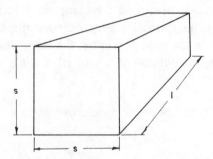

Fig. 14-3. The square prism.

When the end faces are not squares, the prism becomes a *rectangular parallelepiped* or rectangular prism having three unique dimensions: length *l*, width *w*, and height *h* as shown in Fig. 14-4.

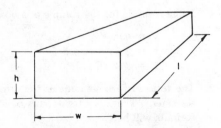

Fig. 14-4. A rectangular parallelepiped.

The total surface area of a cube is the sum of the areas of the six faces. As the measure of the edge of a cube is s, the area of any face is s^2. With six equal faces, the total surface area, A_t, is then:

$$A_t = 6s^2$$

The total surface area of a square prism is the sum of the areas of the two square faces plus the sum of the area of the four rectangular faces. This is stated as:

$$A_t = 2s^2 + 4sl$$

where l is the length of the rectangular faces.

The total surface area of a rectangular prism is the sum of the areas of three pairs of equal size rectangular faces. The edge dimensions, l, w, and h from Fig. 14-4, are used to obtain the formula:

$$A_t = 2wh + 2wl + 2lh$$

Volumes of polyhedrons relate to three-dimensional space. The volume of cube, square prism, or a rectangular prism is calculated as a product of the three mutually perpendicular dimensions: length, width, and height. In the case of a cube, the three dimensions are the same, s. Therefore, the volume of a cube is: $s \cdot s \cdot s$ or:

$$V = s^3$$

The volume of a square prism would be: $s \cdot s \cdot l$, or:

$$V = s^2 l$$

The volume of a rectangular prism would be: $l \cdot w \cdot h$, or:

$$V = lwh$$

Study Examples.

1. A rectangular tank is 9 m long, 6 m wide, and 4.5 m deep. What is its volume? What is its total surface area?

   Step 1. Using formula for the volume of a box:

   $$V = lwh$$
   $$V = 9 \times 6 \times 4.5$$
   $$V = 243 \text{ cu m}$$

   Step 2. The total surface area equals the sum of the areas of the six sides. There are three pairs of identical sides. The formula will be:

$$A_t = 2A_1 + 2A_2 + 2A_3$$

where: A_t = Total surface area of six sides
A_1 = Area of one side of first pair
A_2 = Area of one side of second pair
A_3 = Area of one side of third pair

Step 3. Substituting values from the problem in formula:

$A_t = 2(9 \times 6) + 2(9 \times 4.5) + 2(6 \times 4.5)$
$A_t = 108 + 81 + 54$
$A_t = 243$ sq m

2. Find the volume and the total surface area of a 5-ft cube.

Step 1. A 5-ft cube will have $s = 5$. Using the formula for the volume of a cube:

$V_c = s^3$
$V_c = 5^3 = 125$ cu ft

Step 2. The six faces of the cube will be of equal area. There-fore, the formula, $A_t = 6s^2$ is used:

$A_t = 6s^2$
$A_t = (6)(5)(5) = 6 \times 25 = 150$ sq ft

Study Problems.

1. Calculate the volume of a rectangular tank that is 32 ft long, 40 ft wide and 10 ft deep. If the tank is full of water, how many gallons does it hold?

2. How many square feet of steel plate are necessary to build a rectangular tank that is $7\frac{1}{2}$ ft long, $4\frac{1}{2}$ ft wide and $3\frac{1}{2}$ ft deep? Allow 5% for cutting and welding.

3. A concrete cistern is used to collect rainwater in rural Spain. It is 30 m long, 22 m wide and $11\frac{1}{2}$ m deep (inside dimensions). How many liters of water does it take to fill up the cistern?

4. The cold air return duct on a heating job was 8 in. by 14 in. in cross section. Fifty-five feet of duct was used. How many cubic feet of air will it contain? How many cubic meters?

5. The hydraulic oil storage tank on a machine tool is 250 cm wide by 360 cm long by 150 cm deep. How many liters of oil will it hold? How many gallons?

Prisms. Prisms have been defined as having two parallel faces that are polygons and sides that are parallelograms; the rectangular nature of the cube, the square prism, and the rectangular prism are typical of what is termed a *right* prism. The term "right" refers to the right angle the sides make with the end faces. In Fig. 14-5, a right prism with hexagonal parallel end faces is shown.

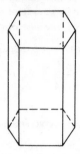

Fig. 14-5. A right prism.

Oblique prisms, such as that shown in Fig. 14-6, are those prisms whose parallel faces are *not* at right angles with all of the sides.

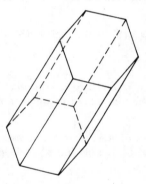

Fig. 14-6. An oblique prism.

The total surface area of any prism is the sum of the areas of the sides plus the sum of the areas of the two end faces. Depending on whether the polygon's parallel faces are regular (hexagonal, octagonal, etc.) or irregular and whether the prism is right or oblique, a formula for the total surface area must be developed for each different configuration. These special formulas will also relate to the area formulas for the regular and irregular plane figures discussed in Chapter 12.

Volume formulas for the various styles of prisms likewise depend on their being right or oblique with regular or irregular parallel faces.

The solution for a volume frequently requires a careful subdivision of the solid into simpler prisms which can be more readily solved for volume. The total volume of the given prism will, of course, equal the sum of the volumes of the separate, simpler prisms (Axiom VIII).

Study Examples.

1. Find the total surface area of a hexagonal right prism 65.0 mm long with an "across-the-flats" dimension of 22.2 mm. See Fig. 14-7.

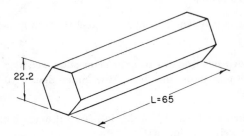

22.2

L=65

Fig. 14-7. Study example 1.

Step 1.   From standard formula for area of a hexagon:

$$\text{Area of base} = (3.4641)\left(\frac{\text{across-the-flats}}{2}\right)^2$$
$$= (3.4641)\left(\frac{22.2}{2}\right)^2$$
$$A_b = (3.4641)(11.1)^2$$
$$= 3.4641 \times 123.21$$
$$A_b = 426.81 \text{ mm}^2$$

Step 2.   Calculate area of a side (rectangle):

$$\text{Area side} = \text{Length} \times \text{side of hexagon}$$
$$\text{Side of hexagon} = 1.155 \times \frac{\text{across-the-flats}}{2}$$
$$= 1.155 \times 11.1 = 12.82 \text{ mm}$$
$$\text{Area of side} = 12.82 \times 65.0$$
$$A_s = 833.3 \text{ mm}^2$$

Step 3.   Total surface area:

$$\text{Total area} = 2A_b + 6A_s$$
$$A_t = 2(426.81) + 6(833.3)$$
$$A_t = 853.62 + 4999.8$$
$$A_t = 5853.42 \text{ mm}^2 \quad \text{say,} \quad 5852 \text{ mm}^2$$

2. Find volume of the prism described in study example 1:

Step 1.   Volume equals area of base times length;

$$V = A_b L$$
$$= 426.81 \times 65.0$$
$$V = 27,742.65 \quad \text{say,} \quad 27,743 \text{ mm}^3$$

Study Problems.

1. Find the total surface area of a hexagon bar that is $2\frac{1}{2}$ in. across the flats and 10 ft long.

2. If the bar in problem 1 is made of steel (0.2833 lb per cu in.), what is its weight?

3. An extruded hexagon bar of aluminum alloy weighing 0.0979 lb per cu in. is 15.9 mm across the flats. How much does a 10-ft piece weigh in kilograms?

4. A flat rolled steel bar, $1\frac{1}{2} \times 7\frac{1}{2}$ in., is 18 ft long. What is its volume in cu in. and weight in pounds?

5. A piece of free cutting brass (0.307 lb per cu in.), hex shaped, is 22.2 mm across the flats. How many kilograms does a piece 5.25 meters long weigh?

The Cylinder. There are two basic types of cylinders. The *right* circular cylinder as shown in Fig. 14-8A is the most common type used in industry. Many storage tanks are of this type and are used in either an upright position resting on the circular base or in a flat position resting on the side. The definition of a right circular cylinder is: "A solid formed by revolving a rectangle about one of its sides as an axis."

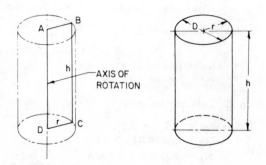

Fig. 14-8A. A right circular cylinder.

The cylindrical surface area of a right circular cylinder, A_s, is calculated by use of the following formula:

$$A_s = 2\pi rh, \quad \text{or} \quad 6.2832rh$$

also: $\quad A_s = \pi Dh, \quad \text{or} \quad 3.1416Dh \quad (D = 2r)$

The total surface area of a right circular cylinder, A_t, is the sum of the cylindrical area plus the areas of the two ends. This is stated in the formulas:

$$A_t = 2\pi r(r + h) \quad \text{or} \quad A_t = 6.2832(r^2 + rh)$$

also: $\quad A_t = \pi D(\frac{1}{2}D + h) \quad \text{or} \quad A_t = 3.1416(\frac{1}{2}D^2 + Dh)$

The volume, V, of a right circular cylinder is found by using the formulas:

$$V = \pi r^2 h, \quad \text{or} \quad V = 3.1416r^2h$$

$$V = \frac{1}{4}\pi D^2 h, \quad \text{or} \quad V = 0.7854D^2h$$

The symbols used in the above formulas are as follows:

A_s = Area of the cylindrical surface only
A_t = Total area of cylinder sides and ends
V = Volume of right circular cylinder
π = 3.1416 (pi)
r = Radius of circular ends
D = Diameter of circular ends
h = Height of cylinder

Study Examples.

1. Find the total surface area of a cylindrical water tank that is 66 in. in diameter and $14\frac{1}{2}$ ft long. Also calculate its volume.

   Step 1. Convert 66 in. to feet:

   $$66 \div 12 = 5.5 \text{ ft}$$

   Step 2. Substitute $D = 5.5$ and $h = 14.5$ in formula:

   $$A_t = 3.1416D(\frac{1}{2}D + h)$$

   $A_t = 3.1416 \cdot 5.5(2.75 + 14.5)$
   $A_t = 17.2788 \cdot (17.25) = 298.06 \quad \text{say,} \quad 298 \text{ sq ft}$

   Step 3. Substitute $D = 5.5$ and $h = 14.5$ in formula:

   $$V = 0.7854D^2h$$
   $$V = 0.7854 \times 30.25 \times 14.5$$
   $$V = 344.496 \quad \text{say} \quad 344.5 \text{ cu ft}$$

2. The diameter of an air cylinder is $3\frac{1}{2}$ in. Its length is $14\frac{1}{2}$ in. Find its volume and surface area of the cylindrical surface.

Step 1. Solve for the volume by using formula:

$$V = 0.7854D^2h$$

Step 2. Substitute values: $D = 3.5$ and $h = 14.5$ in formula:

$$V = 0.7854 \times (3.5)^2 \times 14.5$$
$$V = 0.7854 \times 12.25 \times 14.5$$
$$V = 139.5 \text{ cu in.}$$

Step 3. Solve for surface area using formula:

$$A_s = 3.1416Dh$$

Step 4. Substitute values: $D = 3.5$ and $h = 14.5$ in formula:

$$A_s = 3.1416 \times 3.5 \times 14.5$$
$$A_s = 159.4 \text{ sq in.}$$

The second type of cylinder is the *oblique* circular cylinder as shown in Fig. 14-8B whose axis is not at right angles with its circular base. This type is not found very often in commerce or industry.

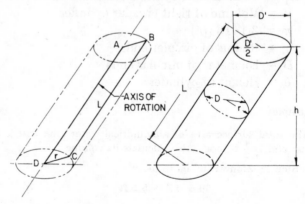

Fig. 14-8B. An oblique circular cylinder.

An oblique circular cylinder with top and bottom cut by parallel planes will have the same cylindrical surface area, A_s, as a right circular cylinder whose height h equals the length L and has the same right circular cross-section. The total surface area, A_t, will be the sum of the cylindrical surface and twice the area of the eliptical base. Expressed as formulas this gives:

$$A_s = 6.2832rL \quad \text{or} \quad A_s = 3.1416DL$$

where: $r =$ Radius of the right circular cross-section

$D =$ Diameter of the right circular cross-section

$L =$ Length of cylinder (see Fig. 14-8B)

The total surface area of an oblique circular cylinder is the sum of the cylindrical surface area plus twice the area of the eliptical base. Expressed as a formula, this gives:

$$A_t = 6.2832rL + 2(3.1416)(\frac{D'}{2})(r)$$

where: $\dfrac{D'}{2} =$ Major radius of the eliptical base

$r =$ Radius of the right circular cross-section.

The volume of an oblique circular cylinder is expressed by the formula:

$$V = 3.1416r^2h \quad \text{or:} \quad V = 0.7854D^2h$$

The above formulas for oblique circular cylinders utilize the symbols shown in Fig. 14-8B.

Study Examples.

1. In Fig. 14-8B, $L = 8$ in., $r = 2$ in., and $\dfrac{D'}{2} = 2.825$ in. Calculate the total surface area, A_t.

Step 1. Substitute the given values in the formula for A_t:

$A_t = 6.2832rL + 2(3.1416 \dfrac{D'}{2}r)$

$A_t = (6.2832)(2)(8) + 2(3.1416)(2.825)(2)$

$A_t = 100.53 + 35.50 = 136.03, \quad \text{say} \quad 136 \text{ sq in.}$

2. Compute the volume of the oblique circular cylinder described in study example 1.

Step 1. Substitute the given values in the formula for V:

$V = 3.1416r^2L = (3.1416)(2^2)(8)$

$V = (3.1416)(4)(8) = 100.53, \quad \text{say,} \quad 100 \text{ cu in.}$

Study Problems.

1. A right circular cylinder is 3.5 in. in diameter and 8.35 in. long. Calculate the volume, the cylindrical surface area, and the total surface area.

2. Calculate the height of a 1-gal paint can whose inside diameter is 6.50 in. (1 gallon = 231 cubic inches).

3. How many square feet of paint are needed to cover the cylindrical surface of a water tank that is 35 ft in diameter and 40 ft high?

4. What is the total displacement of a six-cylinder auto engine with a 3.25 bore (diam) and a 3.5 stroke (h)? Give the displacement (cid) to nearest even cubic inch.

5. A right circular cylinder is cut by two parallel planes that are 9 in. apart. The planes make a 45° angle with the sides of the cylinder whose diameter is 8 in. Calculate the volume and the total surface area of the oblique circular cylinder. See Fig. 14-9.

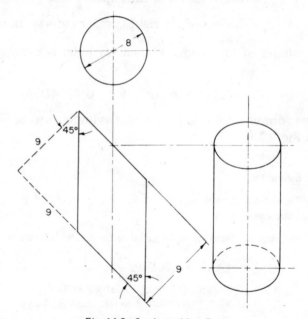

Fig. 14-9. Study problem 5.

6. How many barrels (31.5 gallons per barrel) of hydraulic oil are used in a 3-in. diameter cylinder operating at a 14 in. stroke for 2.5 hours at a speed of 9 strokes per minute?

7. A stand pipe at a waterworks is 35 ft in diameter and 65 ft tall. How many gallons of water does it hold when full? When five-eighths full?

8. A water main is 12 ft inside diameter and runs 5 miles from the reservoir to the pumping station. How many gallons of water does it take to fill the main?

The Cone. There are many types of cones determined by the shape of the base and the relation of the altitude to the base. The most common shape is the *right circular cone*, as shown in Fig. 14-10. This solid is generated by revolving the right triangle, AOB, about the leg AO which becomes the altitude h of the cone or the *axis* of the cone. The other leg, OB, becomes the radius of the base circle.

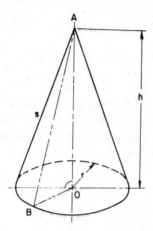

Fig. 14-10. A right circular cone.

The formulas relating to cones, discussed below, relate to the following symbols:

A = Area of the conical surface of a cone
A_t = Total area of a cone (cone and base surface)
r = Radius of the circular base
h = Altitude of cone
s = Length of slant element in cone surface
R = Radius of large base of frustum
r = Radius of small base of frustum
V = Volume of a cone or frustum
D = Diameter of large base of frustum
d = Diameter of small base of frustum

The area of the conical surface, A, of a right circular cone, Fig. 14-10, is found by using the formula:

$$A = \pi r \sqrt{r^2 + h^2}$$

or: $A = 3.1416rs$

The total area of the conical and base surfaces of a right circular cone is found by using the formula:

$$A_t = \pi r^2 + \pi rs = \pi r(r + s)$$

or: $\qquad A_t = 0.7854D(2s + D)$

The volume of a right circular cone, V, is found by the use of the formula:

$$V = \frac{\pi r^2 h}{3}$$

or: $\qquad V = 1.0472 r^2 h$

or: $\qquad V = 0.2618 D^2 h$

When the altitude of a cone does *not* meet the base at its center, the cone is called an *oblique* circular cone. This type is shown in Fig. 14-11. Cones can be formed on elliptical bases and are called *elliptical* cones. These may be *right* or *oblique* in construction.

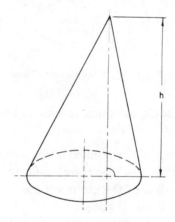

Fig. 14-11. An oblique circular cone.

A frustum of a cone, shown in Fig. 14-12, is a common shape used in funnels or connectors in sheet metal work. A frustum is the part of a cone that remains if the top of the cone is cut off. The cut may be parallel to the base or at an angle.

The conical surface area of a frustum of a cone is calculated by use of the formula:

$$A = 3.1416s(R + r)$$
$$A = 1.5708s(D + d)$$

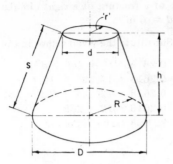

Fig. 14-12. A frustum of a cone.

The volume of a frustum of a cone is given in the formula:

$$V = 1.0472h(R^2 + Rr + r^2)$$

or:
$$V = 0.2618h(D^2 + Dd + d^2)$$

Study Examples.

1. Find the area of the conical surface of a right circular cone 6 in. high with a base circle radius of 3 in.

Step 1. Substitute values in formula:

$$A = 3.1416r \sqrt{r^2 + h^2}$$
$$A = 3.1416 \cdot 3 \sqrt{(3 \cdot 3) + (6 \cdot 6)}$$
$$A = 3.1416 \cdot 3 \sqrt{45}$$
$$A = 3.1416 \cdot 3 \cdot 6.71$$
$$A = 63.24, \quad \text{say} \quad 63.2 \text{ sq in.}$$

2. Find the volume of a right circular cone whose base is a 5-in.-diameter circle and whose height is 10 in.

Step 1. Substitute the values in the formula:

$$V = 0.2618D^2 h$$
$$V = 0.2618 \cdot 5^2 \cdot 10$$
$$V = 0.2618 \cdot 250$$
$$V = 65.45 \quad \text{say} \quad 65 \text{ cu in.}$$

3. Find the conical surface area, A, of a frustum of a cone whose slant height, $s = 14$ in., $D = 12$ in. and $d = 7$ in.

Step 1. Substitute the given values in the formula:

$$A = 1.5708s(D + d)$$
$$A = 1.5708 \cdot 14(12 + 7)$$
$$A = 1.5708 \cdot 14 \cdot 19$$
$$A = 417.8, \quad \text{say} \quad 418 \text{ sq in.}$$

4. Find the volume of a frustum of a right circular cone when $h = 12$ in., $D = 12$ in., and $d = 6$ in.

Step 1.   Substitute the given values in the formula:

$$V = 0.2618h(D^2 + Dd + d^2)$$
$$V = 0.2618 \cdot 12(12^2 + 12 \cdot 6 + 6^2)$$
$$V = 0.2618 \cdot 12(144 + 72 + 36)$$
$$V = 0.2618 \cdot 12 \cdot 252$$
$$V = 791.68 \quad \text{say} \quad 792 \text{ cu in.}$$

Study Problems.

1. Calculate the volume of a right circular cone with a base radius of 2.45 in. and a height of 6.5 in.

2. Compute the surface area of a right circular cone with $r = 4.6$ in. and $h = 6.2$ in. How much will this cone weigh if made from 16 gage copper weighing 2.355 lbs per sq ft?

3. Find the volume of a right circular cone when $r = 2.75$ in. and $h = 1.75$ yds.

4. Compute the surface area of a cone whose base radius equals 25 mm and altitude equals 45 mm.

5. Find the volume of a right frustum of a cone whose major diameter equals 8 m, minor diameter equals 5 m and vertical height equals 135 cm.

6. Find the surface area of a frustum of a right circular cone whose altitude is 46 in., major diameter is 2.4 ft and minor radius is 10 in. How much will this piece weigh if made from 20 gage sheet steel weighing 1.5 lbs per sq ft?

7. Find the volume of the two parts of a right circular cone, $h = 48$ in., base radius $= 18$ in. with the parting plane passing parallel to the base and 1 ft above it.

The Sphere. A sphere is the three-dimensional development of the circle. A sphere is defined as: "A solid that has every point on its surface at an equal distance from the same point, the center." A sphere is formed when a semicircle is revolved about its diameter as an axis.

In Fig. 14-13, the sphere O is formed by the rotation of the semicircle ACB about diameter AB as an axis. During rotation, point C describes a "great circle" while points E and F generate "little circles."

The surface area of a sphere, A, is found by the use of a formula involving the radius or diameter:

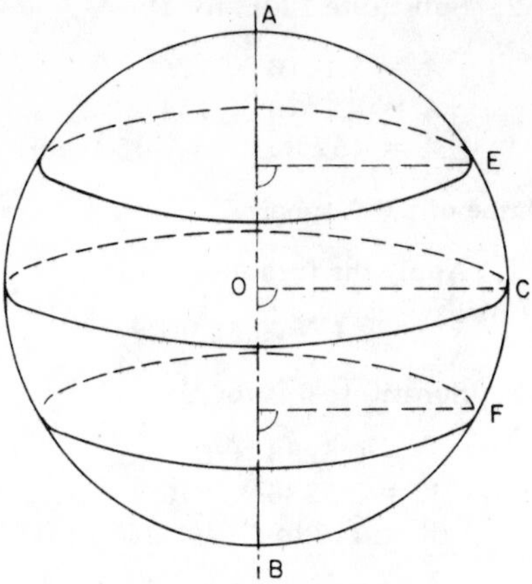

Fig. 14-13. The sphere.

$$A = 4\pi r^2$$
$$= 12.5664r^2$$

or:
$$A = \pi D^2$$
$$= 3.1416D^2$$

The computation of the volume of a sphere, V, is done by the use of the following formulas:

$$V = \frac{4\pi r^3}{3}$$
$$= 4.1888r^3$$

or:
$$V = \frac{\pi D^3}{6}$$
$$= 0.5236D^3$$

The symbols used in the above sphere formulas are defined as follows:

A = Surface area of the sphere
V = Volume of sphere in cubic measure
π = 3.1416 (pi)
r = Radius of the sphere
D = Diameter of the sphere

Study Examples.

1. Find the surface area of a 12-in. sphere.

Step 1. Apply the formula:

$$A = 3.1416D^2$$

Step 2.   Substitute 12 in. for D:

$$A = 3.1416 \times 12^2$$
$$A = 3.1416 \times 144$$
$$A = 452.39, \quad \text{say,} \quad 452 \text{ sq in.}$$

2. Find the volume of a 6-ft sphere.

Step 1.   Apply the formula:

$$V = 0.5236D^3$$

Step 2.   Substitute 6 ft for D:

$$V = 0.5236 \times 6^3$$
$$V = 0.5236 \times 216$$
$$V = 113.09, \quad \text{say} \quad 113 \text{ sq ft}$$

In some problems, the surface area or the volume is known and the diameter of the sphere is asked for. The formulas are then transposed as follows:

$$A = 3.1416D^2$$
$$D^2 = \frac{A}{3.1416}$$

also:
$$V = 0.5236D^3$$
$$D^3 = \frac{V}{0.5236}$$

The solution of the first formula involves the extraction of the square root which is explained in Chapter 4. However, the solution of the second formula involves the extraction of the cubic root which is not described in this book. The reader is referred to *Machinery's Handbook*, pages 124 and 125, where the logarithmic method is described.

Spherical shapes are commonly used in the construction of elevated water tanks used in municipal water systems to provide pressure. The volume of water in such a tank is normally stated in gallons while the tank volume will be calculated from its diameter and given in cubic feet. The conversion from cubic feet to gallons is done by the use of the formula:

$$V_{\text{gallons}} = 7.48 \cdot \text{cu ft}$$

Study Example.

1. An elevated water tank is a sphere 30 ft in diameter. How many gallons of water will it hold?

Step 1.   Calculate volume, using: $V = 0.5236D^3$

$$V = 0.5236 \cdot 30^3$$
$$V = 0.5236 \cdot 27,000$$
$$V = 14,137 \quad \text{say,} \quad 14,140 \text{ cu ft}$$

Step 2.   Calculate volume in gallons: 1 cu ft = 7.48 gal

$$V_g = 7.48 \cdot 14,140$$
$$V_g = 105,767, \text{ say } 105,800 \text{ gals}$$

Study Problems.

1. A hot-air balloon is 35 ft in diameter. How many cubic feet of air will it hold when inflated?

2. A spherical water tank is 25 ft in diameter. What is its total surface area? (Assume no pedestal.)

3. How many cubic feet are there in the tank in problem 2?

4. How many gallons of water will the tank in problem 2 hold?

5. How many liters will the tank in problem 2 hold?

6. A cannonball is 8 in. in diameter. It is made of a material weighing 442 lb/cu ft. How much does the cannonball weigh?

7. What is the volume of a spherical tank that has a diameter of 40 m? How many gallons would this be?

8. What is the surface area of the tank in problem 7?

9. What is the surface area of a sphere whose diameter is 125 ft?

10. How many square yards of cloth are required to cover the balloon described in problem 1? Allow 10% extra for cutting and for seams.

11. Calculate the weight of the water in problem 2 giving the answer in kilograms.

12. How many kilograms will 1235 ball-bearings, 6.35 mm in diameter, weigh when made of alloy steel with a specific weight of 0.2950 lb per cu in.?

13. If the tank in problem 7 is made of $\frac{1}{2}$-in. steel plate, how much will the empty tank weigh allowing 15% added for cutting and reinforcements? One cubic foot of rolled steel weighs 489.6 lbs. Give answer in metric tons.

14. A special reactor tank is fabricated that consists of two hemispherical ends 9.54 m inside diameter. The ends are fastened to a cylinder also 9.54 m I.D. and 22.6 m long. Calculate the volume in cubic meters and the weight of the water the tank will hold when filled to capacity. Give weight in metric tons.

The pyramid. A pyramid is a polyhedron whose base is a polygon and whose other faces are triangles meeting at a common point, the vertex. The base may have any number of sides and be either a *regular* (equal sides and equal interior angles) or an *irregular* polygon.

There are two basic types of pyramids. In the right or regular pyramid, Fig. 14-14, the altitude *h* from the vertex is perpendicular to the center of a regular polygon as its base. In a right pyramid, the angles that the edges make with the base edges are all equal and the triangular sides are isosceles.

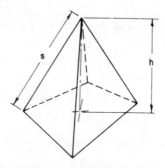

Fig. 14-14. A right (regular) pyramid.

The total surface area of a pyramid can be calculated as the sum of the areas of each triangular side and the area of the polygonal base. No general formula applies.

The volume of any pyramid is given in the formula:

$$V = \frac{1}{3}hA$$

where: h = Height
          A = Area of the base

Study Example.

1. Find the volume of a right pyramid whose base is a regular hexagon with an inscribed radius of 60 ft and whose height is 80 ft.

Step 1.   Substitute values in the formula:

$$V = \frac{1}{3}hA$$

$$V = \frac{80(3.464 \cdot 60^2)}{3}$$

$$V = \frac{80 \cdot 12,470.4}{3}$$

$$V = \frac{997,632}{3}$$

$$V = 332,544, \quad \text{say} \quad 332,500 \text{ cu ft}$$

The *oblique* pyramid shown in Fig. 14-15 is one whose altitude is not perpendicular to the base at its center. When irregular polygons form the base of a pyramid, it is a common practice to refer to it as oblique whether or not the altitude is perpendicular to the base at its center because of the pyramid's nonsymmetrical shape.

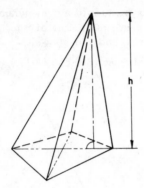

Fig. 14-15. An oblique pyramid.

The frustum of a regular pyramid, shown in Fig. 14-16, is that part of a pyramid remaining after the top portion is cut off by a plane parallel to the base.

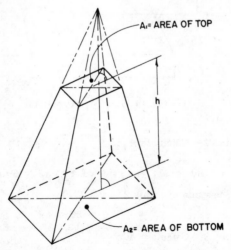

Fig. 14-16. A right frustum of a pyramid.

The volume of a right frustum of a pyramid equals the difference between the volume of the entire pyramid (before cutting off the top) and the small pyramid cut off to make the frustum. The formula is:

$$V = \frac{h}{3}(A_1 + A_2 + \sqrt{A_1 \cdot A_2})$$

Refer to Fig. 14-16 for identification of h, A_1 and A_2.

Study Example.

1. Find the volume of a frustum of a right pyramid with a square base 12 in. on a side, and an original height of 8 in. The pyramid is cut by a plane parallel to the base and 4 in. above it.

   Step 1. Calculate h; (See Fig. 14-17.)

   $$h = 8 - 4 = 4 \text{ in.}$$

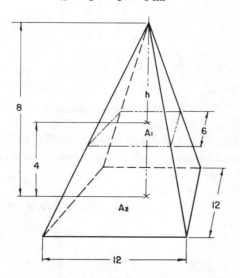

Fig. 14-17. Study example 1.

   Step 2. Calculate A_1:

   $$\text{Side of } A_1 = \frac{1}{2} \text{ side of } A_2. \quad s_1 = 6 \text{ in.}$$

   (See study example 2, page 291)

   $$A_1 = 6 \cdot 6 = 36 \text{ sq in.}$$

   Step 3. Calculate A_2:

   $$A_2 = 12 \cdot 12 = 144 \text{ sq in.}$$

Step 4.   Substitute values in formula:

$$V = \frac{h}{3}(A_1 + A_2 + \sqrt{A_1 \cdot A_2})$$

$$V = \frac{4}{3}(36 + 144 + \sqrt{36 \cdot 144})$$

$$V = \frac{4}{3}(180 + \sqrt{5184})$$

$$V = \frac{4}{3}(180 + 72)$$

$$V = \frac{4}{3}(252)$$

$$V = 336 \text{ cu in.}$$

Regular Polyhedrons

Solids having faces that are regular polygons of the same shape and size are called *regular polyhedrons*. These solids have the same number of edges meeting at each vertex. There are only five regular polyhedrons existing. These are:

Tetrahedron.   A solid with four triangular faces. These triangles are equilateral.

Hexahedron.   This is the cube, a solid with six faces which are all equal squares.

Octahedron.   A solid with eight triangular faces. These triangles are equilateral.

Dodecahedron.   A twelve-sided figure whose faces are equal five-sided pentagons.

Icosahedron.   A twenty-sided figure whose faces are all equal equilateral triangles.

Study Problems.

1. Calculate the volume of a hydraulic cylinder whose I.D. is 3.825 in. and stroke is 3.825 in.:   (a) in the plain end when the rod is fully extended; (b) in the rod end when the piston is fully retracted. Rod diameter = $\frac{15}{16}$ in.

2. Calculate the volume of oil needed in the plain end to move the piston in problem 1 the distance of 3.25 in.

3. Calculate the volume of oil needed to return the piston (rod end volume) in problem 2 to a fully retracted position.

4. Calculate the volume of a rectangular (right prism) oil reservoir whose length is 5 m, width 2500, cm, and depth 1500 cm. Give answer in kiloliters and in gallons.

5. A cylindrical tank, 15 ft in diameter and 10 ft high, is filled with oil. How many gallons are required to fill it?

6. The tank in problem 4 is drained so that 6.25 ft of oil remains. How many gallons were drained?

7. The empty tank in problem 4 is filled to 8 ft depth, drained to 3.75 ft and then filled to capacity. How many gallons were drained? How many kiloliters were then needed to fill it to capacity?

8. A hemispherical dome, 40 ft in diameter, is to be painted. How many square feet of surface will be covered?

9. Find the weight of a solid sphere 75 cm in diameter if the material weighs 450 lbs/cu ft.

10. A sphere is cut from a cube 20 in. on an edge. What is the minimum amount of waste in cubic inches?

11. How many liters of oil will a circular tank hold if its diameter is 10 m and its length is 35 m?

12. What is the volume of a square base pyramid whose base is 35 ft on each side and whose height equals the diagonal of the base?

13. Two spherical tanks, 30 in. in diameter, are packed in a box whose inside measurements are $36'' \times 36'' \times 70''$. How much space (cu in.) is left for packing material?

14. A right circular cone is 14 in. high with a base diameter of 9 in. It is cut by a plane, parallel to the base and 10 in. above the base. What is the volume of: (a) the original cone; (b) the frustum?

15. An ice cream cone is 11.6 cm in length and has a circular opening 4.73 cm in diameter. What is its surface area?

16. A smoke stack for a Madrid powerhouse is 1.5 m in diameter (O.D.) and 60 m high. What is the surface area of the stack in sq cm?

17. A round bar of steel 5.5 in. in diameter and 14 in. long is machined out to have a wall thickness of 0.625 in. What is the volume of the resulting tube, and how many cubic inches of metal were removed?

Calculating Surface Areas and Volumes

There are many more formulas for calculating areas and volumes of solids as well as the areas of plane figures. A comprehensive listing of these formulas will be found in *Machinery's Handbook*. The reader is referred to this reference material not only to find a formula but also to use the examples of solutions that accompany each formula.

CHAPTER 15

Right-Angle Trigonometry

So far in this book, we have discussed: *Arithmetic*, the computation and working with numerals; *Algebra*, a type of generalized arithmetic involving both numerals and letters; and *Geometry*, the study of forms and the relationships of plane figures.

Trigonometry is yet another branch of mathematics. It is a branch that includes arithmetic, algebra, and geometry, but it is primarily concerned with the relationships of lines and angles in triangles. This branch of mathematics forms the basis of measurements used in surveying, engineering, shop mechanics, geodesy and astronomy. The word *trigonometry* comes from the Greek words *trigon*, a triangle, plus *metrein*, to measure. The word trigonometry is often shortened to "trig" in shop talk, and we will use this shortened word in many of our discussions.

In geometry we worked with angles, studying their relationships in various types of triangles. For ordinary purposes, angles can be measured with a protractor to an accuracy of about a half degree (30'). The surveyor uses an instrument called a *transit* to measure angles. This instrument incorporates a protractor for angle measurement. With a transit, he can measure angles to the right or left of a point. By moving the telescope up or down, vertical angles can be measured. A high degree of accuracy can be obtained using a transit, and it is used to set up assembly jigs in the aircraft industry.

The measurement of angles in the machine shop and toolroom is very important, and many precision instruments are employed. Some of the more common angle measuring instruments are:

—Bevel protractors, graduated in degrees.
—Vernier bevel protractors, in 5-minute graduations.
—Precision square, for testing right angles.
—Optical projector, obtaining 1-minute graduations.
—Sine bars and sine plates, accurate to 5 to 10 seconds, or less.

—Circular tables with indexing mechanisms.

—Angle gage blocks, made in steps of 1-second, 1 minute, or 1 degree.

The measurement of angles has traditionally been done in units of degrees, minutes, and seconds. This subject has been discussed in Chapter 5. For everyday use in the shop and in the design office, this system of angle measurement is quite adequate and the reference tables of trig functions are usually written in this system. However, decimal degrees are often used, particularly in electrical engineering and in computer calculations.

The *right*, or 90-degree, triangle is the basis of trig calculations, and a special vocabulary must be learned by the tradesman to better understand and correctly use the special properties of the right triangle.

In the right triangle shown in Fig. 15-1, the vertex angles are designated by the capital letters A, B, and C. The three sides of the triangle are designated by the small letters a, b, and c. Note that a is opposite $\angle A$, b is opposite $\angle B$, and c is opposite $\angle C$.

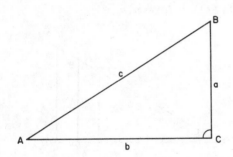

Fig. 15-1. Right-triangle designations.

The angle at A is referred to as $\angle A$ or $\angle BAC$. In naming a triangle, the letters of the three vertices are used. In a right triangle it is helpful to identify the right angle. This can be done by placing the letter of the right angle between the other two letters of the triangle and underscoring it, as in; $\triangle A\underline{C}B$.

The sides of a right triangle adjacent to the right angle are referred to as *sides* or *legs*. The sides are further identified in respect to one of the acute angles in the triangle as the *opposite side* (that is, lying opposite the specified angle), or *adjacent side* (that is, lying adjacent to the specified angle). The side opposite the right angle is always referred to as the *hypotenuse*. The hypotenuse is the longest side in a right triangle, and the shortest side always lies opposite the smaller acute angle.

Definitions Used in Trigonometry

Acute Angle: Any angle less than 90°.
Right Angle: An angle of 90°.
Obtuse Angle: Any angle greater than 90° (never found in a right triangle).
Hypotenuse: The side opposite the 90° angle.
Opposite Side: The side opposite the specified angle.
Adjacent Side: The side touching the angle being used.

Functions of Angles

The study of trigonometry is based on certain ratios called trigonometric *functions*. Angles are formed by the counterclockwise rotation of a line such as *OP* in Fig. 15-2. As line *OP* rotates away from line *OX*, angle *POX* is formed.

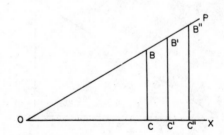

Fig. 15-2. Formation of an angle and proportional sides.

If parallel lines *BC*, *B′C′* and *B″C″* are drawn perpendicular to line *OX*, a series of similar right triangles will be formed. From our study of geometry we know that the corresponding sides of similar triangles are proportional; that is:

$$\frac{BC}{OB} = \frac{B'C'}{O'B'} = \frac{B''C''}{O''B''}$$

and:
$$\frac{BC}{OC} = \frac{B'C'}{O'C'} = \frac{B''C''}{O''C''}$$

and:
$$\frac{OB}{OC} = \frac{O'B'}{O'C'} = \frac{O''B''}{O''C''}$$

These ratios remain unchanged as long as the angle *POX* remains unchanged, but they change as the angle changes. Each of the above ratios is therefore a *function* of angle *POX*.

The term *function* describes a relationship. Thus a circle is a "function" of the radius; the area of a square is a "function" of a side; the surface of a sphere is a "function" of its diameter, and the volume of a pyramid is a "function" of the base and the altitude.

The Six Trig Functions

In trig we make use of the various functions of an angle, but we give them special names and symbols. There are six trig functions and they are stated in reference to a standard right triangle as shown in Fig. 15-3.

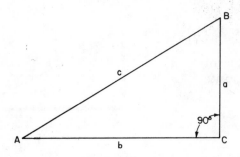

Fig. 15-3. Standard right triangle.

It has long been the custom to letter the right angle *C* and its opposite side, the hypotenuse, as *c*. The legs or the shorter sides of the triangle then are labeled *a* and *b* and are opposite the acute angles *A* and *B* respectively. The six trig functions are: the *sine*, abbreviated *sin*; *cosine*, abbreviated *cos*; *tangent*, abbreviated *tan*; *cotangent*, abbreviated *cot*; *secant*, abbreviated *sec*; and *cosecant*, abbreviated *csc*.

The functions of angle *A* are ratios involving two of the three sides of the triangle; $\frac{a}{c}, \frac{b}{c}, \frac{a}{b}, \frac{b}{a}, \frac{c}{b}$, and $\frac{c}{a}$. These ratios are called:

Sine of $\angle A$ (Fig. 15-4):

$$\sin A = \frac{\text{side opposite}}{\text{hypotenuse}} = \frac{a}{c}$$

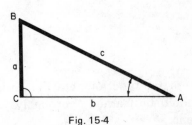

Fig. 15-4

Cosine of ∠A (Fig. 15-5):

$$\cos A = \frac{\text{side adjacent}}{\text{hypotenuse}} = \frac{b}{c}$$

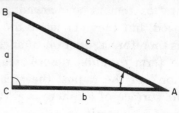

Fig. 15-5.

Tangent ∠A (Fig. 15-6):

$$\tan A = \frac{\text{side opposite}}{\text{side adjacent}} = \frac{a}{b}$$

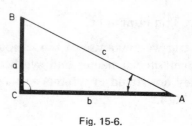

Fig. 15-6.

Cotangent ∠A (Fig. 15-7):

$$\cot A = \frac{\text{side adjacent}}{\text{side opposite}} = \frac{b}{a}$$

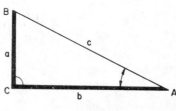

Fig. 15-7.

Secant ∠A (Fig. 15-8):

$$\sec A = \frac{\text{hypotenuse}}{\text{side adjacent}} = \frac{c}{b}$$

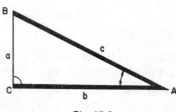

Fig. 15-8.

Cosecant ∠A (Fig. 15-9):

$$\csc A = \frac{\text{hypotenuse}}{\text{side opposite}} = \frac{c}{a}$$

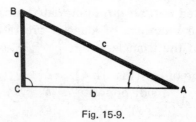

Fig. 15-9.

These six functions are the basis for all work in trigonometry and should be learned thoroughly to facilitate solving future shop problems.

Reading the Trig Tables

Numerical values for all the trigonometric functions have been accurately calculated for all angles in degrees and minutes. These values are given to five decimal figures in the tables at the back of the book, pages 486 to 500.

These tables of trigonometric functions list the functions of all angles between $0°$ and $180°$ ($180° = 179°60'$). For the present we are only concerned with the values of the functions for angles between $0°$ and $90°$. Furthermore, the tables of versed sine and cosine functions will not be treated in this book.*

The table lists all of the angles between $0°$ and $45°$ at the upper left-hand corner of the page; the minutes for angles in this range are listed in the "M" column at the left and are read from top to the bottom of the page. Note that $44°60' = 45°$.

For angles from $45°$ to $90°$, the angle is printed in the lower right-hand corner of the page and the minutes for angles in this range are listed in the "M" column at the right and are read from bottom to top.

For angles between $0°$ and $45°$, the trigonometric functions (sine, cosine, etc.) are named at the top of each column and the values of the function for the given angle and minute are found in the column below.

For angles between $45°$ and $90°$, the trigonometric functions are named at the bottom of each column and the value of a given angle and minute is found by reading up the column.

To find the value of a trig function of a given angle:

1. Find the page on which the degree part of the angle appears.

2. Locate the column which lists the required trigonometric function such as sin, cos, tan, etc.

3. Locate the required minutes by going down the "M" column for angles between $0°$ and $45°$, where $45° = 44°60'$; and going up the right-hand "M" column for angles between $45°$ and $90°$. Read the numerical value of the function at the intersection of the function column and the row corresponding to the required minutes.

---

*(Information on versed sines and cosines can be found in *Mathematics at Work* by Holbrook L. Horton, Industrial Press Inc.)

Study Examples.

Check the values given to practice for the problems.

1. The sine of $36°45'$ is read: 0.59832

2. The tangent of $24°40'$ is read: 0.45924

3. The cosine of $55°10'$ is read: 0.57119

4. The cotangent of $70°30'$ is read: 0.35412

5. The tangent of $66°35'$ is read: 2.3090

Study Problems.

Determine the values for the following angle functions:

1. $\sin 30°$ 5. $\sin 5°22'$ 9. $\csc 22°22'$

2. $\tan 45°$ 6. $\tan 66°66'$ 10. $\sin 60°45'$

3. $\cos 45°$ 7. $\cot 45°$ 11. $\cos 3°50'$

4. $\sec 15°30'$ 8. $\tan 46°$ 12. $\tan 19°22'$

To find an angle measurement from a given trig value:

1. Scan down the columns of the function used in the trig tables until the numerical value is located;

   such as: tangent $A = 0.79070$.

2. Look at the upper left corner of the table to find the angle for angles from $0°$ to $44°59'$ and at the lower right corner for angles from $45°$ to $90°$.

   tangent $A = 0.79070$, then A is on the $38°$ page.

3. The minutes are found in the left column opposite the numerical value for angles from $0°$ to $44°59'$ and in the right column for angles over $45°$. The value 0.79070 in the tangent column reads $20'$ in the left column.

4. Therefore, the angle measurement is: $38°20'$.

Care must be taken in reading the proper function at the top or the bottom of the page. Note that the value 0.79070, above, is also the "cotangent" of $51°40'$ reading up from the bottom in the right column above $51°$.

In any right triangle, ACB (shown in Fig. 15-10): $\sin A = \cos B$, $\cos A = \sin B$, $\tan A = \cot B$, $\cot A = \tan B$, $\sec A = \csc B$ and $\csc A = \sec B$.

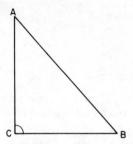

Fig. 15-10. Function relationship in a right triangle.

Study Problems.

1. $\sin A = 0.03490$, find angle measure of $\angle A$

2. $\tan A = 0.14945, \angle A = ?$　　　　6. $\csc B = 1.6034, \angle B = ?$

3. $\cos A = 0.92388, \angle A = ?$　　　　7. $\sin B = 0.78801, \angle B = ?$

4. $\cot B = 2.412, \angle B = ?$　　　　8. $\cos A = 0.36650, \angle A = ?$

5. $\sec A = 1.1879, \angle A = ?$

Since one angle in a right triangle is always equal to $90°$ and the sum of all the interior angles in the triangle is $180°$, the sum of the two acute angles must be $180° - 90°$ or $90°$. Therefore, if one acute angle in a right triangle is known, the other angle can easily be found as follows:

$$\text{unknown angle} = 90° - \text{known angle}$$

For example, if one acute angle is 55 degrees, the other angle is found as follows:

$$\text{unknown angle} = 90° - 55° = 35°$$

All of the dimensions of a right triangle can be found by trigonometry when one of the following conditions exists:

1. When one acute angle and the length of one of the sides or of the hypotenuse of the right triangle is known.

2. When the lengths of two sides or of one side and the hypotenuse of the right triangle are known.

Use of the Tangent Function

Study Example.

1. Determine $\angle A$ in the right triangle in Fig. 15-11:

   Step 1.   Determine the trig ratio (function) to use that will involve known sides, a and b.

$$\text{tangent } A = \frac{\text{side opposite}}{\text{side adjacent}} = \frac{a}{b}$$

   Step 2.   Substitute known values for a and b:

$$\text{tangent } A = \frac{3.8}{6.5}$$

   Step 3.   Convert ratio to a decimal:

$$\tan A = 0.58462$$

   Step 4.   Find nearest angle in trig tables:

$$\angle A = 30°19'$$

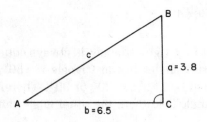

Fig. 15-11. Study example 1.

The actual value of tan 30°19′ is 0.58474. To get the exact value of angle A which lies between 30°19′ and 30°20′, a method called interpolation is used. This will be explained later.

Use of the Cotangent Function

Study Example.

2. Solve for $\angle B$ in the figure of example 1.

   Step 1.   Use cotangent function and values of a and b:

$$\text{cotangent } B = \frac{\text{side adjacent}}{\text{side opposite}} = \frac{a}{b}$$

Step 2. Substitute known values for a and b:

$$\cot B = \frac{3.8}{6.5}$$

Step 3. Convert ratio to decimal:

$$\cot B = 0.58462$$

Step 4. Find nearest angle in trig table:

$$\angle B = 59°41'$$

An alternate method of finding $\angle B$ when $\angle A$ is known, as in example 1, is to subtract $\angle A$ from $90°$:

$$\angle B = 90° - \angle A$$
$$\angle B = 90° - 30°19' \; (90° = 89°60')$$
$$\angle B = 89°60' - 30°19'$$
$$\angle B = 59°41'$$

Use of the Sine Function

Study Example.

3. In the right triangle shown in Fig. 15-12, solve for hypotenuse c:

Step 1. Use the sine function involving sides a and c:

$$\sin A = \frac{\text{side opposite}}{\text{hypotenuse}} = \frac{a}{c}$$

Step 2. Transpose equation in step 1 to solve for c:

$$c = \frac{a}{\sin A}$$

Step 3. Determine trig value of $\sin 30°30'$:

$$\sin 30°30' = 0.50754$$

Step 4. Substitute values of $\sin 30°30'$ and side c:

$$c = \frac{3.8}{0.50754} = 7.49, \quad \text{say} \quad 7.5$$

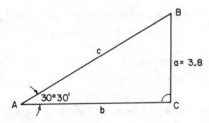

Fig. 15-12. Study example 3.

Use of the Cosine Function

Study Example.

4. Determine the length of the hypotenuse of a right triangle when $\angle A = 40°45'$ and side $b = 32$.

> Step 1. Determine a trig function that involves side b and the hypotenuse c:
>
> $$\cos A = \frac{\text{side adjacent}}{\text{hypotenuse}} = \frac{b}{c}$$
>
> Step 2. Transpose equation in step 1 to solve for c:
>
> $$c = \frac{b}{\cos A} = \frac{b}{\cos 40°45'}$$
>
> Step 3. Determine trig value of $\cos 40°45'$:
>
> $$\cos 40°45' = 0.75756$$
>
> Step 4. Substitute values for $\cos 40°45'$ and side b:
>
> $$c = \frac{32}{0.75756} = 42.24$$

Use of the Secant Function

Study Example.

5. Determine the length of the hypotenuse of a right triangle when $\angle A = 40°45'$ and side $b = 32$.

> Step 1. Using the secant function, state the ratio involving the hypotenuse and the adjacent side:
>
> $$\sec A = \frac{\text{hypotenuse}}{\text{side adjacent}} = \frac{c}{b}$$
>
> Step 2. Transpose equation in step 1 to solve for c:
>
> $$c = \sec A \cdot \text{side adjacent} = \sec A \cdot b$$
>
> Step 3. Determine trig value of $\sec 40°45'$:
>
> $$\sec 40°45' = 1.3200$$
>
> Step 4. Substitute values for $\sec 40°45'$ and side b:
>
> $$c = 1.3200 \times 32 = 42.24$$

Use of the Cosecant Function

Study Example.

6. Determine the length of side a of a right triangle when the hypotenuse = 42.24 and $\angle A = 40°45'$.

Step 1. Using the cosecant function, state the ratio that involves the hypotenuse and the side a:

$$\csc A = \frac{\text{hypotenuse}}{\text{side opposite}} = \frac{c}{a}$$

Step 2. Transpose equation 1 to solve for a:

$$a = \frac{c}{\csc A} = \frac{c}{\csc 40°45'}$$

Step 3. Determine trig value for $\csc 40°45'$:

$$\csc 40°45' = 1.5319$$

Step 4. Substitute values for $\csc 40°45'$ and side c:

$$a = \frac{42.24}{1.5319} = 27.57$$

In some cases when an acute angle and two legs of a right triangle are known, the use of the Pythagorean Law will be a shorter method. For example, suppose hypotenuse c is to be found in problem 1:

Step 1. State Pythagorean Law in terms of the triangle sides:

$$c^2 = a^2 + b^2$$

Step 2. Substitute values of a and b:

$$c^2 = (3.8)^2 + (6.5)^2$$
$$c^2 = 14.44 + 42.25$$
$$c^2 = 56.69$$
$$c = \sqrt{56.69} = 7.53$$

Study Problems.

Finding an angle in a right triangle given two sides:

1. Using the sine function, solve for $\angle x$ when the side opposite equals 25 ft and the hypotenuse equals 50 ft. See Fig. 15-13.

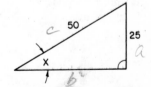

Fig. 15-13. Study problem 1.

2. A flagstaff breaks off 22 ft from the top and, with the top and bottom still holding together, the top of the staff touches the ground 11 ft from the foot of the flagstaff. What angle does the upper portion (hypotenuse) make with the ground?

3. An A-shaped roof has a span AA' of 24 ft. The ridgepole is 12 ft above AA'. What angle does side AR make with AA'? (See Fig. 15-14.)

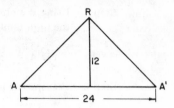

Fig. 15-14. Study problem 3.

4. A rafter on one side of a roof covers a horizontal run of 12 ft and a vertical rise of 10 ft. What angle does it make with the horizontal and vertical? (See Fig. 15-15.)

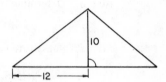

Fig. 15-15. Study problem 4.

Finding sides in a right triangle given an angle and one side.

5. Using the cosine function, solve for side b in the right triangle whose hypotenuse, c, is 28 in. and $\angle A = 46°$. See sketch in Fig. 15-16.

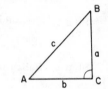

Fig. 15-16. Study problem 5.

6. A builder wishes to measure a pond. He constructs a right triangle ACB, as seen in Fig. 15-17. If $AB = 928$ ft and $\angle A$ equals $29°$, find length of BC.

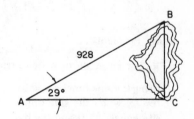

Fig. 15-17. Study problem 6.

7. A regular hexagon is inscribed in a circle of radius 9 in. How far is it from the center to a side? See sketch in Fig. 15-18.

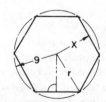

Fig. 15-18. Study problem 7.

8. An equilateral triangle is inscribed in a circle of 12-in. radius. How far is it from the center to a side?

9. Using the tangent function, solve for side a when $b = 12$ in. and $\angle A = 35°$. (See Fig. 15-19.)

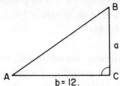

Fig. 15-19. Study problem 9.

Finding the angles and sides in right triangles.

10. Calculate $\angle \theta$ which is the angle between the geometrical tangent AB to the 3 in. circle and the line AO passing through the center of the circle, as seen in Fig. 15-20. The length of $AB = 10.625$ in.

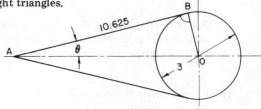

Fig. 15-20. Study problem 10.

11. A man standing 120 ft from the base of a smokestack finds that the angle of elevation (the angle a line from the eye to the object makes with the horizontal) is $50°$. If his eye is 5 ft 8 in. from the ground, what is the height of the stack?

12. The principal dimensions of the truss are shown in Fig. 15-21. Calculate $\angle CHG$ and the length of HC. All dimensions are in feet.

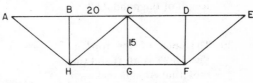

Fig. 15-21. Study problem 12.

13. What is the angle of inclination of a stairway from the floor if the steps have a tread of 8 in. and a rise of $6\frac{1}{2}$ in.?

14. A grade of 1% in a roadbed is a rise of 1 ft in a horizontal distance of 100 ft. What is the angle of slope of a roadbed that has a grade of 6%? Of another roadbed with a grade of 0.64%?

15. From the top of a smoke-stack 300 ft high, a point on the ground has an angle of depression of $35°$. (See Fig. 15-22.) How far is the point from the base of the stack? Angle of depression is the angle a line from the eye to object below makes with the horizontal.

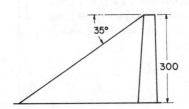

Fig. 15-22. Study problem 15.

16. A tree known to be 50 ft high, standing on the far bank of a stream is observed from the opposite bank to have an angle of elevation of 20°. The angle is measured on a line 5 ft above the ground. How wide is the stream?

17. A building, standing on level ground, has a fire-escape platform 20 ft from the ground. From a point O on the ground, the angle of elevation to the platform is 38°. The angle of elevation from the same point O to the top of the building is 75°. What is the building height?

18. A ladder rests against the side of a building, and makes an angle of 28° with the ground. The foot of the ladder is 20 ft from the building wall. How long is the ladder?

19. In measuring the distance AB across a pond, a surveyor ran a line AC, making an angle of 50° with AB, and a line BC perpendicular to AC. See Fig. 15-23. He measured AC and found it was 880 ft. What was the length of the pond AB?

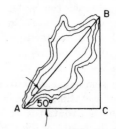

Fig. 15-23. Study problem 19.

20. The span of the roof in Fig. 15-24 is 40 ft and the roof timbers AB make an angle of 40° with the horizontal. Find the length of AB.

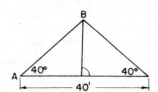

Fig. 15-24. Study problem 20.

21. A typical rigging to load and unload ships at a pier is shown in Fig. 15-25. Solve for $\angle EWD$ (α) when DW is parallel to BC and AF. All dimensions are in feet.

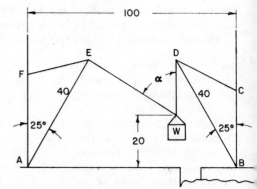

Fig. 15-25. Study problem 21.

22. From the top of a rock, a
cord is stretched to a point
on the ground, making an
angle of 40° with the horizontal plane.   The cord
measures 84 ft. Assuming
the cord to be straight,
how high is the rock? See
sketch in Fig. 15-26.

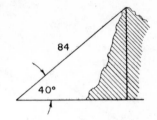

Fig. 15.26. Study problem 22.

23. The edge of the Great Pyramid is 609 ft and makes an
angle of 52° with the base.
What is the height of the
pyramid?

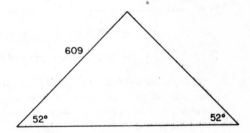

Fig. 15-27. Study problem 23.

Interpolation

As the tables of trigonometric functions give values for degrees
and minutes only, the method used to determine the value for an
angle given in degrees, minutes, *and seconds* is a process called *interpolation*.

The precise dimensions commonly used in machine shop practice
require interpolation.   While in most instances the angles are sufficiently precise when measured in minutes, interpolation is usually
necessary when calculating precise distances. The following example
illustrates the concept of the procedure used in determining the decimal function value when an angle measurement is given:

Study Example.

Determine the decimal trig value for the tangent of $22°10'22''$.
Figure 15-28 shows an enlarged angle of $22°10'$ upon which is drawn,
greatly enlarged, an angle of $1'$. The $1'$ angle is divided by a line (not to
scale) representing $22''$ added to the $22°10'$ line.
To the right of the figure, the vertical representation of the tangent is
shown. The horizontal length of the base of the angle is unity (1). The
height of the vertical line to the $22°10'$ intersection is indicated as 0.40741.
This is the value given in the trig table for the tangent of $22°10'$. The tangent of an angle of $22°11' = 0.40775$. The difference between these two
tangent values is $0.40775 - 0.40741$, or 0.00034.

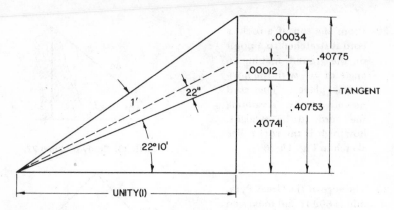

Fig. 15-28. Interpolation of the tangent function.

Hence, between $22°10'$ and $22°11'$, 0.00034 is the increase in value of the tangent of that particular minute or 60 seconds. $(1' = 60'')$. What is wanted is the increase in value of the tangent of a 22 second angle in this specific region.

This increase can be obtained by taking a proportional part of the difference of 0.00034; or, by taking $\frac{22}{60}$ ths of 0.00034. Thus:

$$0.00034 \times \frac{22}{60} = 0.00012$$

Since the value of the tangent *increases* in going from $22°10'$ to $22°11'$, 0.00012 must be *added* to the value of the tangent of $22°10'$ in order to obtain the tangent of $22°10'22''$. Therefore:

$$\tan 22°10'22'' = 0.40741 + 0.00012 = 0.40753$$

The difference of the value of a function for a 1-minute change in angle will not be the same for different angles and must be calculated for each specific angle, although sometimes the difference in the value of the function will be so small that it will not be evident in a five- or six-place trig table. Moreover, when the angle is increasing, the value of some functions of the angle will *increase*, but those of other functions will *decrease*; the procedure for handling this will be explained later in this chapter.

It is important at this point to understand that interpolation involves ratio and proportion; i.e., finding the proportional part of the difference between the value of a function or the value of an angle.

Rules for Interpolation of Trigonometric Functions

Interpolating Angles into Trig Functions. Note from the trig tables that the numerical values of the trig functions for sine, tangent, and secant *increase* as the angle size *increases*; however, for cosines, cotangents and cosecants they *decrease* as the angle size *increases*. It is important that this be understood and remembered for accurate interpolation.

1. When an angle is given in degrees, minutes, *and* seconds, determine the nearest smaller and larger angles in degrees and minutes.

2. Subtract the nearest smaller angle (in degrees and minutes) from the given angle. The difference will be in seconds.

3. As the difference between the nearest smaller and larger angle in degrees and minutes is 1 minute, this difference is also 60 seconds.

4. Form a ratio from the number of seconds found from step 2 and the 60 seconds from step 3.

5. Find the decimal value from the trig table for the given function of the nearest smaller angle and the nearest larger angle that was determined in step 1. Calculate the decimal difference.

6. Multiply the decimal difference found in step 5 by the ratio formed in step 4.

7. For sine, tangent, and secant functions, *add* the decimal product of step 6 to the decimal value of the nearest smaller angle. For cosine, cotangent and cosecant functions *subtract* the decimal product from the decimal value of the nearest smaller angle.

Study Example.

Find the tangent value for an angle of $25° \, 39' 15''$.

Step 1. (From above rules)

Nearest smaller angle is $25° \, 39'$

Nearest larger angle is $25° \, 40'$

Step 2. Subtract the nearest smaller angle from the given angle; the difference will be in seconds:

$$
\begin{array}{r}
25° 39' 15'' \\
-25° 39' 00'' \\
\hline
15''
\end{array}
$$

Step 3. Note that the difference between the nearest smaller angle and the nearest larger angle is 1 minute, or 60 seconds.

Step 4. Form a ratio, step 2 to step 3:

$$\frac{15}{60} = \frac{1}{4}$$

Step 5. Find numerical decimal value of trig functions of the smaller and larger angles and calculate decimal difference:

$$\tan 25°40' = 0.48055$$
$$\tan 25°39' = \underline{0.48019}$$
$$\text{Difference} = 0.00036$$

Step 6. Multiply the difference of step 5 by ratio of step 4:

$$0.00036 \times \frac{1}{4} = 0.00009$$

Step 7. For a tangent, add the difference in step 6 to the trig value of the smaller angle:

$$\tan 25°39' = 0.48019$$
$$+ \text{ Difference} = \underline{0.00009}$$
$$0.48028 = \tan 25°39'15''$$

Study Example.

Find the cosine value for an angle of $16°48'45''$.

Step 1. Find the nearest smaller angle to the given angle and also the nearest larger angle, in degrees and minutes.
Nearest smaller angle is $16°48'$
Nearest larger angle is $16°49'$

Step 2. Subtract the nearest smaller angle from the given angle, the difference will be in seconds.

$$16°48'45''$$
$$\underline{-16°48'00''}$$
$$45''$$

Step 3. Note that the difference between the nearest smaller angle and the nearest larger angle is 1 minute, or 60 seconds.

Step 4. Form a ratio step 2 to step 3:

$$\frac{45}{60} = \frac{3}{4}$$

Step 5. Find the difference between the decimal value of the trig function of the nearest smaller angle and the nearest larger angle:

$$\text{cosine } 16°48' = 0.95732$$
$$\text{cosine } 16°49' = \underline{0.95723}$$
$$0.00009$$

Step 6. Multiply difference of step 5 by ratio of step 4:

$$0.00009 \times \frac{3}{4} = 0.00007$$

Step 7. For a cosine, subtract the result of step 6 from the decimal value of the function of the nearest smaller angle:

$$0.95732$$
$$-\,\underline{0.00007}$$
$$0.95725 = \text{ cosine } 16° \; 48' \; 47''$$

Interpolating Functions into Their Angles. Note from the trig tables that the numerical values of the trig functions for sine, tangent, and secant *increase* as the angle size *increases*; but for cosines, cotangents, and cosecants, values *decrease* as the angle size *increases*.

1. When the decimal trig function is given for a function of an unknown angle, locate this decimal in the trig tables for that function. If the specific value does not appear, find the nearest smaller decimal that appears.

2. Determine the angle size in degrees and minutes for this nearest smaller decimal value.

3. Determine the decimal value for the next larger angle size (the next minute entry).

4. Calculate the difference between the decimal trig values of the nearest smaller angle and the nearest larger angle.

5. Calculate the difference between the decimal trig value of the given trig function and the decimal trig value of the nearest smaller angle.

6. Find the ratio of the smaller difference to the larger difference, that is, step 5 to step 4.

7. As the large difference represents 1 minute of angle or 60 seconds, multiply 60 seconds by the ratio from step 6.

8. For sine, tangent, and secant functions, *add* the product of step 7 to the nearest smaller angle. For cosine, cotangent and

cosecant functions also, *add* the product of step 7 to the nearest smaller angle.

Study Example.

Determine the angle size whose sine is 0.54714:

Step 1.   In the trig tables, locate the page and column where 0.54714 would be expected to be found under the "sine" designation. This value does not appear, but the nearest smaller value, 0.54708, is listed and is opposite $33°10'$. This means that the unknown angle must be $33°10'$ plus some seconds.

Step 2.   Calculate the difference in the decimal values for the nearest smaller angle, $33°10'$ and the nearest larger angle, $33°11'$:

$$\begin{array}{rl} \sin 33°11' = & 0.54732 \\ -\sin 33°10' = & \underline{0.54708} \\ \text{Difference} = & 0.00024 \end{array}$$

Step 3.   Calculate the difference between the decimal value for sine $33°10'$ and the given value of the unknown angle:

$$\begin{array}{rl} \sin \text{ of unknown angle} = & 0.54714 \\ -\sin 33°10' \quad\quad\quad = & \underline{0.54708} \\ \text{Difference} \quad\quad\quad = & 0.00006 \end{array}$$

Step 4.   Form a ratio (fraction) of the small difference to the big difference, step 3 to step 2:

$$\frac{0.00006}{0.00024} = \frac{1}{4}$$

Step 5.   Multiply 60 seconds (1 minute difference between $33°10'$ and $33°11'$) by the ratio in step 4:

$$60 \times \frac{1}{4} = 15 \text{ seconds}$$

Step 6.   Add the 15 seconds difference to the nearest smaller angle ($33°10'$):

$$33°10' + 15'' = 33°10'15''$$

Study Example.

Determine the angle size whose cosine is 0.80380.

Step 1.   Locate the nearest smallest angle in degrees and minutes from trig tables:

0.80380 lies between $36°30'$ and $36°31'$

Step 2. Determine decimal difference between values for the two angles in step 1:

$$\cos 36°30' = 0.80386$$
$$\cos 36°31' = 0.80368$$
$$\text{Difference} = 0.00018$$

Step 3. Calculate decimal difference between unknown angle and the nearest smaller angle:

$$\cos 36°30' \qquad = 0.80386$$
$$-\cos \text{unknown angle} = 0.80380$$
$$\text{Difference} \qquad = 0.00006$$

Step 4. Form a ratio of the small difference in step 3 to the larger difference in step 2:

$$\frac{0.00006}{0.00018} = \frac{1}{3}$$

Step 5. Multiply 60 seconds (one minute difference between nearest smaller and greater angles) by ratio in step 4:

$$60 \times \frac{1}{3} = 20 \text{ seconds}$$

Step 6. *Add* 20 seconds to nearest smaller angle:

$$36°30'00'' + 20'' = 36°30'20''$$

Study Problems.

Find the decimal trig value for the given functions:

1. $\sin 27°10'30''$

2. $\tan 52°10'45''$

3. $\sin 42°15'30''$

4. $\cos 36°14'30''$

5. $\cos 43°12'20''$

6. $\sin 65°29'40''$

7. $\cos 64°18'45''$

8. $\tan 28°32'20''$

9. $\cot 6°32'45''$

10. $\cot 9°20'10''$

11. $\sec 36°47'50''$

12. $\csc 15°25'38''$

13. Given; $\sin x = 0.6391$, find $\angle x$ and then find $\cos x$.

14. Given; $\tan x = 2.8649$, find $\angle x$ and then find $\cot x$.

Find the angle from a given trig function decimal:

15. $\sin x = 0.52250$

16. $\cos x = 0.78534$

17. $\tan x = 2.63954$ 21. $\sin x = 0.70613$

18. $\cot x = 3.79833$ 22. $\cos x = 0.48359$

19. $\sec x = 1.2147$ 23. $\csc x = 1.4140$

20. $\csc x = 1.2206$ 24. $\sec x = 1.2209$

Solutions of Right Triangles Using Interpolation

In a right triangle where an acute interior angle and one side are known, the other side and the hypotenuse may be determined by the use of the trig functions.

Figure 15-29 shows a right triangle, BCA with $\angle A = 42°22'16''$, side b is 7.6 inches. Solve for side a and the hypotenuse c.

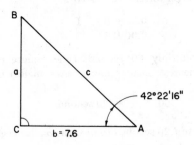

Fig. 15-29. Study example 1.

Study Examples.

1. Solve for side a.

    Step 1. Choose a function of $\angle A$ that will include the known side b and the side a.

$$\tan A = \frac{a}{b} \quad \text{or} \quad a = b \tan A$$

    Step 2. Determine value for $\tan 42°22'16''$

$$\tan 42°23' = 0.91259$$
$$\tan 42°22' = 0.91206$$
$$\text{Difference} = 0.00053$$

$$16'' = \frac{16}{60} \text{ minute}; \quad 0.00053 \times \frac{16}{60} = 0.00014$$

$$\tan 42°22'16'' = 0.91206 + 0.00014 = 0.91220$$

Step 3. Substitute values for *b* and tan *A*;

$$a = b \tan A$$
$$a = 7.6 \times 0.91220 = 6.93 \text{ in.}$$

2. Solve for hypotenuse *c*.

Step 1. Choose function of ∠*A* that includes *c* and *b*.

$$\sec A = \frac{c}{b} \quad \text{or;} \quad c = b \sec A$$

Step 2. Determine value for sec 42°22′16″.

$$\sec 42°23' = 1.3538$$
$$\sec 42°22' = \underline{1.3534}$$
$$\text{Difference} = 0.0004$$

$$16'' = \frac{16}{60} \text{ minute;} \quad 0.0004 \times \frac{16}{60} = 0.0001$$

$$\sec 42°22'16'' = 1.3534 + 0.0001 = 1.3535$$

Step 3. Substitute values for *b* and sec *A*;

$$c = b \sec A$$
$$c = 7.6 \times 1.3535 = 10.29 \text{ in.}$$

In a right triangle where the lengths of the two sides are known, the acute interior angles may be determined by the use of trig functions.

In Fig. 15-30, right triangle *BCA* is shown with side *a* equal to 6.3 cm and side *b* = 8.2 cm.

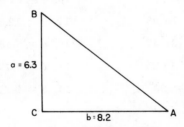

Fig. 15-30. Study example 1.

Study Examples.

1. Solve for angle *BAC* (∠*A*) in Fig. 15-30.

Step 1. Choose a function of ∠*A* that includes known sides *a* and *b*.

$$\tan A = \frac{a}{b}$$

Step 2. Substitute values for a and b;

$$\tan A = \frac{6.3}{8.2} = 0.76829$$

Step 3. Determine the angle from trig tables.

$$\angle A = 37°32'5''$$

Solve for angle ABC ($\angle B$) in Fig. 15-30.

Step 1. Subtract $\angle A$ from $90°$;

$$\angle B = 90° - 37°32'5''$$
$$\angle B = 89°59'60'' - 37°32'5'' = 52°27'55''$$

Study Problems.

The following problems relate to the standard right triangle shown in Fig. 15-31. Solve for all the unknown sides, hypotenuses and angles using interpolation.

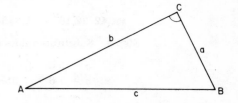

Fig. 15-31. Study problem 1.

1. $a = 6, c = 12$

2. $b = 20, \angle B = 3°38'$

3. $b = 4, \angle A = 60°$

4. $a = 992, \angle B = 76°38'$

5. $c = 200, \angle B = 21°47'$

6. $c = 30.69, b = 18.25$

7. $c = 72.15, \angle A = 39°34'$

8. $c = 2194, b = 1312.7$

9. $b = 50.94, \angle B = 43°48'$

10. $c = 91.92, a = 2.19$

11. $b = 12, \angle A = 29°8'$

12. $c = 8.462, \angle B = 86°4'$

13. $a = 7, \angle A = 18°14'$

14. $b = 9, \angle B = 34°44'$

Reciprocal Functions and Identities

In solving problems involving trig functions, it is sometimes desirable to be able to express a given function in terms of another function. In many trigonometric tables and tables of logarithms of trig functions, the secant and cosecant functions are missing. Formulas stated in secants or cosecants would be unsolvable unless one of the other four functions could be substituted. Also, in cases of small angles in the region of $0°$ and $90°$, the difference in the values for the secant and cosecant functions are so small that accurate interpolation is impossible.

Reciprocals

In Figure 15-32, the sine of angle A is the ratio $\frac{a}{c}$ and the cosecant of angle A is the ratio $\frac{c}{a}$.

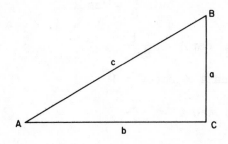

Fig. 15-32. Right triangle for reciprocal functions.

The ratio $\frac{c}{a}$ can be expressed as: $1 \times \frac{c}{a}$ and also: $1 \div \frac{a}{c}$ therefore:

$$\sin A = \frac{a}{c} \quad \text{and:} \quad \csc A = \frac{c}{a} = 1 \div \frac{a}{c} = \frac{1}{\frac{a}{c}}$$

$$\text{but:} \quad \frac{a}{c} = \sin A$$

1. therefore: $\csc A = \dfrac{1}{\sin A}$

   hence: $\sin A \times \csc A = 1$

2. therefore: $\sin A = \dfrac{1}{\csc A}$

also: $\quad \cos A = \dfrac{b}{c} \quad \text{and:} \quad \sec A = \dfrac{c}{b} = \dfrac{1}{\frac{b}{c}}$

$$\text{but:} \quad \frac{b}{c} = \cos A$$

3. therefore: $\sec A = \dfrac{1}{\cos A}$

hence: $\cos A \times \sec A = 1$

4. therefore: $\cos A = \dfrac{1}{\sec A}$

also: $\tan A = \dfrac{a}{b}$ and: $\cot A = \dfrac{b}{a} = \dfrac{1}{\dfrac{a}{b}}$

but: $\dfrac{a}{b} = \tan A$

5. therefore: $\cot A = \dfrac{1}{\tan A}$

hence: $\tan A \times \cot A = 1$

6. therefore: $\tan A = \dfrac{1}{\cot A}$

Identities

In Fig. 15-32, $\sin A = \dfrac{a}{c}$ and: $\cos A = \dfrac{b}{c}$

Then: $a = c \sin A$ and: $b = c \cos A$

If side c in $\triangle ACB$ is equated to one (unity):

then: $a = \sin A$ and: $b = \cos A$

and: $\tan A = \dfrac{a}{b} = \dfrac{\sin A}{\cos A}$

also: $\cot A = \dfrac{b}{a} = \dfrac{\cos A}{\sin A}$

In Fig. 15-32, $\sin A = \dfrac{a}{c}$ and $\cos B = \dfrac{a}{c}$,

therefore: $\sin A = \cos B$

The foregoing identities and reciprocals are shown in Table 15-1 for your reference in the solution of problems involving trigonometric functions.

Study Examples.

1. Determine the sine of $\angle A$ when the cosine of $\angle B$ equals 0.89101.

Table 15-1. Reciprocal Functions and Right-Triangle Formulas.

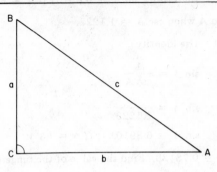

$$\sin A = \frac{a}{c} = \cos B \qquad\qquad \cos A = \frac{b}{c} = \sin B$$

$$\tan A = \frac{a}{b} = \cot B \qquad\qquad \cot A = \frac{b}{a} = \tan B$$

$$\sec A = \frac{c}{b} = \csc B \qquad\qquad \csc A = \frac{c}{a} = \sec B$$

$$\sin A = \frac{1}{\csc A} \qquad \cos A = \frac{1}{\sec A} \qquad \tan A = \frac{1}{\cot A}$$

$$\csc A = \frac{1}{\sin A} \qquad \sec A = \frac{1}{\cos A} \qquad \cot A = \frac{1}{\tan A}$$

$$a = c \sin A \qquad\qquad a = b \tan A$$

$$b = c \cos A \qquad\qquad b = a \cot A$$

$$c = a \csc A \qquad\qquad c = b \sec A$$

$$c^2 = a^2 + b^2 \qquad\qquad c = \sqrt{a^2 - b^2}$$

$$a^2 = c^2 - b^2 \qquad\qquad a = \sqrt{c^2 - b^2}$$

$$b^2 = c^2 - a^2 \qquad\qquad b = \sqrt{c^2 - a^2}$$

Step 1. Use identity: $\sin A = \cos B$;

therefore: $\sin A = 0.89101$ $(\angle A = 63°)$

2. Determine $\sin A$ when $\csc A = 1.1223$.

Step 1. Use identity:

$$\sin A = \frac{1}{\csc A}$$

$$\sin A = \frac{1}{1.1223}$$

$$\sin A = 0.89101 \quad (\angle A = 63°)$$

3. Cotangent $A = 0.78128$. Find the value of the tangent.

Step 1. Use identity:

$$\tan A = \frac{1}{\cot A}$$

$$\tan A = \frac{1}{0.78128}$$

$$\tan A = 1.2799 \quad (\angle A = 52°)$$

Study Problems.

1. Demonstrate by substituting identities that the trig formula, $\dfrac{\cos A \csc A}{\cot A} = 1$, is true.

2. Show that: $\sin B \csc B = 1$.

3. Show that: $\cos w \sec w = 1$.

4. Prove that: $\tan A = \dfrac{\sec A}{\csc A}$

Solving Trigonometric Shop Problems

Most trig problems may be reduced by constructions to a series of right triangles. The drawing of the necessary construction lines is the secret of solving complicated problems.

The solution of a problem is most easily approached by working *backwards* from the answer to be found. Using the unknown as your starting point, construct triangles that relate to the known data. The use of the geometric axioms and propositions is helpful in establishing relations of known and unknown angles and lines. Auxiliary lines are necessary to construct triangles to which the trig functions and laws can be applied.

Study Example.

1. Calculate the dimension X on the gage shown in Fig. 15-33. All dimensions are in inches.

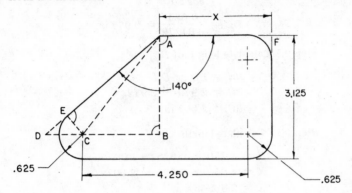

Fig. 15-33. Study example 1.

Step 1. Construct line AB from point A, parallel to the right edge of the gage.

Step 2. Draw CB from C, parallel to the base. Angle $B = 90°$.

Step 3. Extend lines BC and AE to D. Triangle ABD is a right triangle.

Step 4. Draw line CE from point C perpendicular to line AD. Triangle DEC is a right triangle.

Step 5. Calculate length of AB.

$$AB = 3.125 - 0.625 \text{ (radius)} = 2.500 \text{ in.}$$

Step 6. Calculate $\angle BAD$.

$$\angle BAD = 140° - \angle BAF$$
$$\angle BAD = 140° - 90° = 50°$$

Step 7. Calculate length of BD in triangle ABD;

$$\tan \angle BAD = \frac{BD}{AB}$$
$$BD = AB \tan \angle BAD$$
$$BD = 2.50 \times 1.1917$$
$$BD = 2.979 \text{ in.}$$

Step 8. In triangle DEC, calculate length of DC;

$$EC = 0.625$$
$$\angle ECD = \angle BAD \quad \text{(Proposition 26)}$$
$$\text{hence: } \angle ECD = 50°$$

$$\cos \angle ECD = \frac{EC}{DC}$$

$$DC = \frac{EC}{\cos 50°}$$

$$DC = \frac{0.625}{0.64279} = 0.972 \text{ in.}$$

Step 9.   Calculate length of BC:

$$BC = BD - DC$$
$$BC = 2.979 - 0.972$$
$$BC = 2.007 \text{ in.}$$

Step 10.   Calculate length X;

$$X = 4.250 + 0.625 - BC$$
$$X = 4.250 + 0.625 - 2.007$$
$$X = 2.868 \text{ in.}$$

Study Problems.

1. In Fig. 15-34 solve for diameter D. The dimension shown is in inches.

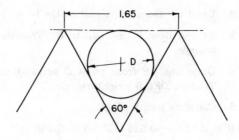

Fig. 15-34. Study problem 1.

2. In Fig. 15-35 determine angle A. The dimensions shown are in inches.

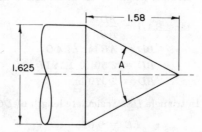

Fig. 15-35. Study problem 2.

3. In Fig. 15-36 solve for length L. The dimensions shown are in millimeters.

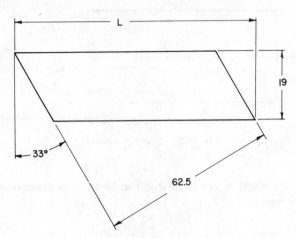

Fig. 15-36. Study problem 3.

4. Determine angle A in Fig. 15-37. The dimensions shown are in inches.

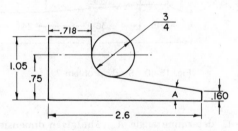

Fig. 15-37. Study problem 4.

5. In Fig. 15-38 solve for distance Z. The dimensions shown are in inches.

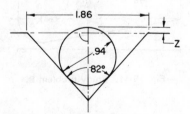

Fig. 15-38. Study problem 5.

6. In Fig. 15-39 determine radius *r*. The given dimensions are in inches.

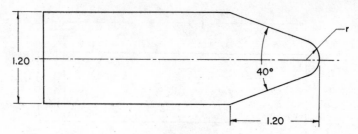

Fig. 15-39. Study problem 6.

7. Solve for length *L* and angle *A* in Fig. 15-40. The dimensions shown are in millimeters.

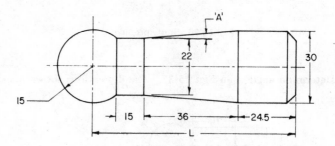

Fig. 15-40. Study problem 7.

8. In Fig. 15-41 determine angle *A*. The given dimensions are in inches.

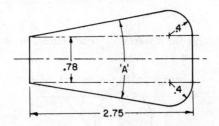

Fig. 15-41. Study problem 8.

9. In Fig. 15-42 determine the height *H* (mm). The dimensions shown are in millimeters.

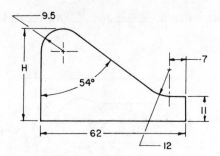

Fig. 15-42. Study problem 9.

10. In Fig. 15-43 solve for dimension A.   The dimensions shown are in inches.

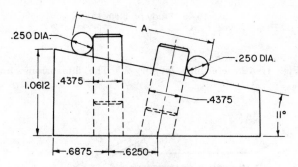

Fig. 15-43. Study problem 10.

11. In Fig. 15-44 solve for dimension A.   All dimensions are in inches.

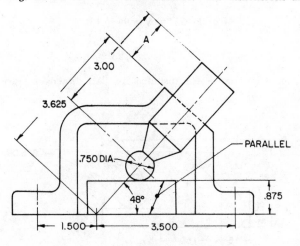

Fig. 15-44. Study problem 11.

12. In Fig. 15-45 solve for dimension X. All dimensions are in inches.

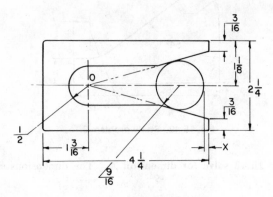

Fig. 15-45. Study problem 12.

13. In Fig. 15-46 determine length Z. The dimensions shown are in inches.

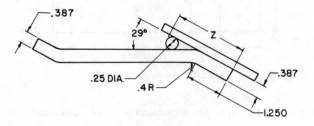

Fig. 15-46. Study problem 13.

14. In Fig. 15-47 solve for distance A. All dimensions are in inches.

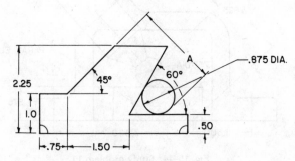

Fig. 15-47. Study problem 14.

15. In Fig. 15-48 solve for angle A and length L (see problem 7). All dimensions are in inches.

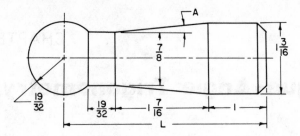

Fig. 15-48. Study problem 15.

16. In Fig. 15-49 solve for dimension H (see problem 9). All dimensions are in inches.

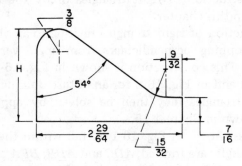

Fig. 15-49. Study problem 16.

17. In Fig. 15-50 solve for dimension X (see problem 12). All dimensions are in millimeters.

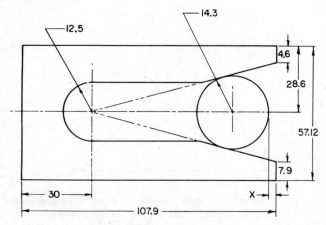

Fig. 15-50. Study problem 17.

CHAPTER 16

Oblique-Angle Trigonometry

Oblique triangles are often encountered in shop problems. Distances between holes, lengths of sides, and angles between center lines are usually required in the layout of machine parts. The oblique triangles usually formed by sides and center lines may be solved by further construction of right triangles or by the use of special formulas given in this chapter.

The construction of right triangles relating to an oblique triangle is done by dropping perpendiculars from angle vertices to their opposite sides. This construction is shown in Fig. 16-1 for an acute oblique triangle and in Fig. 16-2 for an obtuse oblique triangle. The resulting right triangles may then be solved by applying the trig functions as shown in Chapter 15.

The altitudes shown in Fig. 16-1 all fall within the triangle ABC and six right triangles are formed; $A\underline{D}C$ and $A\underline{D}B$; $B\underline{F}A$ and $B\underline{F}C$; $C\underline{E}A$ and $C\underline{E}B$. Two of the altitudes shown in Fig. 16-2 fall outside the triangle $A'B'C'$ and one, $A'E'$, falls within the triangle. Here, too, six right triangles are formed; $A'\underline{F}'B'$ and $A'\underline{D}'C'$; $B'\underline{D}'C'$ and $B'\underline{E}'A'$; $C'\underline{E}'A'$ and $C'\underline{F}'B'$.

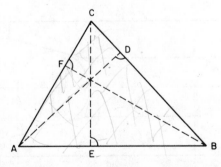

Fig. 16-1. Constructing right triangles by dropping perpendiculars to the opposite sides in acute oblique triangles.

426

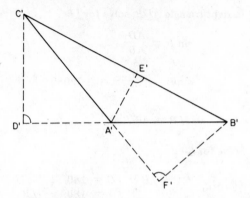

Fig. 16-2. Constructing right triangles by dropping perpendiculars
to the opposite sides in obtuse oblique triangles.

The choice of which triangle to use in a solution depends on
which angles and sides are known. In the following two study examples an acute and an obtuse oblique triangle are solved.

Study example.

1. In the acute oblique triangle ABC (Fig. 16-3), side $AB = 22.50$ mm,
   side $AC = 19.80$ mm and angle $C = 55°$. Solve for $\angle A$, $\angle B$ and side BC.

   Step 1.   Draw altitude AD from $\angle A$ to side BC forming the right triangle
   $A\underline{D}C$ where $\angle C$ is known. Solve for AD.

$$AD = AC \sin 55°$$
$$AD = 19.80 \times 0.81915$$
$$AD = 16.22 \text{ mm}.$$

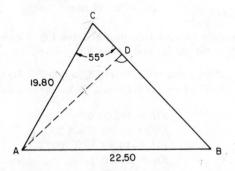

Fig. 16-3. Study example 1.

Step 2.   In right triangle $A\underline{D}B$, solve for $\angle B$:

$$\sin B = \frac{AD}{AB}$$

$$\sin B = \frac{16.22}{22.50} = 0.72089$$

$$\angle B = 46°7'42''$$

Step 3.   Solve for $\angle A$:

$$\angle A + \angle B + \angle C = 180°$$
$$\angle A = 180° - (\angle B + \angle C)$$
$$\angle A = 180° - (46°7'42'' + 55°)$$
$$\angle A = 180° - 101°7'42''$$
$$\angle A = 78°52'18''$$

Step 4.   Side $BC = BD + CD$. In right triangle $A\underline{D}C$, solve for side DC:

$$DC = AC \cos 55°$$
$$DC = 19.8 \times 0.57358$$
$$DC = 11.36 \text{ mm}$$

Step 5.   Solve for BD in right triangle $A\underline{D}B$:

$$BD = AB \cos B$$
$$BD = AB \cos 46°7'42''$$
$$BD = 22.5 \times 0.69305$$
$$BD = 15.59 \text{ mm}$$

Step 6.   Solve for BC:

$$BC = BD + CD$$
$$BC = 15.59 + 11.36$$
$$BC = 26.95 \text{ mm}$$

2. In the obtuse oblique triangle EFG (see Fig. 16-4), side $EF = 38.00$ in. side $EG = 44.50$ in. and $\angle G = 23°$. Solve for $\angle E$, $\angle F$, and side FG.

Step 1.   Draw altitude EH from $\angle E$ to side FG, forming the right triangle $E\underline{H}G$ where $\angle G$ is known. Solve for side EH:

$$EH = EG \sin G$$
$$EH = 44.50 \times \sin 23°$$
$$EH = 44.50 \times 0.39073$$
$$EH = 17.39 \text{ in.}$$

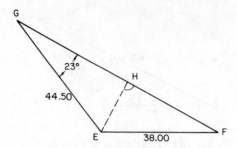

Fig. 16-4. Study example 2.

Step 2.   In right triangle $E\underline{H}F$, solve for $\angle F$:

$$\sin F = \frac{EH}{EF}$$

$$\sin F = \frac{17.39}{38.0} = 0.45763$$

$$F = 27°14'4''$$

Step 3. $FG = FH + GH$. In right triangle $E\underline{H}G$, solve for GH:

$$GH = EG \cos \angle G$$
$$GH = 44.50 \times \cos 23°$$
$$GH = 44.50 \times 0.92050$$
$$GH = 40.96 \text{ in.}$$

Step 4.   In right triangle $E\underline{H}F$, solve for side HF:

$$HF = EF \cos \angle F$$
$$HF = 38.00 \times \cos 27°14'4''$$
$$HF = 38.00 \times 0.88914$$
$$HF = 33.79 \text{ in.}$$

Step 5. Solve for FG:

$$FG = HF + GH$$
$$FG = 33.79 + 40.96$$
$$FG = 74.75 \text{ in.}$$

Study Problems.

1. Using the altitude method, solve for angle A in $\triangle ABC$ in Fig. 16-5.

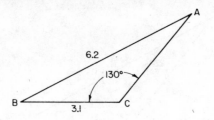

Fig. 16-5. Study problems 1 and 2.

2. Solve for the length of AC in $\triangle ABC$.

3. In $\triangle EFG$ (Fig. 16-6), solve for the length of side g.

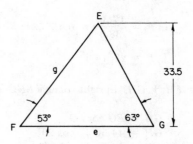

Fig. 16-6. Study problems 3 and 4.

4. In $\triangle EFG$ (Fig. 16-6), solve for the length of side e.

5. In $\triangle ABC$ (Fig. 16-7), determine the length of the altitude from point C.

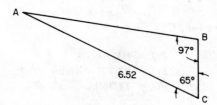

Fig. 16-7. Study problems 5 and 6.

6. Solve for the length of side AB in $\triangle ABC$ (Fig. 16-7).

7. In Fig. 16-8 determine the radius R.

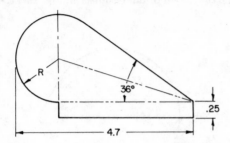

Fig. 16-8. Study problem 7.

8. Determine the length of radius R in Fig. 16-9.

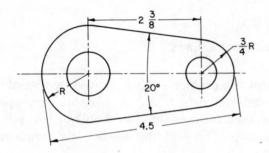

Fig. 16-9. Study problem 8.

9. Determine the length of side AC in Fig. 16-10.

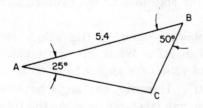

Fig. 16-10. Study problems 9 and 10.

10. Determine the length of side *BC* in Fig. 16-10.

11. Determine the length of dimension *X* in Fig. 16-11.

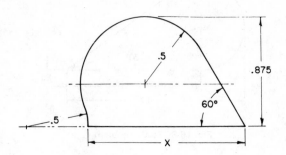

Fig. 16-11. Study problem 11.

Special Trigonometric Laws

While most trig problems can be solved by special constructions leading to a series of right triangles, there are some special formulas which can be used for directly solving oblique triangles when certain data are known.   Three of the special trig formulas for oblique triangles are:

> The Sine Law
>
> The Cosine Law
>
> The Cotangent Law

In many shop problems, much calculation can be avoided by using these laws, and the solutions to the problems are often quicker and easier to handle.

In solving oblique triangles, trig values are often required for angles greater than 90°.   While the trig tables in this book permit direct readings of values for angles from 0° to 180°, many tables do not go above 90°.   However, any angle between 90° and 180° can be evaluated from any trig table by using the following table:

Where an angle, θ, is greater than 90° and less than 180°:

$$\sin \theta = \sin (180° - \theta)$$
$$\cos \theta = -\cos (180° - \theta)$$
$$\tan \theta = -\tan (180° - \theta)$$
$$\cot \theta = -\cot (180° - \theta)$$
$$\sec \theta = -\sec (180° - \theta)$$
$$\csc \theta = \csc (180° - \theta)$$

for example

$$\sin 135° = \sin (180° - 135)$$
$$= \sin 45° = 0.70711$$
$$\cos 122° = -\cos (180° - 122)$$
$$= -\cos 58° = -0.52992$$
$$\tan 167° = -\tan (180° - 167)$$
$$= -\tan 13° = -0.23087$$

Study Problems.

Determine the decimal trig value for the following:

1. $\sin 95°$　　　　　5. $\sec 111°$　　　　　9. $\cot 163°$

2. $\cos 132°$　　　　6. $\cos 100°$　　　　10. $\sin 159°$

3. $\tan 103°$　　　　7. $\csc 159°$　　　　11. $\sec 108°$

4. $\cot 98°$　　　　　8. $\tan 92°$　　　　　12. $\cos 105°$

The Sine Law

If any triangle—right, acute, or obtuse—is examined, it is readily apparent that the longest side always lies opposite the largest angle and the shortest side always lies opposite the smallest angle. In trigonometry, this relationship is developed into the *Sine Law*:

> *In any triangle, the sines of the angles are pro-*
> *portional to the lengths of the opposite sides.*

In Fig. 16-12, an acute oblique triangle ABC is shown. An altitude BD is drawn from angle B to the opposite side AC. Then by definition of the trig function:

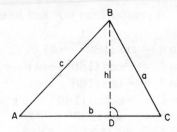

Fig. 16-12. An oblique triangle: the Sine Law.

$$\sin A = \frac{h}{c}, \qquad \text{and} \qquad \sin C = \frac{h}{a}$$

also $\qquad\qquad h = c \sin A, \qquad$ and $\qquad h = a \sin C$

however $\qquad\qquad h = h$

then $\qquad\qquad\qquad c \sin A = a \sin C$

or $\qquad\qquad\qquad\qquad \dfrac{\sin A}{a} = \dfrac{\sin C}{c}$

In similar fashion we can develop the Sine Law:

$$\frac{\sin B}{b} = \frac{\sin C}{c}$$

and $\qquad\qquad\qquad \dfrac{\sin A}{a} = \dfrac{\sin B}{b}$

These relationships may also be inverted, thus:

$$\frac{a}{\sin A} = \frac{b}{\sin B}$$

$$\frac{a}{\sin A} = \frac{c}{\sin C}$$

$$\frac{b}{\sin B} = \frac{c}{\sin C}$$

The use of the Sine Law is in the solving of a proportion where three of the four terms are known. If an oblique triangle, either acute or obtuse, has two sides and one angle opposite one of the known sides, or, two angles and a side opposite one of the known angles, the Sine Law may be used.

Study Example.

1. In triangle ABC (Fig. 16-13) side $AB = 12.56$ in., side $BC = 10.48$ in. and angle $A = 22°33'$. Solve for angle C and side AC using the Sine Law.

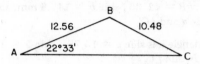

Fig. 16-13. Study example 1.

Step 1. Determine a known side that has a known angle opposite it.

$$\text{side } BC \text{ and } \angle A$$

Step 2. Set up the Sine Law with the unknown in the numerator of the first ratio and equate to the ratio of the knowns. Solving for angle C:

$$\frac{\sin C}{AB} = \frac{\sin A}{BC}$$

$$\frac{\sin C}{12.56} = \frac{\sin 22°33'}{10.48}$$

$$\sin C = \frac{12.56 \times \sin 22°33'}{10.48}$$

$$\sin C = \frac{12.56 \times 0.38349}{10.48} = 0.45960$$

$$\angle C = 27°21'41''$$

Step 3. Solving for side AC:

$$\frac{AC}{\sin B} = \frac{BC}{\sin A}$$

$$AC = \frac{BC \times \sin B}{\sin A}$$

$$AC = \frac{10.48 \times \sin [180° - (22°33' + 27°21'41'')]}{\sin 22°33'}$$

$$AC = \frac{10.48 \times \sin 130°5'19''}{\sin 22°33'}$$

$$AC = \frac{10.48 \times 0.76505}{0.38349} = 20.91$$

Study Problems.

Solve by the Sine Law.

1. An acute oblique triangle has two sides and two angles known. $\angle A = 36°25'$, $\angle B = 42°36'$, side $b = 147.6$ mm, and side $c = 222.4$ mm. Determine side a and $\angle C$.

2. An oblique triangle has side $a = 15.75''$, side $b = 21''$, and $\angle A = 43°15'$. Find side c, $\angle B$, and $\angle C$.

3. A vee-shaped cut is made in a steel plate, as shown in Fig. 16-14. The angle of the vee is $23°15'$ and the width of the vee on the surface of the plate is 0.478 in. The sides of the vee are of equal length. What is their length?

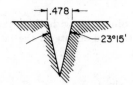

Fig. 16-14. Study problem 3.

4. An oblique triangle has two angles and a side opposite one of the angles known. If $\angle A = 55°$, $\angle B = 51°$ and side $b = 5.64''$, what is the length of side c?

5. An extrusion die as shown in Fig. 16-15 has 12 points symmetrically placed around a base circle of 2.75 in. Find the diameter D.

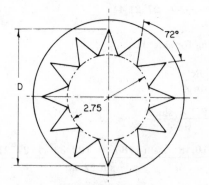

Fig. 16-15. Study problem 5.

6. An internal cam is shown in Fig. 16-16. Compute the angle ϕ.

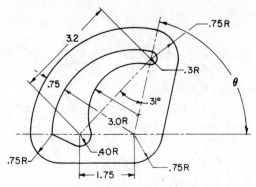

Fig. 16-16. Study problem 6.

7. Three gears mesh as shown in Fig. 16-17. Solve for dimension AC.

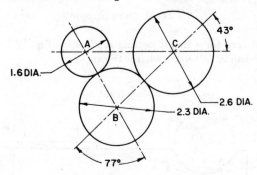

Fig. 16-17. Study problem 7.

8. Two oblique triangles have a common side AC as shown in Fig. 16-18. Sides AB and CD are equal in length. Solve for the length of AC, $\angle ABC$, and side BC.

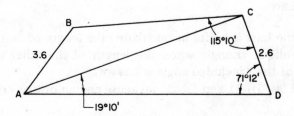

Fig. 16-18. Study problem 8.

9. The broach shown in Fig. 16-19 has ten teeth. A special diameter ball contacts the sides of the teeth at the pitch diameter. What is the diameter of the special ball? What is the distance M across two special balls of calculated diameter?

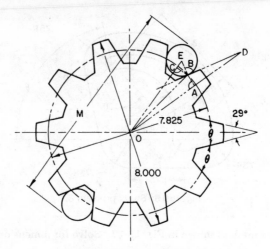

Fig. 16-19. Study problem 9.

10. A ball race for a thrust bearing is shown in Fig. 16-20. If the diameter of the ball is 1.000 in., what is the angle ϕ?

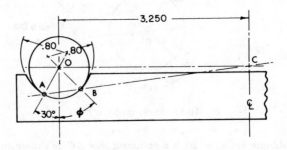

Fig. 16-20. Study problem 10.

The Cosine Law

The Cosine Law is useful in calculating the length of an unknown side of an oblique triangle when the length of the other two sides and the size of their included angle is known.

In both Figs. 16-21 and 16-22, an acute and an obtuse triangle:

$$a^2 = h^2 + BD^2 \tag{1}$$

and
$$b^2 = h^2 + AD^2 \tag{2}$$

also
$$BD = c - AD \text{ (in acute } \triangle)$$
$$\tag{3}$$
and
$$BD = AD - c \text{ (in obtuse } \triangle)$$

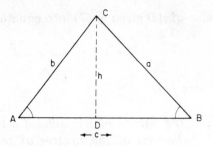

Fig. 16-21. An acute oblique triangle: the Cosine Law.

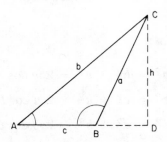

Fig. 16-22. An obtuse oblique triangle: the Cosine Law.

Substitute the values of BD given in (3) into equation (1):

$$a^2 = h^2 + (c - AD)^2 \quad \text{and} \quad a^2 = h^2 + (AD - c)^2$$

giving $a^2 = h^2 + c^2 - 2cAD + AD^2$ \hfill (4)

Transposing equation (2) gives:

$$h^2 = b^2 - AD^2 \tag{5}$$

Substitute the value of h^2 given in (5) into equation (4):

$$a^2 = b^2 + c^2 - 2cAD + AD^2 - AD^2$$

or $a^2 = b^2 + c^2 - 2cAD$ \hfill (6)

In both triangles (Fig. 16-21 and Fig. 16-22):

$$\cos A = \frac{AD}{b}$$

or $AD = b \cos A$ \hfill (7)

Substitute the value of AD given in (7) into equation (6):

$$a^2 = b^2 + c^2 - 2bc \cos A$$

The Cosine Law:

> *The square of any side of a triangle is equal to the sum of the squares of the other two sides diminished by twice their product times the cosine of the included angle.*

Expressed as equations:

$$a^2 = b^2 + c^2 - 2bc \cos A$$
$$b^2 = c^2 + a^2 - 2ca \cos B$$
$$c^2 = a^2 + b^2 - 2ab \cos C$$

This law is sometimes called the *Generalized Pythagorean Theorem.*

Study Examples.

1. In the $\triangle ABC$ shown in Fig. 16-21, side $a = 91$ mm, side $b = 82.5$ mm, and $\angle C = 82°14'10''$. Solve for side c using the Cosine Law.

Step 1.   State the Cosine Law for c (c^2) being the unknown:

$$c^2 = a^2 + b^2 - 2ab \cos C$$

Step 2. Substitute values given in problem and solve for c:

$$c^2 = (91)^2 + (82.5)^2 - 2(91)(82.5)(\cos 82°14'10'')$$
$$c^2 = 8,281 + 6,806.25 - (15,015)(0.13509)$$
$$c^2 = 15,087.25 - 2,028.38$$
$$c^2 = 13,058.87$$
$$c = 114.275 \text{ mm, say; } 114.28$$

2. In $\triangle ABC$ (Fig. 16-21) solve for $\angle A$ using the Cosine Law.

Step 1.   State the Cosine Law with $\angle A$ as the unknown:

$$a^2 = b^2 + c^2 - 2bc \cos A$$

$$\cos A = \frac{b^2 + c^2 - a^2}{2bc}$$

Step 2. Substitute known values from example 1 and solve:

$$\cos A = \frac{(82.5)^2 + (114.275)^2 - (91)^2}{2(82.5)(114.275)}$$

$$\cos A = \frac{6806.25 + 13058.78 - 8281}{18855.4}$$

$$\cos A = \frac{11584.0}{18855.4} = .61436$$

$$\angle A = 52°5'40''$$

3. In $\triangle ABC$ (Fig. 16-22), solve for $\angle B$ using the Cosine Law when side $a = 7.85$, $b = 11.00$ and $c = 5.15$.

Step 1.   State the Cosine Law with $\angle B$ as unknown angle:

$$\cos B = \frac{a^2 + c^2 - b^2}{2ac}$$

Step 2.   Substitute known values for sides a, b, and c:

$$\cos B = \frac{(7.85)^2 + (5.15)^2 - (11.00)^2}{2(7.85)(5.15)}$$

$$= \frac{61.6225 + 26.5225 - 121}{80.855} = \frac{-32.855}{80.855}$$

$$= -0.40634$$

$$\angle B = 113°58'31''$$

Study Problems.

1. Solve for angle BAC in Fig. 16-23.

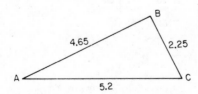

Fig. 16-23. Study problems 1 and 2.

2. Solve for angle ABC in Fig. 16-23.

3. Solve for angle *BAC* in Fig. 16-24.

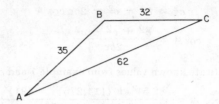

Fig. 16-24. Study problems 3 and 4.

4. Solve for angle *ABC* in Fig. 16-24.

5. Determine the distance *D* in Fig. 16-25.

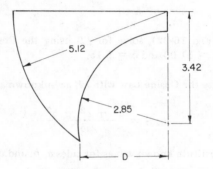

Fig. 16-25. Study problem 5.

6. Solve for dimension *A* in Fig. 16-26.

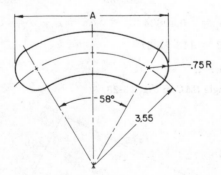

Fig. 16-26. Study problem 6.

7. Solve for line *BD* in Fig. 16-27.

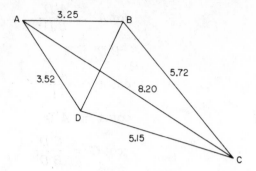

Fig. 16-27. Study problem 7.

The Cotangent Law

An oblique triangle with two angles and the included side given may be solved directly for the length of the perpendicular from the apex of the unknown angle to the known side by the use of the Cotangent Law.

The oblique triangle ABC shown in Fig. 16-28 has side b and the adjacent angles A and C given. To develop the cotangent formula, a similar triangle $A'B'C'$ is constructed whose altitude $B'D'$ is equal to one (unity).

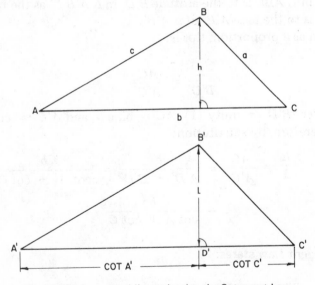

Fig. 16-28. An acute oblique triangle: the Cotangent Law.

In $\triangle A'B'D'$; $\qquad\qquad\qquad\qquad \cot A' = \dfrac{A'D'}{B'D'}$

but $\qquad\qquad\qquad\qquad\qquad\qquad B'D' = 1$

hence $\qquad\qquad\qquad\qquad\qquad \cot A' = \dfrac{A'D'}{1}$

or $\qquad\qquad\qquad\qquad\qquad\qquad \cot A' = A'D'$

likewise $\qquad\qquad\qquad\qquad\quad \cot C' = \dfrac{C'D'}{B'D'}$

but $\qquad\qquad\qquad\qquad\qquad\qquad B'D' = 1$

hence $\qquad\qquad\qquad\qquad\qquad \cot C = C'D'$

however $\qquad\qquad\qquad\qquad A'C' = A'D' + D'C'$

Substitute the values for $A'D'$ and $D'C'$ in the equation above:

$$A'C' = \cot A' + \cot C'$$

But $\triangle A'B'C'$ is similar to $\triangle ABC$ by construction. Hence $\angle A' = \angle A$, $\angle B' = \angle B$, and $\angle C' = \angle C$. Therefore, $\cot A' + \cot C'$ equals $\cot A + \cot C$ and $A'C' = \cot A + \cot C$.

Since ABC and $A'B'C'$ are similar triangles by construction, the altitude h in $\triangle ABC$ is to the altitude $B'D'$ in $\triangle A'B'C'$ as the base AC of $\triangle ABC$, is to the base $A'C'$ of $\triangle A'B'C'$.

Written as a proportion, this is:

$$\frac{h}{B'D'} = \frac{AC}{A'C'}$$

However, $B'D' =$ unity (1), $AC =$ base b, and $A'C' = \cot A' + \cot C$. Therefore, by sutstitution:

$$\frac{h}{1} = \frac{AC}{A'C'} = \frac{b}{A'D' + D'C'} = \frac{b}{\cot A' + \cot C'}$$

or $\qquad\qquad\qquad\qquad h = \dfrac{b}{\cot A + \cot C}$

The Cotangent Law states:

When a side and the two adjacent angles
are given, the altitude to the given side is

equal to the length of that side divided by the sum of the cotangents of the two known, adjacent angles.

In the case of an obtuse triangle where the desired altitude lies *outside* the triangle as in Fig. 16-29, the cotangent formula becomes:

$$h = \frac{b}{\cot A - \cot (180° - C)}$$

$$h = \frac{b}{\cot A - \cot \omega}$$

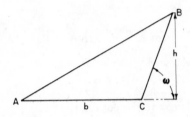

Fig. 16-29. The Cotangent Law in an obtuse oblique triangle.

Study Examples.

1. In $\triangle ABC$ shown at right, AC = 1225 mm, $\angle A = 70°$, and $\angle C = 35°$. Solve for the altitude BD.

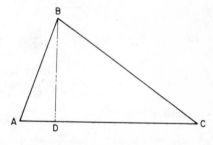

Fig. 16-30. Study example 1.

    Step 1. Use the Cotangent Formula:

$$BD = \frac{AC}{\cot A + \cot C}$$

    Step 2. Substitute known values in formula:

$$BD = \frac{1225}{\cot 70° + \cot 35°}$$

$$BD = \frac{1225}{0.36397 + 1.4281}$$

$$BD = \frac{1225}{1.792}$$

$$BD = 683.57 \text{ mm, say; } 683.6 \text{ mm}$$

2. In an obtuse $\triangle ABC$, $AC =$ 4.62 in., $\angle A = 45°$, and $\angle C = 110°$. Solve for the altitude h.

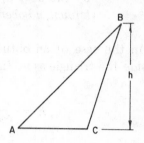

Step 1. Use the Cotangent Formula for an obtuse triangle:

Fig. 16-31. Study example 2.

$$h = \frac{AC}{\cot A - \cot (180° - C)}$$

Step 2. Substitute known values in the formula:

$$h = \frac{4.62}{\cot 45° - \cot (180° - 110°)}$$

$$h = \frac{4.62}{\cot 45° - \cot 70°}$$

$$h = \frac{4.62}{1.000 - 0.36397}$$

$$h = \frac{4.62}{0.63603}$$

$$h = 7.2638, \text{ say}; 7.26 \text{ in.}$$

Study Problems.

1. Determine the altitude in triangle ABC shown in Fig. 16-32.

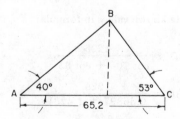

Fig. 16-32. Study problem 1.

2. Solve for the distance h in obtuse triangle ABC in Fig. 16-33.

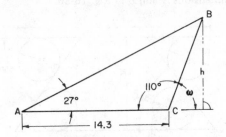

Fig. 16-33. Study problem 2.

3. Determine the diameter D in gage shown in Fig. 16-34.

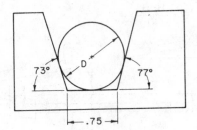

Fig. 16-34. Study problem 3.

4. Determine distance P in Fig. 16-35.

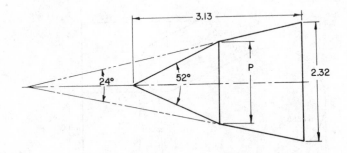

Fig. 16-35. Study problem 4.

Review Problems.

1. Solve for dimensions A and B in Fig. 16-36.

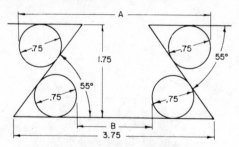

Fig. 16-36. Review problem 1.

2. The vernier height gage reading obtained for H_1 is 4.678 in. and for H_2 it is 6.122 in. Calculate the included angle of the taper and the taper per foot of the arbor. See Fig. 16-37.

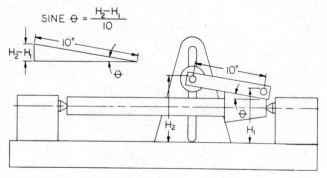

Fig. 16-37. Review problem 2.

3. Solve for angle α in Fig. 16-38.

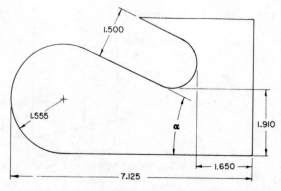

Fig. 16-38. Review problem 3.

4. Solve for dimension X in Fig. 16-39.

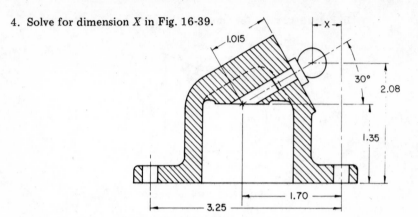

Fig. 16-39. Review problem 4.

5. Solve for dimension X in Fig. 16-40.

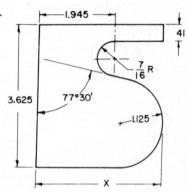

Fig. 16-40. Review problem 5.

6. Solve for distance X in Fig. 16-41.

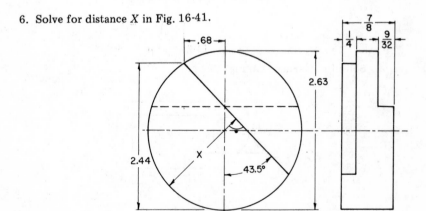

Fig. 16-41. Review problem 6.

7. In Fig. 16-42 solve for dimension D.

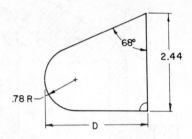

Fig. 16-42. Review problem 7.

8. Solve for dimension A in Fig. 16-43.

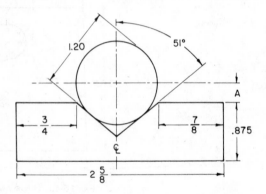

Fig. 16-43. Review problem 8.

9. Solve for distance D in Fig. 16-44.

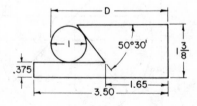

Fig. 16-44. Review problem 9.

10. Solve for dimension X and angle α, in Fig. 16-45.

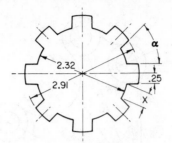

Fig. 16-45. Review problem 10.

11. Solve for distance D and angle α in Fig. 16-46.

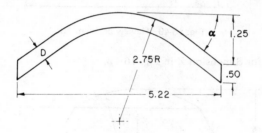

Fig. 16-46. Review problem 11.

12. Solve for diameter X and angles α and β in Fig. 16-47.

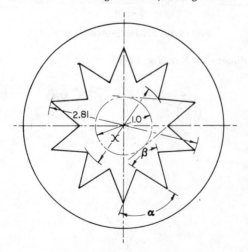

Fig. 16-47. Review problem 12.

13. Solve for radii X and Y in Fig. 16-48.

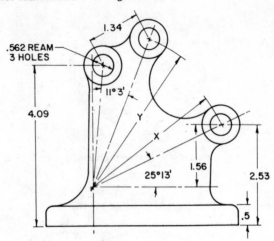

Fig. 16-49. Review problem 13.

14. Solve for dimension X in Fig. 16-49.

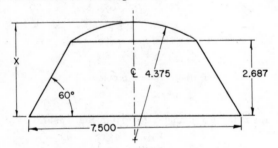

Fig. 16-49. Review problem 14.

15. Solve for dimension X in Fig. 16-50.

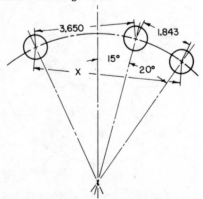

Fig. 16-50. Review problem 15.

16. Solve for dimension X in Fig. 16-51.

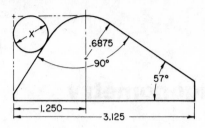

Fig. 16-51. Review problem 16.

Shop Trigonometry

Trigonometry is a most useful tool in the shop and in the drafting room. Many parts that are designed and made, involve angles or angular relationships. Trigonometry is useful in designing these parts, setting them up on machine tools, designing jigs and fixtures on which they are to be made, and measuring them after machining. The objective of this chapter is to provide examples and problems of typical shop and drafting room situations where trigonometry is used.

Sine Bars and Sine Plates

Sine bars and sine plates, shown in Figs. 17-1 and 17-2, are precision tools used in tool and die shops to make precise angular measurements and to hold workpieces in a precise angular relationship on machine tools such as jig borers, jig grinders, and surface grinders. When the angle to be machined or measured has a limit of accuracy of 5 minutes or less, sine bars and sine plates are almost indispensable tools.

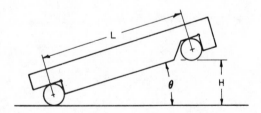

Fig. 17-1. Angular relationship on a sine bar.

These tools are precision ground on all surfaces. Two cylindrical rolls, having equal diameters, are attached to each end of these tools so that they are parallel and held at a precise distance, L, apart. This distance is usually 5, 10, or 20 inches, and 100 mm or 200 mm in the metric system.

The angular relationship is obtained by the distance, L, at which the rolls are held apart and by the distance, H, at which one of the rolls is elevated above the reference surface by a stack of precision gage blocks.* Thus, in Fig. 17-1 the following simple trigonometric relationship is obtained:

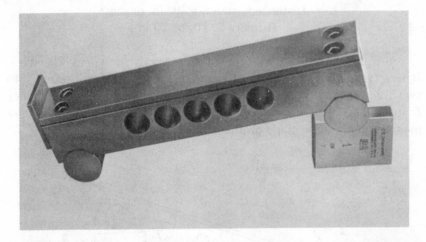

Courtesy of Taft-Pierce

Fig. 17-2. (Upper) The sine bar. (Lower) The sine plate.

---

*See Chapter 5 for a description of precision gage blocks.

$$\sin \theta = \frac{H}{L}$$

or $H = L \sin \theta$

where H = Height of gage blocks stack; in. or mm

L = Distance between rolls; in. or mm
(usually 5, 10, or 20 in.)

A somewhat different type of sine bar is shown in Fig. 17-3. It has two cylindrical plugs of equal diameter that are pressed into two holes in the sine bar which have been machined a precise distance L apart. The two plugs project out from this sine bar enabling precise measurements to be made on the projecting ends.

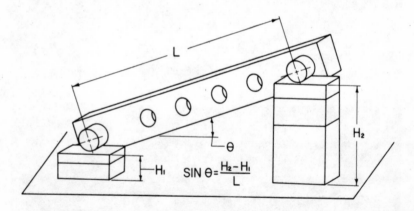

$$SIN \theta = \frac{H_2 - H_1}{L}$$

Fig. 17-3. Plug-type sine bar supported on gage blocks.

In practice, these measurements are made over the top of the plugs with a vernier height gage, usually in combination with a test type dial indicator. This sine bar may also be set at a given angle by resting the bar on the angle of the part to be measured as in Fig. 17-4. While in this position, the H_1 and H_2 dimensions are determined by using the vernier height gage set-up and measuring "over" the plugs.

In either case, the angular relationship depends upon the distance L and the difference in the heights H_1 and H_2, as shown by the simple formula:

$$\sin \theta = \frac{H_2 - H_1}{L}$$

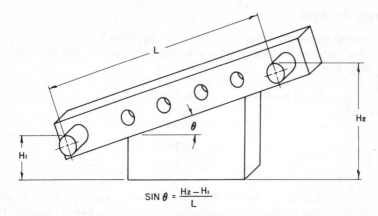

$$SIN\ \theta = \frac{H_2 - H_1}{L}$$

Fig. 17-4. Set-up for measuring angle of part with plug type sine bar.

In Fig. 17-5, a small (5-inch) sine bar is shown measuring the angle of a precision gage block scriber attachment. The sine bar is tipped to the required angle by the gage blocks placed under one of the rolls.

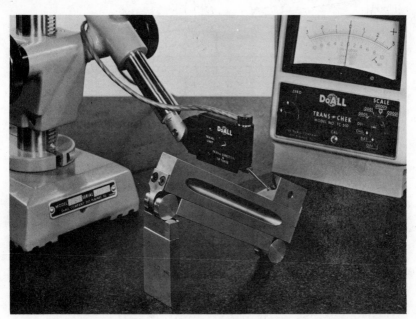

Courtesy of the DoALL Co.

Fig. 17-5. Measuring an angle on a surface plate using a sine bar, gage blocks, and an electronic height gage.

Study Examples.

1. The angle of the scriber point shown in Fig. 17-5 is to be 35°. Calculate the height of the gage block stack required to obtain this angle using a 5-in. sine bar.

Step 1. Using formula: $H = L \sin \theta$

$$H = 5 \times \sin 35°$$

$$= 5 \times 0.57358$$

$$H = 2.8679 \text{ in.}$$

2. A 10-in. sine bar, shown in Fig. 17-6, is used to measure the angularity of the taper on the shaft positioned between two bench centers. The difference in height over the buttons, $H_2 - H_1$, is measured, using a vernier height gage and a dial test indicator. Calculate the half-angle of taper, θ, when $H_2 = 6.122$ in. and $H_1 = 4.678$ in.

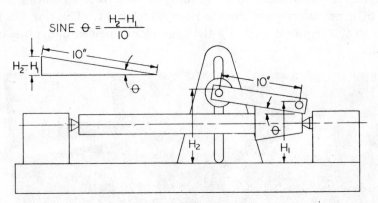

Fig. 17-6. Study example 2.

Step 1. Using the formula: $\sin \theta = \dfrac{H}{L} = \dfrac{H_2 - H_1}{10}$

Step 2. Substitute values for H_2 and H_1:

$$\sin \theta = \frac{6.122 - 4.678}{10} = \frac{1.444}{10}$$

$$\sin \theta = 0.1444$$

Half-angle $\theta = 8°18'$ (From trig table)

The included angle of the taper is:

$$2 \times \theta = 2(8°18') = 16°36'$$

Study Problems.

1. When a sine bar was in position to measure an angle, the gage block stack height was 1.99368 in. What is the angle if a 5-in. sine bar is used? A 10-in. bar?

2. In checking a taper shaft similar to the one shown in Fig. 17-6, it was found that dimension H_2 equaled 8.326 and $H_1 = 6.168$ in. Calculate the total included angle of the taper if a 10-in. sine bar was used.

3. A toolmaker requires an angle of $39°38'10''$. Using a 5-in. sine plate, what height H of gage blocks was required?

4. If the stack of gage blocks calculated in problem 3 were used with a 10-in. sine plate, what would be the angle measurement?

5. In checking an angle on a machined casting, a 5-in. sine bar was used that had buttons (rolls) fastened to the side. The lower button was 2.1672 above the surface plate and the upper button was 4.8634 above the surface plate. What is the angle measurement?

6. A 10-in. sine bar with buttons on the side has the following two measurements: lower, 5.6729; upper, 9.2173. What is the measure of the angle?

7. The measurement to the buttons of a 10-in. sine bar from the flat were taken as 3.2763 and 8.4976 in. What was the measure of the angle?

8. Determine the stack height (gage blocks to the upper button) of a 5-in. sine bar with two buttons when the lower button is 0.7500 in. above the flat and the angle is $22°22'22''$.

9. Calculate the stack height of gage blocks to set a 10-in. sine plate to an angle of $37°10'30''$.

10. The stack height for a 10-in. sine plate in measuring an angle is 3.5837 in. What is the angle measurement? If a 5-in. sine plate is used on the same stack, what is the angle measurement?

11. A 10-in. sine bar with buttons is placed on a taper shaft whose included angle of taper is $19°30'$. If the short stack measures 1.350 in., what is the height of the upper button stack?

Hole Circle Spacing

It is often necessary to space holes in a circular pattern such as around a flange of a pipe, etc. The circle on which the holes are located is called the *hole circle*. For even spacing of the holes, it is necessary to determine the chordal distance between the holes in order to "lay out" the holes. In the unique case of six, equally spaced holes, used quite often in industry, the chordal distance is equal to the radius of the hole circle. For other spacings, a trigonometric solution is necessary.

Study Examples.

1. In Fig. 17-7, twenty equally spaced holes are to be placed along a 10-in. diameter hole circle. Calculate the chordal distance, AB, between the adjacent holes.

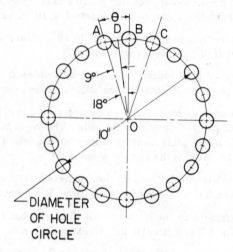

Fig. 17-7. Study example 1.

Step 1. Draw radii, AO and BO. Calculate $\angle AOB$:

$$\angle AOB = 360° \div \text{number of holes}$$

$$\angle AOB = \frac{360}{20} = 18°$$

Step 2. Bisect $\angle AOB$ with line OD:

$$\angle AOD = \frac{18}{2} = 9°$$

$$AD = \frac{AB}{2} = \frac{1}{2}AB$$

Step 3. In rt $\triangle ADO$:

$$\sin AOD = \frac{AD}{AO}$$

$$AD = AO \sin 9°$$

$$AO = \frac{1}{2}\text{ diameter} = \frac{10}{2} = 5$$

$$AD = 5 \times 0.15643 = 0.78215$$

Step 4.   Calculate chordal distance AB:

$$AB = 2AD$$

$$AB = 2 \times 0.78215 = 1.5643 \text{ in.}$$

An alternate formula for solving chordal distances of equally spaced holes along a known diameter hole circle is:

$$AB = \text{Diameter of hole circle} \times \sin\left[\frac{1}{2}\left(\frac{360^\circ}{\text{no. holes}}\right)\right]$$

or $AB = D \sin\left(\dfrac{360^\circ}{2 \times \text{no. holes}}\right)$

2. Solve study example 1 by the alternate formula:

Step 1.   Using above formula:

$$AB = D \sin\left(\frac{360^\circ}{2 \times \text{no. holes}}\right)$$

$$AB = 10 \sin\frac{360}{40}$$

$$AB = 10 \sin 9^\circ$$

$$AB = 10 \times 0.156434 = 1.5643 \text{ in.}$$

Study Problems.

Find the chordal distances between adjacent holes for the number of equally spaced holes and hole circle diameter shown in the accompanying chart:

No.	Number of Holes	Diameter Hole-Circle	No.	Number of Holes	Diameter Hole-Circle
1.	5	8-inch	10.	5	6-inch
2.	9	6-inch	11.	8	10-inch
3.	14	12-inch	12.	7	9-inch
4.	17	10-inch	13.	11	10-inch
5.	22	20-inch	14.	22	22-inch
6.	8	7-inch	15.	14	9-inch
7.	4	9-inch	16.	9	6-inch
8.	7	7-inch	17.	10	10-inch
9.	3	3-inch	18.	18	12-inch

Coordinate Distances

The precise location of holes is a problem that is frequently encountered in tool and die work. The *coordinate* dimensioning system has been developed to accurately measure the location of holes. This system is used on drawings for jig-bore work and is designed to be compatible with the coordinate axes of movement available on the jig-boring machine. For jig-boring work, the coordinates are generally referred from the intersections of lines that are tangent to the hole-circle at the *top* and at the left, as shown in Fig. 17-8. Thus, the horizontal coordinate dimensions are taken from a vertical line passing through the center of the hole at the extreme left and the vertical coordinate dimensions are taken from a horizontal line passing through the center of the hole at the top.

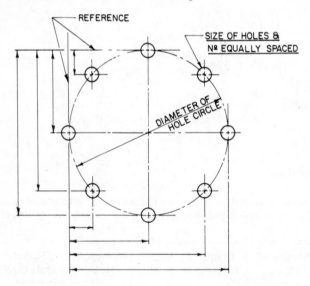

Fig. 17-8. Jig bore coordinates.

This problem is also frequently encountered on numerically controlled machines. In the system used on these machines, the table movements must be compatible with the coordinate axes of the dimensioning system. The base or reference tangents in this system are usually to the left and *below* the hole-circle as shown in Fig. 17-9.

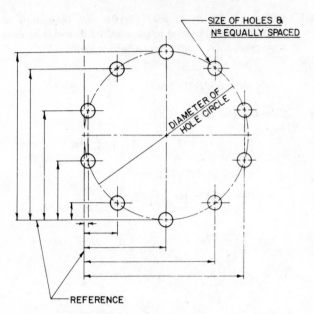

Fig. 17-9. Numerical control coordinates.

Study Examples.

1. Calculate the coordinate distances for the eight holes equally spaced around a 10-in. diameter hole-circle to be machined on a jig-borer. See Fig. 17-10A.

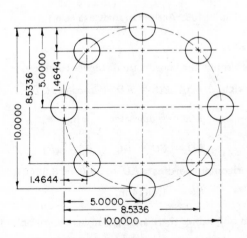

Fig. 17-10A. Answers to study example 1.

Step 1. Select a hole, A, in Fig. 17-10B and then draw the related reference lines, central angle, and the diameter as shown. Draw a perpendicular from A to OC giving point B.

$$\text{In right } \triangle O\underline{B}A:\ \angle AOB = \frac{360^{\circ}}{8} = 45^{\circ}$$

Step 2. Calculate the length of AB:

$$AB = AO \sin 45^{\circ} \qquad AO = \frac{1}{2}\text{ dia.}$$

$$AB = 5 \times 0.70711$$

$$AB = 3.5356 \text{ in.}$$

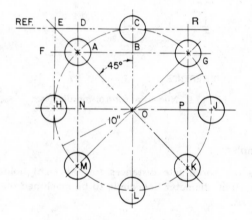

Fig. 17-10B. Analysis of study example 1.

Step 3. Determine the coordinate distances to hole A;

In right $\triangle E\underline{D}A;\ ED = AD$ (isosceles rt. $\triangle$)

$$EC = \frac{1}{2}\text{ diameter} = 5 \text{ in.}$$

$$ED = EC - DC \quad (DC = AB)$$

Horizontal coordinate: $ED = 5.00 - 3.5356$

$$ED = 1.4644 \text{ in.}$$

In this problem only, since there are eight holes on the bolt circle:

Vertical coordinate: $EF = AD = ED = 1.4644$ in.

Step 4.   Determine the coordinate distance to hole C:

In right $\triangle ECO$:

$$EC = OC = \frac{1}{2}\text{ diameter}$$

also $OC = EH = \frac{1}{2}\text{ diameter}$

Horizontal coordinate:

$$EC = EH = 5.000 \text{ in.}$$

Vertical coordinate in this case is zero (0).

Step 5.   Determine the coordinate distances to hole G:

In right $\triangle AOG$;

$$BG = AB = 3.5356 \text{ in.}$$

$$FG = BG + FB \quad (FB = \frac{1}{2}\text{ dia.})$$

Horizontal coordinate:

$$FG = 5.000 + 3.5356$$
$$FG = 8.5356$$

Vertical coordinate:

$$EF = 1.4644 \text{ in.}$$

Step 6.   Determine the coordinate distances to hole J:

Horizontal coordinate:

$$HJ = \text{diameter} = 10.000 \text{ in.}$$

Vertical coordinate:

$$EH = \frac{1}{2}\text{ diameter} = 5.000 \text{ in.}$$

Step 7.   Determine the coordinate distances to hole H:

Horizontal coordinate $= 0$

Vertical coordinate:

$$EH = \frac{1}{2}\text{ diameter} = 5.000 \text{ in.}$$

Step 8.   Determine the coordinate distances to hole K:

$$KP = PG = BG = AB = 3.5356 \text{ in.}$$

Horizontal coordinate:

$$FG = 5.0000 + 3.5356 = 8.5356 \text{ in.}$$

Vertical coordinate:

$$RK = 5.0000 + 3.5356 = 8.5356 \text{ in.}$$

Step 9.   Determine the coordinate distances to hole L:

Horizontal coordiante:

$\dfrac{1}{2}$ diameter = 5.000 in.

Vertical coordinate:

diameter = 10.000 in.

Step 10.   Determine the coordinate distances to hole M:

Horizontal distance:

ED = 1.4644 in.

Vertical distance:

$$DM = KP + \dfrac{1}{2}\text{ diameter}$$

$$DM = 3.5356 + 5.000 = 8.5356 \text{ in.}$$

2. Calculate the coordinate distances for the ten equally spaced holes on an 8-in. hole-circle to be machined on a numerically controlled (NC) boring machine as shown in Fig. 17-11.

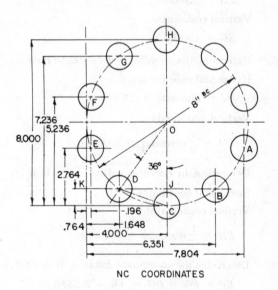

NC   COORDINATES

Fig. 17-11. Study example 2.

Step 1.   Select two adjacent holes, C and D, and construct related reference lines, central angle, and given dimensions.

Step 2. In right triangle OJD, solve for length of DJ;

$$DJ = OD \sin DOJ$$

$$OD = \frac{1}{2} \text{ diameter} = \frac{8}{2} = 4$$

$$\angle DOJ = \frac{360^\circ}{10} = 36^\circ$$

$$DJ = 4 \times \sin 36^\circ = 4 \times 0.58778$$

$$DJ = 2.3511 \text{ in.}$$

Step 3. Calculate length of line KD;

$$KD = JK - DJ$$

$$KD = 4.00 - 2.3511$$

$$KD = 1.6489 \text{ in. (horizontal coordinate)}$$

Step 4. Calculate vertical coordinate distance CJ.

In right $\triangle CJD$, $\angle JCD = \frac{1}{2}(180^\circ - 36^\circ) = 72^\circ$

$$CJ = \frac{DJ}{\tan 72^\circ} = \frac{2.3511}{3.0777} =$$

$$CJ = 0.7639 \text{ in. (Vertical coordinate)}$$

Continue solution for the remaining holes. See Fig. 17-11.

When an odd number of holes is involved, or when the holes are not spaced symmetrically with respect to the reference axes, a separate X-Y coordinate calculation is necessary for each hole.

Study Problems.

Solve for the "jig-bore" coordinate distances in the following problems for the number of equally spaced holes and the hole-circle diameters shown:

1. Coordinate dimensions for five holes on an 8-in. diameter hole-circle.

2. Coordinate dimensions for ten holes on a 12-in. diameter hole-circle.

3. Coordinate dimensions for six holes on an 8-in. diameter hole-circle.

4. Coordinate dimensions for seven holes on a 10-in. diameter hole-circle.

5. Coordinate dimensions for twelve holes on a 12-in. diameter hole-circle.

Solve for the numerical control coordinate distances in the following problems for the number of equally spaced holes indicated for the specified hole-circle diameters (one or more holes on a center line):

6. Seven holes on a 10-in. diameter hole-circle.

7. Three holes on a 5-in. diameter hole-circle.

8. Six holes on an 8-in. diameter hole-circle.

9. Twelve holes on a 12-in. diameter hole-circle.

10. Ten holes on a 6-in. diameter hole-circle.

Solving Practical Shop Problems

The solutions of the following problems are based on the use of geometric construction and trigonometric calculations. In general, the solution for linear dimensions can be approached by the construction of lines parallel to the linear dimension to be solved for: lines parallel to lines in the figure of the problem, lines connecting the centers of circles or arcs, and/or lines drawn perpendicular to known or constructed lines.

When problems have rolls (circles) shown contacting two lines in a "V" construction, remember that a line drawn from the center of the circle to the apex of the angle of the "V" bisects that angle. Also radii drawn to the point of tangency are perpendicular to that side of the "V" at the point of tangency.

Study Examples.

1. The included angle of the "V" block is $90°$. Solve for dimension M when $L = 1.2500$ and the roll diameter is 1.7500. See Fig. 17-12.

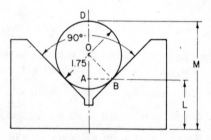

Fig. 17-12. Study example 1.

Step 1. Draw line DA through center, O, of roll and perpendicular to the base.

Step 2. Draw line OB perpendicular to side of "V." Draw AB parallel to the base.

Step 3. In right $\triangle OAB$:

$$\angle OBA = \angle AOB = 45°$$

$$OA = OB \sin 45°$$

$$OB = \frac{1}{2} \text{ roll diameter} = \frac{1.7500}{2} = 0.8750$$

$$OA = 0.8750 \times 0.70711$$

$$OA = 0.6187$$

Step 4. Dimension $M = DO + OA + L$

$$M = 0.8750 + 0.6187 + 1.2500$$

$$M = 2.7437$$

2. A detail shown in Fig. 17-13 is being checked for accuracy of angular displacement of the two holes using the tooling-ball and pin technique. The top of the ball and the top surface of the pin are measured from a surface plate. Dimension M is the difference of the two measurements. What should dimension M be if the angle between the two holes is exactly $20°$?

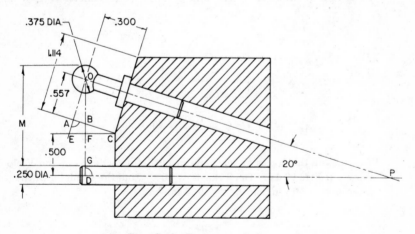

Fig. 17-13. Study example 2.

Step 1. Draw line $OD \perp$ to center line DP and $OE \perp$ to center line OP. In rt. $\triangle EAC$: $AC = 0.300$ and $\angle ACE = 20°$. Solve for AE:

$$AE = AC \tan 20°$$

$$AE = 0.300 \times 0.36397$$

$$AE = 0.10919$$

Step 2. In rt. $\triangle OFE$: $AO = 0.557$ and $\angle AOF = 20°$. Solve for OF:

$$OF = OE \cos 20°$$

$$OE = AO + AE = 0.557 + 0.10919$$

$$OE = 0.66619$$

$$OF = 0.66619 \times 0.93969$$

$$OF = 0.6260$$

Step 3. Distance $M = \dfrac{\text{ball dia.}}{2} + OF + FG$ $\left(FG = 0.500 - \dfrac{0.250}{2}\right)$

$$M = 0.1875 + 0.6260 + 0.3750$$

$$M = 1.1885 \text{ or } 1.188 \text{ to three decimal places.}$$

Study Problems.

Practical shop problems using trigonometry in their solution:

1. The included angle of the V-block shown in Fig. 17-14 is $90°$. Solve for dimension M when $L = 2.45$ mm and the roll diameter is 1.75 mm.

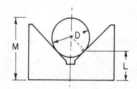

Fig. 17-14. Study problem 1.

2. The top of the roll is at the same height as the side, K, of the V-block shown in Fig. 17-15. The V angle is $90°$. Find the diameter of the roll when $L = 1.563$ in. and $K = 2.575$ in.

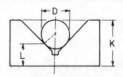

Fig. 17-15. Study problem 2.

3. In the drill jig shown in Fig. 17-16, the large circle, D_1, is the diameter of the part to be drilled. D_2 is the diameter of the setting roll. Dimension L locates the center of the drill bushing from the locating surface.

The V angle is $90°$. Determine M so that the center of the part will be on the center of the jig bushing. $D_1 = 3.75$ in., $D_2 = 1.00$ in., $L = 4.25$ in.

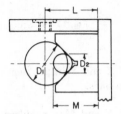

Fig. 17-16. Study problem 3.

4. An Acme thread is shown in Fig. 17-17. Solve for dimension A.

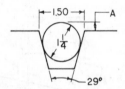

Fig. 17-17. Study problem 4.

5. An 8.000-in. disc is positioned vertically on a surface plate as shown in Fig. 17-18. The vertical distance to the hole is measured by a height gage to be 1.625 in. Calculate angle α.

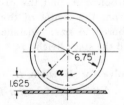

Fig. 17-18. Study problem 5.

6. Dovetails are common in machine design. The usual method of measurement is to measure across a pair of rolls and compare the measurement to a known dimension such as X. Calculate the dimensions X and Y when the angle of the dovetail $= 60°$ and the rolls are 0.75 in. diameter. See Fig. 17-19.

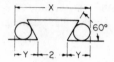

Fig. 17-19. Study problem 6.

7. Find the values of X and Y in problem 6 if 0.500-in. rolls were used in place of the 0.75-in ones.

8. Find the value of X and Y in problem 6 if 0.625-in. rolls were used.

9. A workpiece is placed on a simple sine plate as shown in Fig. 17-20. Calculate the stack height H of required gage blocks to raise the upper plate to an angle of $31°50'$, when $L = 100$ mm.

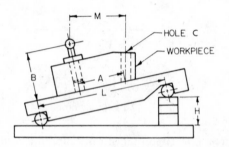

Fig. 17-20. Study problem 9.

10. In the set-up in problem 9, find the distance M that a jig-borer table must be moved from the center of the tooling ball to drill hole C. Hole C is to be bored to size. Distance $L = 5$ in., $A = 1.125$ in., $B = 1.1875$ in., and the stack height $= 1.4201$ in.

11. In problem 10, solve for M when $L = 127$ mm, $H = 82.672$ mm, $A = 28.575$ mm, and $B = 30.162$ mm. Give answer to 0.001 mm.

12. The L distance on a metric sine bar is 100 mm. Find the angle of a set-up when the gage block stack equals 14.925 mm.

13. An angle of $47°32'47''$ is to be set up on a sine bar whose L dimension is 200 mm. Calculate the stack height of the required gage-blocks.

14. Solve for dimension M in Fig. 17-21. Dimensions are in mm.

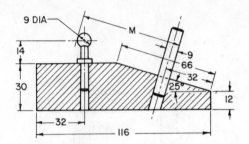

Fig. 17-21. Study problem 14.

15. Solve for angle β in Fig. 17-22. Dimensions are in mm.

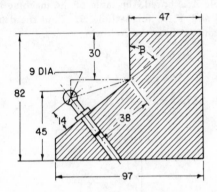

Fig. 17-22. Study problem 15.

16. Solve for dimension M in Fig. 17-23. Dimensions are in mm.

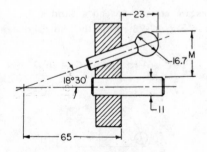

Fig. 17-23. Study problem 16.

17. Solve for dimension M in Fig. 17-24. Dimensions are in mm.

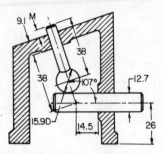

Fig. 17-24. Study problem 17.

18. A plate is to have two holes bored on a boring machine. See Fig. 17-25. After hole A is bored, the table of the machine is moved in two directions, X and Y, to locate hole B. Find the distances X and Y when line AB and $\angle A$ have the following values.

a. $AB = $ 8.75 in. $\angle A = 45°$

b. $AB = $ 13.125 in. $\angle A = 9°$

c. $AB = 237.45$ mm $\angle A = 66°54'$

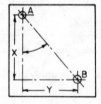

Fig. 17-25. Study problem 18.

19. Nine holes are equally spaced around a 7.125-in. diameter bolt-circle as shown in Fig. 17-26. Find the chordal distance between the following holes:

A to B; A to C; A to D.

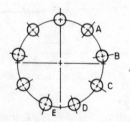

Fig. 17-26. Study problem 19.

20. Seven holes are equally spaced on a 114.30 mm bolt-circle as shown in Fig. 17-27. They are to be checked by measuring the distances AB and AC. Determine these distances.

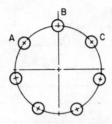

Fig. 17-27. Study problem 20.

21. Five equally spaced holes are on a 6.435 in. bolt-circle as shown in Fig. 17-28. Find the chordal distance from hole A to hole B and from hole A to hole C.

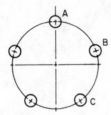

Fig. 17-28. Study problem 21.

22. Solve for dimensions X and Y, as seen in Fig. 17-29.

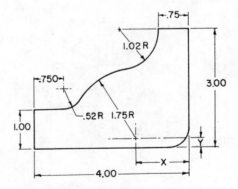

Fig. 17-29. Study problem 22.

23. Solve for distances X and Y, as seen in Fig. 17-30.

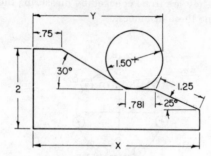

Fig. 17-30. Study problem 23.

24. Solve for dimension D in Fig. 17-31.

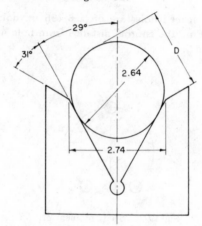

Fig. 17-31. Study problem 24.

25. Solve for A in Fig. 17-32.

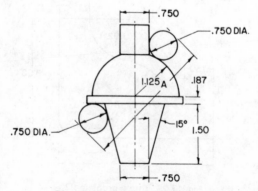

Fig. 17-32. Study problem 25.

26. Solve for dimension X in Fig. 17-33.

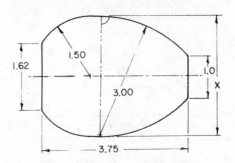

Fig. 17-33. Study problem 26.

27. Solve for dimensions X and Y in Fig. 17-34.

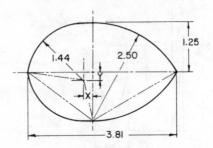

Fig. 17-34. Study problem 27.

28. Solve for distance X in Fig. 17-35.

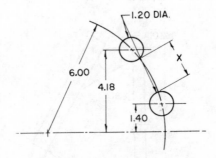

Fig. 17-35. Study problem 28.

29. Solve for dimensions X, Z, and radius Y in Fig. 17-36.

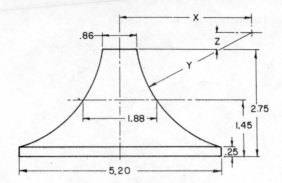

Fig. 17-36. Study problem 29.

30. Solve for dimension X in Fig. 17-37.

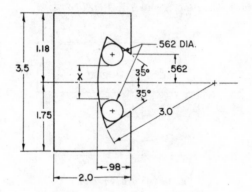

Fig. 17-37. Study problem 30.

31. Solve for distance X in Fig. 17-38.

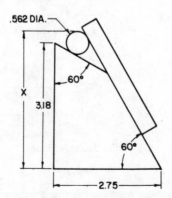

Fig. 17-38. Study problem 31.

32. Solve for dimensions A, B, and C in Fig. 17-39.

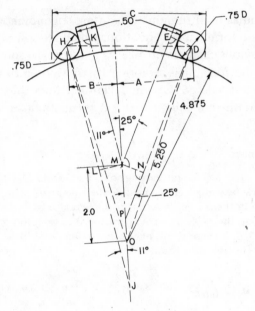

Fig. 17-39. Study problem 32.

33. A typical construction hole and tooling ball set-up is shown in Fig. 17-40. Solve for dimension M_1 and M_2 when $\angle A = 117°$ and dimension $B = 2.264$ in.

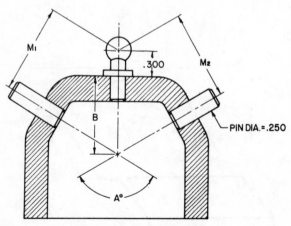

Fig. 17-40. Study problem 34.

34. In the above figure, solve for M_1 and M_2 when $\angle A = 121°$ and dimension $B = 51.275$ mm. (Convert height of tooling ball, 0.300 in. to millimeters.) Answers to be to 0.01 mm.

Trigonometric Shop Formulas

Many formulas used in industry contain trigonometric functions. Essentially, these formulas are equations which are solved by methods given in the previous chapters on algebra and trig. To solve the following problems, the formula is algebraically transposed to place the unknown on the left side of the equal sign and then the known values for the literal terms are substituted. Trig functions become numerical terms.

Study Examples.

1. The face of a single-point cutting tool is shown in Fig. 17-41. Angle α is the back rake angle and angle β is the side rake angle. The angle ϕ is any angle in the XY plane and ρ is the actual rake angle on the face of the tool at the angle ϕ. There will be one value of the angle ϕ, designated ϕ_{max}, where ρ will have a maximum value. These angles can be found by the following formulas:

$$\tan \phi_{max} = \frac{\tan \beta}{\tan \alpha} \text{ ; and: } \tan \rho = \sin \phi \tan \beta + \cos \phi \tan \alpha$$

Find the maximum rake angle when the back rake angle is $5°$ and the side rake angle is $15°$.

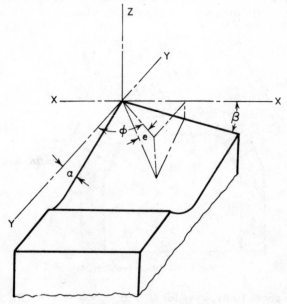

Fig. 17-41. Study example 1.

Step 1. Solving for ϕ_{max}:

$$\tan \phi_{max} = \frac{\tan \beta}{\tan \alpha} = \frac{\tan 15°}{\tan 5°} = \frac{0.26795}{0.08749} = 3.06264$$

$$\angle\phi = 71°55'2''$$

Step 2. Solving for angle ρ:

$$\tan \rho = \sin \phi \tan \beta + \cos \phi \tan \alpha$$

$$\tan \rho = (\sin 71°55'2'')(\tan 15°) + (\cos 71°55'2'')(\tan 5°)$$

$$\tan \rho = (0.95061)(0.26795) + (0.31039)(0.08749)$$

$$= 0.25472 + 0.02716$$

$$= 0.28188$$

$$\angle\rho = 15°44'31''$$

2. The pitch diameter, d, of a worm can be calculated by the following formula:

$$d = \frac{L \cot a}{\pi}$$

where: L is the lead of the worm
$\angle a$ is the lead angle.

Calculate the pitch diameter of a worm having a lead of 0.750 in. and a lead angle equal to $10°$.

$$d = \frac{L \cot a}{\pi} = \frac{0.750 \cot 10°}{3.1416} = \frac{(0.750)(5.67128)}{3.1416}$$

$$d = \frac{4.25346}{3.1416} = 1.3539 \text{ in.}$$

Study Problems.

1. Using the formula given in study example 1, calculate the rake angle at $45°$ from the side cutting edge ($\phi = 45°$) when the back rake angle and the side rake angle are both $10°$.

2. Calculate the maximum rake angle of a single-point cutting tool when the back rake angle and the side rake angle are both $10°$.

3. The efficiency of a worm gear drive, E, can be calculated by the following formula:

$$E = \frac{\cos \phi - f \tan \theta}{\cos \phi + f \cotan \theta}$$

where: ϕ = Normal pressure angle

θ = Lead angle of the worm

f = Coefficient of friction

Calculate the efficiency of a worm gear drive when:

$$\phi = 20°, \theta = 15°, \text{ and } f = 0.05$$

4. A formula for checking the accuracy of threads by the three-wire method is given in *Machinery's Handbook* as;

$$M = E - T \cot A + W (1 + \csc A)$$

where: M = Dimension over the wires, inches (by micrometer)

E = Pitch dia. of thread to be checked, inches

$T = \dfrac{1}{2}$ pitch or width of thread at dia. E, inches

$A = \dfrac{1}{2}$ included thread angle in axial plane.

W = Dia. of wires used in measurements, inch.

Solve for M when: $E = 0.450, T = 0.03846, A = 30°$, and $W = 0.07013$ in.

5. Calculate the transverse offset required to cut a circular flute in a $\frac{1}{2}$ in., 13 threads per inch tap using a circular form milling cutter having a radius of 0.150 in. The tap is to be made with a tangential hook angle of 5° Use the formula:

$$X = \frac{D}{2} (1 - \sin \phi) + r[1 - \cos (\phi - \alpha)]$$

where: X = Tangential offset

D = Actual outside dia. of the tap

r = Radius of form milling cutter.

ϕ = One-half of angle between flutes, in this case, 45°

α = Tangential hook angle.

6. The wire diameter for obtaining pitch-line contact at the back of a buttress thread may be determined by the following formula:

where: $P =$ Pitch of the thread $\left(\dfrac{1}{\text{No. thds./in.}}\right)$

$A =$ Included angle of thread

$a =$ Angle of front face

$W =$ Diameter of wire.

Solve for the wire diameter, W, when $P = 0.0625$, $a = 7°$, and $A = 52°$.

7. In gearing work it is sometimes necessary to check involute teeth by measuring the distance across two or more teeth. This distance is related to the pressure angle of the gear, the pitch radius, tooth thickness and the number of teeth on the gear. The checking formula is given as:

$$M = R \times \cos A \times \left(\frac{T}{R} + \frac{6.2832 \times S}{N} + F\right)$$

where: $N =$ Number of teeth on the gear

$R =$ Pitch radius of gear

$A =$ Pressure angle of gear

$T =$ Tooth thickness along pitch circle

$S =$ Number of tooth spaces within caliper jaws.

$F =$ Factor relating to pressure angle; .01109 for $14\frac{1}{2}°$

Determine the distance M over three teeth of a spur gear having a pressure angle of $14°30'$ with 30 teeth of 6 diametral pitch. Pitch radius $= 2.5$ in. and T for 6 diametral pitch is 0.2618 in. (no allowance for backlash). $F = 0.01109$.

Appendix

Multiply	By	To Obtain
LENGTH		
centimetre	0.03280840	foot
centimetre	0.3937008	inch
fathom	1.8288*	metre (m)
foot	0.3048*	metre (m)
foot	30.48*	centimetre (cm)
foot	304.8*	millimetre (mm)
inch	0.0254*	metre (m)
inch	2.54*	centimetre (cm)
inch	25.4*	millimetre (mm)
kilometre	0.6213712	mile [U. S. statute]
metre	39.37008	inch
metre	0.5468066	fathom
metre	3.280840	foot
metre	0.1988388	rod
metre	1.093613	yard
metre	0.0006213712	mile [U. S. statute]
microinch	0.0254*	micrometre [micron] (μm)
micrometre [micron]	39.37008	microinch
mile [U. S. statute]	1609.344*	metre (m)
mile [U. S. statute]	1.609344*	kilometre (km)
millimetre	0.003280840	foot
millimetre	0.03937008	inch
rod	5.0292*	metre (m)
yard	0.9144*	metre (m)
AREA		
acre	4046.856	metre^2 (m^2)
acre	0.4046856	hectare
centimetre^2	0.1550003	inch^2
centimetre^2	0.001076391	foot^2
foot^2	0.09290304*	metre^2 (m^2)
foot^2	929.0304*	centimetre^2 (cm^2)
foot^2	92,903.04*	millimetre^2 (mm^2)
hectare	2.471054	acre
inch^2	645.16*	millimetre^2 (mm^2)
inch^2	6.4516*	centimetre^2 (cm^2)
inch^2	0.00064516*	metre^2 (m^2)
metre^2	1550.003	inch^2
metre^2	10.763910	foot^2
metre^2	1.195990	yard^2
metre^2	0.0002471054	acre
millimetre^2	0.0001076387	foot^2
millimetre^2	0.001550003	inch^2
yard^2	0.8361274	metre^2 (m^2)

* Where an asterisk is shown, the figure is exact.

Metric Conversion Factors (*Continued*)

Multiply	By	To Obtain
VOLUME (including CAPACITY)		
centimetre³	0.06102376	inch³
foot³	0.02831685	metre³ (m³)
foot³	28.31685	litre
gallon [U. K. liquid]	0.004546092	metre³ (m³)
gallon [U. K. liquid]	4.546092	litre
gallon [U. S. liquid]	0.003785412	metre³ (m³)
gallon [U. S. liquid]	3.785412	litre
inch³	16,387.06	millimetre³ (mm³)
inch³	16.38706	centimetre³ (cm³)
inch³	0.00001638706	metre³ (m³)
litre	0.001*	metre³ (m³)
litre	0.2199692	gallon [U. K. liquid]
litre	0.2641720	gallon [U. S. liquid]
litre	0.03531466	foot³
metre³	219.9692	gallon [U. K. liquid]
metre³	264.1720	gallon [U. S. liquid]
metre³	35.31466	foot³
metre³	1.307951	yard³
metre³	1000.*	litre
metre³	61,023.76	inch³
millimetre³	0.00006102376	inch³
yard³	0.7645549	metre³ (m³)
PRESSURE and STRESS		
bar	100,000.*	pascal (Pa)
bar	14.50377	pound/inch²
bar	100,000.*	newton/metre² (N/m²)
kilogram/centimetre²	14.22334	pound/inch²
kilogram/metre²	9.806650*	newton/metre² (N/m²)
kilogram/metre²	9.806650*	pascal (Pa)
kilogram/metre²	0.2048161	pound/foot²
kilonewton/metre²	0.1450377	pound/inch²
newton/centimetre²	1.450377	pound/inch²
newton/metre²	0.00001*	bar
newton/metre²	1.0*	pascal (Pa)
newton/metre²	0.0001450377	pound/inch²
newton/metre²	0.1019716	kilogram/metre²
newton/millimetre²	145.0377	pound/inch²
pascal	0.00001*	bar
pascal	0.1019716	kilogram/metre²
pascal	1.0*	newton/metre² (N/m²)
pascal	0.02088543	pound/foot²
pascal	0.0001450377	pound/inch²
pound/foot²	4.882429	kilogram/metre²
pound/foot²	47.88026	pascal (Pa)
pound/inch²	0.06894757	bar
pound/inch²	0.07030697	kilogram/centimetre²
pound/inch²	0.6894757	newton/centimetre²
pound/inch²	6.894757	kilonewton/metre²
pound/inch²	6894.757	newton/metre² (N/m²)
pound/inch²	0.006894757	newton/millimetre² (N/mm²)
pound/inch²	6894.757	pascal (Pa)

* Where an asterisk is shown, the figure is exact.

0° Natural Trigonometric Functions 179°

M	Sine	Cosine	Tan.	Cotan.	Secant	Cosec.	M
0	0.00000	1.0000	0.00000	Infinite	1.0000	Infinite	60
1	.00029	.0000	.00029	3437.7	.0000	3437.7	59
2	.00058	.0000	.00058	1718.9	.0000	1718.9	58
3	.00087	.0000	.00087	1145.9	.0000	1145.9	57
4	.00116	.0000	.00116	859.44	.0000	859.44	56
5	0.00145	1.0000	0.00145	687.55	1.0000	687.55	55
6	.00174	.0000	.00174	572.96	.0000	572.96	54
7	.00204	.0000	.00204	491.11	.0000	491.11	53
8	.00233	.0000	.00233	429.72	.0000	429.72	52
9	.00262	.0000	.00262	381.97	.0000	381.97	51
10	0.00291	0.99999	0.00291	343.77	1.0000	343.77	50
11	.00320	.99999	.00320	312.52	.0000	312.52	49
12	.00349	.99999	.00349	286.48	.0000	286.48	48
13	.00378	.99999	.00378	264.44	.0000	264.44	47
14	.00407	.99999	.00407	245.55	.0000	245.55	46
15	0.00436	0.99999	0.00436	229.18	1.0000	229.18	45
16	.00465	.99999	.00465	214.86	.0000	214.86	44
17	.00494	.99999	.00494	202.22	.0000	202.22	43
18	.00524	.99999	.00524	190.98	.0000	190.99	42
19	.00553	.99998	.00553	180.93	.0000	180.93	41
20	0.00582	0.99998	0.00582	171.88	1.0000	171.89	40
21	.00611	.99998	.00611	163.70	.0000	163.70	39
22	.00640	.99998	.00640	156.26	.0000	156.26	38
23	.00669	.99998	.00669	149.46	.0000	149.47	37
24	.00698	.99997	.00698	143.24	.0000	143.24	36
25	0.00727	0.99997	0.00727	137.51	1.0000	137.51	35
26	.00756	.99997	.00756	132.22	.0000	132.22	34
27	.00785	.99997	.00785	127.32	.0000	127.32	33
28	.00814	.99997	.00814	122.77	.0000	122.78	32
29	.00843	.99996	.00844	118.54	.0000	118.54	31
30	0.00873	0.99996	0.00873	114.59	1.0000	114.59	30
31	.00902	.99996	.00902	110.89	.0000	110.90	29
32	.00931	.99996	.00931	107.43	.0000	107.43	28
33	.00960	.99995	.00960	104.17	.0000	104.17	27
34	.00989	.99995	.00989	101.11	.0000	101.11	26
35	0.01018	0.99995	0.01018	98.218	1.0000	98.223	25
36	.01047	.99994	.01047	95.489	.0000	95.495	24
37	.01076	.99994	.01076	92.908	.0000	92.914	23
38	.01105	.99994	.01105	90.463	.0001	90.469	22
39	.01134	.99993	.01134	88.143	.0001	88.149	21
40	0.01163	0.99993	0.01164	85.940	1.0001	85.946	20
41	.01193	.99993	.01193	83.843	.0001	83.849	19
42	.01222	.99992	.01222	81.847	.0001	81.853	18
43	.01251	.99992	.01251	79.943	.0001	79.950	17
44	.01280	.99992	.01280	78.126	.0001	78.133	16
45	0.01309	0.99991	0.01309	76.390	1.0001	76.396	15
46	.01338	.99991	.01338	74.729	.0001	74.736	14
47	.01367	.99991	.01367	73.139	.0001	73.146	13
48	.01396	.99990	.01396	71.615	.0001	71.622	12
49	.01425	.99990	.01425	70.153	.0001	70.160	11
50	0.01454	0.99989	0.01454	68.750	1.0001	68.757	10
51	.01483	.99989	.01484	67.402	.0001	67.409	9
52	.01512	.99988	.01513	66.105	.0001	66.113	8
53	.01542	.99988	.01542	64.858	.0001	64.866	7
54	.01571	.99988	.01571	63.657	.0001	63.664	6
55	0.01600	0.99987	0.01600	62.499	1.0001	62.507	5
56	.01629	.99987	.01629	61.383	.0001	61.391	4
57	.01658	.99987	.01658	60.306	.0001	60.314	3
58	.01687	.99986	.01687	59.266	.0001	59.274	2
59	.01716	.99985	.01716	58.261	.0001	58.270	1
60	0.01745	0.99985	0.01745	57.290	1.0001	57.299	0
M	Cosine	Sine	Cotan.	Tan.	Cosec.	Secant	M

90° 89°

1° Natural Trigonometric Functions 178°

M	Sine	Cosine	Tan.	Cotan.	Secant	Cosec.	M
0	0.01745	0.99985	0.01745	57.290	1.0001	57.299	60
1	.01774	.99984	.01775	56.350	.0001	56.359	59
2	.01803	.99984	.01804	55.441	.0001	55.450	58
3	.01832	.99983	.01833	54.561	.0002	54.570	57
4	.01861	.99983	.01862	53.708	.0002	53.718	56
5	0.01891	0.99982	0.01891	52.882	1.0002	52.891	55
6	.01920	.99981	.01920	52.081	.0002	52.090	54
7	.01949	.99981	.01949	51.303	.0002	51.313	53
8	.01978	.99980	.01978	50.548	.0002	50.558	52
9	.02007	.99980	.02007	49.816	.0002	49.826	51
10	0.02036	0.99979	0.02036	49.104	1.0002	49.114	50
11	.02065	.99979	.02066	48.412	.0002	48.422	49
12	.02094	.99978	.02095	47.739	.0002	47.750	48
13	.02123	.99977	.02124	47.085	.0002	47.096	47
14	.02152	.99977	.02153	46.449	.0002	46.460	46
15	0.02181	0.99976	0.02182	45.829	1.0002	45.840	45
16	.02210	.99975	.02211	45.226	.0002	45.237	44
17	.02240	.99975	.02240	44.638	.0002	44.650	43
18	.02269	.99974	.02269	44.066	.0002	44.077	42
19	.02298	.99974	.02298	43.508	.0003	43.520	41
20	0.02326	0.99973	0.02327	42.964	1.0003	42.976	40
21	.02356	.99972	.02357	42.433	.0003	42.445	39
22	.02385	.99971	.02386	41.916	.0003	41.928	38
23	.02414	.99971	.02415	41.410	.0003	41.423	37
24	.02443	.99970	.02444	40.917	.0003	40.930	36
25	0.02472	0.99969	0.02473	40.436	1.0003	40.448	35
26	.02501	.99969	.02502	39.965	.0003	39.978	34
27	.02530	.99968	.02531	39.506	.0003	39.518	33
28	.02559	.99967	.02560	39.057	.0003	39.069	32
29	.02589	.99966	.02589	38.618	.0003	38.631	31
30	0.02618	0.99966	0.02618	38.188	1.0003	38.201	30
31	.02647	.99965	.02648	37.769	.0003	37.782	29
32	.02676	.99964	.02677	37.358	.0003	37.371	28
33	.02705	.99963	.02706	36.956	.0004	36.969	27
34	.02734	.99963	.02735	36.563	.0004	36.576	26
35	0.02763	0.99962	0.02764	36.177	1.0004	36.191	25
36	.02792	.99961	.02793	35.800	.0004	35.814	24
37	.02821	.99960	.02822	35.431	.0004	35.445	23
38	.02850	.99959	.02851	35.069	.0004	35.084	22
39	.02879	.99958	.02881	34.715	.0004	34.729	21
40	0.02908	0.99958	0.02910	34.368	1.0004	34.382	20
41	.02937	.99957	.02939	34.027	.0004	34.042	19
42	.02967	.99956	.02968	33.693	.0004	33.708	18
43	.02996	.99955	.02997	33.366	.0004	33.381	17
44	.03025	.99954	.03026	33.045	.0004	33.060	16
45	0.03054	0.99953	0.03055	32.730	1.0005	32.745	15
46	.03083	.99952	.03084	32.421	.0005	32.437	14
47	.03112	.99951	.03113	32.118	.0005	32.134	13
48	.03141	.99951	.03143	31.820	.0005	31.836	12
49	.03170	.99950	.03172	31.528	.0005	31.544	11
50	0.03199	0.99949	0.03201	31.241	1.0005	31.257	10
51	.03228	.99948	.03230	30.960	.0005	30.976	9
52	.03257	.99947	.03259	30.683	.0005	30.699	8
53	.03286	.99946	.03288	30.411	.0005	30.428	7
54	.03315	.99945	.03317	30.145	.0005	30.161	6
55	0.03344	0.99944	0.03346	29.882	1.0005	29.899	5
56	.03374	.99943	.03375	29.624	.0006	29.641	4
57	.03403	.99942	.03405	29.371	.0006	29.388	3
58	.03432	.99941	.03434	29.122	.0006	29.139	2
59	.03461	.99940	.03463	28.877	.0006	28.894	1
60	0.03490	0.99939	0.03492	28.636	1.0006	28.654	0
M	Cosine	Sine	Cotan.	Tan.	Cosec.	Secant	M

91° 88°

2° Natural Trigonometric Functions 177°

M	Sine	Cosine	Tan.	Cotan.	Secant	Cosec.	M
0	0.03490	0.99939	0.03492	28.636	1.0006	28.654	60
1	.03519	.99938	.03521	28.399	.0006	28.417	59
2	.03548	.99937	.03550	28.166	.0006	28.184	58
3	.03577	.99936	.03579	27.937	.0006	27.955	57
4	.03606	.99935	.03608	27.712	.0006	27.730	56
5	0.03635	0.99934	0.03638	27.490	1.0007	27.508	55
6	.03664	.99933	.03667	27.271	.0007	27.290	54
7	.03693	.99932	.03696	27.056	.0007	27.075	53
8	.03722	.99931	.03725	26.845	.0007	26.864	52
9	.03751	.99930	.03754	26.637	.0007	26.655	51
10	0.03781	0.99928	0.03783	26.432	1.0007	26.450	50
11	.03810	.99927	.03812	26.230	.0007	26.249	49
12	.03839	.99926	.03842	26.031	.0007	26.050	48
13	.03868	.99925	.03871	25.835	.0007	25.854	47
14	.03897	.99924	.03900	25.642	.0008	25.661	46
15	0.03926	0.99923	0.03929	25.452	1.0008	25.471	45
16	.03955	.99922	.03958	25.264	.0008	25.284	44
17	.03984	.99921	.03987	25.080	.0008	25.100	43
18	.04013	.99919	.04016	24.898	.0008	24.918	42
19	.04042	.99918	.04045	24.718	.0008	24.739	41
20	0.04071	0.99917	0.04075	24.542	1.0008	24.562	40
21	.04100	.99916	.04104	24.367	.0008	24.388	39
22	.04129	.99915	.04133	24.196	.0008	24.216	38
23	.04158	.99913	.04162	24.026	.0009	24.047	37
24	.04187	.99912	.04191	23.859	.0009	23.880	36
25	0.04217	0.99911	0.04220	23.694	1.0009	23.716	35
26	.04246	.99910	.04249	23.532	.0009	23.553	34
27	.04275	.99908	.04279	23.372	.0009	23.393	33
28	.04304	.99907	.04308	23.214	.0009	23.235	32
29	.04333	.99906	.04337	23.058	.0009	23.079	31
30	0.04362	0.99905	0.04366	22.904	1.0009	22.925	30
31	.04391	.99903	.04395	22.752	.0010	22.774	29
32	.04420	.99902	.04424	22.602	.0010	22.624	28
33	.04449	.99901	.04453	22.454	.0010	22.476	27
34	.04478	.99900	.04483	22.308	.0010	22.330	26
35	0.04507	0.99898	0.04512	22.164	1.0010	22.186	25
36	.04536	.99897	.04541	22.022	.0010	22.044	24
37	.04565	.99896	.04570	21.881	.0010	21.904	23
38	.04594	.99894	.04599	21.742	.0010	21.765	22
39	.04623	.99893	.04628	21.606	.0011	21.629	21
40	0.04652	0.99892	0.04657	21.470	1.0011	21.494	20
41	.04681	.99890	.04687	21.337	.0011	21.360	19
42	.04711	.99889	.04716	21.205	.0011	21.228	18
43	.04740	.99888	.04745	21.075	.0011	21.098	17
44	.04769	.99886	.04774	20.946	.0011	20.970	16
45	0.04798	0.99885	0.04803	20.819	1.0011	20.843	15
46	.04827	.99883	.04832	20.693	.0012	20.717	14
47	.04856	.99882	.04862	20.569	.0012	20.593	13
48	.04885	.99881	.04891	20.446	.0012	20.471	12
49	.04914	.99879	.04920	20.325	.0012	20.350	11
50	0.04943	0.99878	0.04949	20.205	1.0012	20.230	10
51	.04972	.99876	.04978	20.087	.0012	20.112	9
52	.05001	.99875	.05007	19.970	.0012	19.995	8
53	.05030	.99873	.05037	19.854	.0013	19.880	7
54	.05059	.99872	.05066	19.740	.0013	19.766	6
55	0.05088	0.99870	0.05095	19.627	1.0013	19.653	5
56	.05117	.99869	.05124	19.515	.0013	19.541	4
57	.05146	.99867	.05153	19.405	.0013	19.431	3
58	.05175	.99866	.05182	19.296	.0013	19.322	2
59	.05204	.99864	.05212	19.188	.0013	19.214	1
60	0.05234	0.99863	0.05241	19.081	1.0014	19.107	0
M	Cosine	Sine	Cotan.	Tan.	Cosec.	Secant	M

92° 87°

3° Natural Trigonometric Functions 176°

M	Sine	Cosine	Tan.	Cotan.	Secant	Cosec.	M
0	0.05234	0.99863	0.05241	19.081	1.0014	19.107	60
1	.05263	.99861	.05270	18.975	.0014	19.002	59
2	.05292	.99860	.05299	18.871	.0014	18.897	58
3	.05321	.99858	.05328	18.768	.0014	18.794	57
4	.05350	.99857	.05357	18.665	.0014	18.692	56
5	0.05379	0.99855	0.05387	18.564	1.0014	18.591	55
6	.05408	.99854	.05416	18.464	.0015	18.491	54
7	.05437	.99852	.05445	18.365	.0015	18.393	53
8	.05466	.99850	.05474	18.268	.0015	18.295	52
9	.05495	.99849	.05503	18.171	.0015	18.198	51
10	0.05524	0.99847	0.05532	18.075	1.0015	18.103	50
11	.05553	.99846	.05562	17.980	.0015	18.008	49
12	.05582	.99844	.05591	17.886	.0016	17.914	48
13	.05611	.99842	.05620	17.793	.0016	17.821	47
14	.05640	.99841	.05649	17.701	.0016	17.730	46
15	0.05669	0.99839	0.05678	17.610	1.0016	17.639	45
16	.05698	.99837	.05707	17.520	.0016	17.549	44
17	.05727	.99836	.05737	17.431	.0016	17.460	43
18	.05756	.99834	.05766	17.343	.0017	17.372	42
19	.05785	.99832	.05795	17.256	.0017	17.285	41
20	0.05814	0.99831	0.05824	17.169	1.0017	17.198	40
21	.05843	.99829	.05853	17.084	.0017	17.113	39
22	.05872	.99827	.05883	16.999	.0017	17.028	38
23	.05902	.99826	.05912	16.915	.0017	16.944	37
24	.05931	.99824	.05941	16.832	.0018	16.861	36
25	0.05960	0.99822	0.05970	16.750	1.0018	16.779	35
26	.05989	.99820	.05999	16.668	.0018	16.698	34
27	.06018	.99819	.06029	16.587	.0018	16.617	33
28	.06047	.99817	.06058	16.507	.0018	16.538	32
29	.06076	.99815	.06087	16.428	.0018	16.459	31
30	0.06105	0.99813	0.06116	16.350	1.0019	16.380	30
31	.06134	.99812	.06145	16.272	.0019	16.303	29
32	.06163	.99810	.06175	16.195	.0019	16.226	28
33	.06192	.99808	.06204	16.119	.0019	16.150	27
34	.06221	.99806	.06233	16.043	.0019	16.075	26
35	0.06250	0.99804	0.06262	15.969	1.0019	16.000	25
36	.06279	.99803	.06291	15.894	.0020	15.926	24
37	.06308	.99801	.06321	15.821	.0020	15.853	23
38	.06337	.99799	.06350	15.748	.0020	15.780	22
39	.06366	.99797	.06379	15.676	.0020	15.708	21
40	0.06395	0.99795	0.06408	15.605	1.0020	15.637	20
41	.06424	.99793	.06437	15.534	.0021	15.566	19
42	.06453	.99791	.06467	15.464	.0021	15.496	18
43	.06482	.99790	.06496	15.394	.0021	15.427	17
44	.06511	.99788	.06525	15.325	.0021	15.358	16
45	0.06540	0.99786	0.06554	15.257	1.0021	15.290	15
46	.06569	.99784	.06583	15.189	.0022	15.222	14
47	.06598	.99782	.06613	15.122	.0022	15.155	13
48	.06627	.99780	.06642	15.056	.0022	15.089	12
49	.06656	.99778	.06671	14.990	.0022	15.023	11
50	0.06685	0.99776	0.06700	14.924	1.0022	14.958	10
51	.06714	.99774	.06730	14.860	.0023	14.893	9
52	.06743	.99772	.06759	14.795	.0023	14.829	8
53	.06772	.99770	.06788	14.732	.0023	14.765	7
54	.06801	.99768	.06817	14.668	.0023	14.702	6
55	0.06830	0.99766	0.06846	14.606	1.0023	14.640	5
56	.06859	.99764	.06876	14.544	.0024	14.578	4
57	.06888	.99762	.06905	14.482	.0024	14.517	3
58	.06918	.99760	.06934	14.421	.0024	14.456	2
59	.06947	.99758	.06963	14.361	.0024	14.395	1
60	0.06976	0.99756	0.06993	14.301	1.0024	14.335	0
M	Cosine	Sine	Cotan.	Tan.	Cosec.	Secant	M

93° 86°

4° Natural Trigonometric Functions 175°

M	Sine	Cosine	Tan.	Cotan.	Secant	Cosec.	M
0	0.06976	0.99756	0.06993	14.301	1.0024	14.335	60
1	.07005	.99754	.07022	14.241	.0025	14.276	59
2	.07034	.99752	.07051	14.182	.0025	14.217	58
3	.07063	.99750	.07080	14.123	.0025	14.159	57
4	.07092	.99748	.07110	14.065	.0025	14.101	56
5	0.07121	0.99746	0.07139	14.008	1.0025	14.043	55
6	.07150	.99744	.07168	13.951	.0026	13.986	54
7	.07179	.99742	.07197	13.894	.0026	13.930	53
8	.07208	.99740	.07226	13.838	.0026	13.874	52
9	.07237	.99738	.07256	13.782	.0026	13.818	51
10	0.07266	0.99736	0.07285	13.727	1.0026	13.763	50
11	.07295	.99733	.07314	13.672	.0027	13.708	49
12	.07324	.99731	.07343	13.617	.0027	13.654	48
13	.07353	.99729	.07373	13.563	.0027	13.600	47
14	.07382	.99727	.07402	13.510	.0027	13.547	46
15	0.07411	0.99725	0.07431	13.457	1.0027	13.494	45
16	.07440	.99723	.07460	13.404	.0028	13.441	44
17	.07469	.99721	.07490	13.351	.0028	13.389	43
18	.07498	.99718	.07519	13.299	.0028	13.337	42
19	.07527	.99716	.07548	13.248	.0028	13.286	41
20	0.07556	0.99714	0.07577	13.197	1.0029	13.235	40
21	.07585	.99712	.07607	13.146	.0029	13.184	39
22	.07614	.99710	.07636	13.096	.0029	13.134	38
23	.07643	.99707	.07665	13.046	.0029	13.084	37
24	.07672	.99705	.07694	12.996	.0029	13.034	36
25	0.07701	0.99703	0.07724	12.947	1.0030	12.985	35
26	.07730	.99701	.07753	12.898	.0030	12.937	34
27	.07759	.99698	.07782	12.849	.0030	12.888	33
28	.07788	.99696	.07812	12.801	.0030	12.840	32
29	.07817	.99694	.07841	12.754	.0031	12.793	31
30	0.07846	0.99692	0.07870	12.706	1.0031	12.745	30
31	.07875	.99689	.07899	12.659	.0031	12.698	29
32	.07904	.99687	.07929	12.612	.0031	12.652	28
33	.07933	.99685	.07958	12.566	.0032	12.606	27
34	.07962	.99682	.07987	12.520	.0032	12.560	26
35	0.07991	0.99680	0.08016	12.474	1.0032	12.514	25
36	.08020	.99678	.08046	12.429	.0032	12.469	24
37	.08049	.99675	.08075	12.384	.0032	12.424	23
38	.08078	.99673	.08104	12.339	.0033	12.379	22
39	.08107	.99671	.08134	12.295	.0033	12.335	21
40	0.08136	0.99668	0.08163	12.250	1.0033	12.291	20
41	.08165	.99666	.08192	12.207	.0033	12.248	19
42	.08194	.99664	.08221	12.163	.0034	12.204	18
43	.08223	.99661	.08251	12.120	.0034	12.161	17
44	.08252	.99659	.08280	12.077	.0034	12.118	16
45	0.08281	0.99656	0.08309	12.035	1.0034	12.076	15
46	.08310	.99654	.08339	11.992	.0035	12.034	14
47	.08339	.99652	.08368	11.950	.0035	11.992	13
48	.08368	.99649	.08397	11.909	.0035	11.950	12
49	.08397	.99647	.08426	11.867	.0035	11.909	11
50	0.08426	0.99644	0.08456	11.826	1.0036	11.868	10
51	.08455	.99642	.08485	11.785	.0036	11.828	9
52	.08484	.99639	.08514	11.745	.0036	11.787	8
53	.08513	.99637	.08544	11.704	.0036	11.747	7
54	.08542	.99634	.08573	11.664	.0037	11.707	6
55	0.08571	0.99632	0.08602	11.625	1.0037	11.668	5
56	.08600	.99629	.08632	11.585	.0037	11.628	4
57	.08629	.99627	.08661	11.546	.0037	11.589	3
58	.08658	.99624	.08690	11.507	.0038	11.550	2
59	.08687	.99622	.08719	11.468	.0038	11.512	1
60	0.08715	0.99619	0.08749	11.430	1.0038	11.474	0
M	Cosine	Sine	Cotan.	Tan.	Cosec.	Secant	M

94° 85°

5° Natural Trigonometric Functions 174°

M	Sine	Cosine	Tan.	Cotan.	Secant	Cosec.	M
0	0.08715	0.99619	0.08749	11.430	1.0038	11.474	60
1	.08744	.99617	.08778	11.392	.0038	11.436	59
2	.08773	.99614	.08807	11.354	.0039	11.398	58
3	.08802	.99612	.08837	11.316	.0039	11.360	57
4	.08831	.99609	.08866	11.279	.0039	11.323	56
5	0.08860	0.99607	0.08895	11.242	1.0039	11.286	55
6	.08889	.99604	.08925	11.205	.0040	11.249	54
7	.08918	.99601	.08954	11.168	.0040	11.213	53
8	.08947	.99599	.08983	11.132	.0040	11.176	52
9	.08976	.99596	.09013	11.095	.0040	11.140	51
10	0.09005	0.99594	0.09042	11.059	1.0041	11.104	50
11	.09034	.99591	.09071	11.024	.0041	11.069	49
12	.09063	.99588	.09101	10.988	.0041	11.033	48
13	.09092	.99586	.09130	10.953	.0041	10.998	47
14	.09121	.99583	.09159	10.918	.0042	10.963	46
15	0.09150	0.99580	0.09189	10.883	1.0042	10.929	45
16	.09179	.99578	.09218	10.848	.0042	10.894	44
17	.09208	.99575	.09247	10.814	.0043	10.860	43
18	.09237	.99572	.09277	10.780	.0043	10.826	42
19	.09266	.99570	.09306	10.746	.0043	10.792	41
20	0.09295	0.99567	0.09335	10.712	1.0043	10.758	40
21	.09324	.99564	.09365	10.678	.0044	10.725	39
22	.09353	.99562	.09394	10.645	.0044	10.692	38
23	.09382	.99559	.09423	10.612	.0044	10.659	37
24	.09411	.99556	.09453	10.579	.0044	10.626	36
25	0.09440	0.99553	0.09482	10.546	1.0045	10.593	35
26	.09469	.99551	.09511	10.514	.0045	10.561	34
27	.09498	.99548	.09541	10.481	.0045	10.529	33
28	.09527	.99545	.09570	10.449	.0046	10.497	32
29	.09556	.99542	.09599	10.417	.0046	10.465	31
30	0.09584	0.99540	0.09629	10.385	1.0046	10.433	30
31	.09613	.99537	.09658	10.354	.0046	10.402	29
32	.09642	.99534	.09688	10.322	.0047	10.371	28
33	.09671	.99531	.09717	10.291	.0047	10.340	27
34	.09700	.99528	.09746	10.260	.0047	10.309	26
35	0.09729	0.99525	0.09776	10.229	1.0048	10.278	25
36	.09758	.99523	.09805	10.199	.0048	10.248	24
37	.09787	.99520	.09834	10.168	.0048	10.217	23
38	.09816	.99517	.09864	10.138	.0048	10.187	22
39	.09845	.99514	.09893	10.108	.0049	10.157	21
40	0.09874	0.99511	0.09922	10.078	1.0049	10.127	20
41	.09903	.99508	.09952	10.048	.0049	10.098	19
42	.09932	.99505	.09981	10.019	.0050	10.068	18
43	.09961	.99503	.10011	9.9893	.0050	10.039	17
44	.09990	.99500	.10040	9.9601	.0050	10.010	16
45	0.10019	0.99497	0.10069	9.9310	1.0050	9.9812	15
46	.10048	.99494	.10099	9.9021	.0051	9.9525	14
47	.10077	.99491	.10128	9.8734	.0051	9.9239	13
48	.10106	.99488	.10158	9.8448	.0051	9.8955	12
49	.10134	.99485	.10187	9.8164	.0052	9.8672	11
50	0.10163	0.99482	0.10216	9.7882	1.0052	9.8391	10
51	.10192	.99479	.10246	9.7601	.0052	9.8112	9
52	.10221	.99476	.10275	9.7322	.0053	9.7834	8
53	.10250	.99473	.10305	9.7044	.0053	9.7558	7
54	.10279	.99470	.10334	9.6768	.0053	9.7283	6
55	0.10308	0.99467	0.10363	9.6493	1.0053	9.7010	5
56	.10337	.99464	.10393	9.6220	.0054	9.6739	4
57	.10366	.99461	.10422	9.5949	.0054	9.6469	3
58	.10395	.99458	.10452	9.5679	.0054	9.6200	2
59	.10424	.99455	.10481	9.5411	.0055	9.5933	1
60	0.10453	0.99452	0.10510	9.5144	1.0055	9.5668	0
M	Cosine	Sine	Cotan.	Tan.	Cosec.	Secant	M

95° 84°

6° Natural Trigonometric Functions 173°

M	Sine	Cosine	Tan.	Cotan.	Secant	Cosec.	M
0	0.10453	0.99452	0.10510	9.5144	1.0055	9.5668	60
1	.10482	.99449	.10540	.4878	.0055	.5404	59
2	.10511	.99446	.10569	.4614	.0056	.5141	58
3	.10540	.99443	.10599	.4351	.0056	.4880	57
4	.10568	.99440	.10628	.4090	.0056	.4620	56
5	0.10597	0.99437	0.10657	9.3831	1.0057	9.4362	55
6	.10626	.99434	.10687	.3572	.0057	.4105	54
7	.10655	.99431	.10716	.3315	.0057	.3850	53
8	.10684	.99428	.10746	.3060	.0057	.3596	52
9	.10713	.99424	.10775	.2806	.0058	.3343	51
10	0.10742	0.99421	0.10805	9.2553	1.0058	9.3092	50
11	.10771	.99418	.10834	.2302	.0058	.2842	49
12	.10800	.99415	.10863	.2051	.0059	.2593	48
13	.10829	.99412	.10893	.1803	.0059	.2346	47
14	.10858	.99409	.10922	.1555	.0059	.2100	46
15	0.10887	0.99406	0.10952	9.1309	1.0060	9.1855	45
16	.10916	.99402	.10981	.1064	.0060	.1612	44
17	.10944	.99399	.11011	.0821	.0060	.1370	43
18	.10973	.99396	.11040	.0579	.0061	.1129	42
19	.11002	.99393	.11069	.0338	.0061	.0890	41
20	0.11031	0.99390	0.11099	9.0098	1.0061	9.0651	40
21	.11060	.99386	.11128	8.9860	.0062	.0414	39
22	.11089	.99383	.11158	.9623	.0062	.0179	38
23	.11118	.99380	.11187	.9387	.0062	8.9944	37
24	.11147	.99377	.11217	.9152	.0063	.9711	36
25	0.11176	0.99373	0.11246	8.8918	1.0063	8.9479	35
26	.11205	.99370	.11276	.8686	.0063	.9248	34
27	.11234	.99367	.11305	.8455	.0064	.9018	33
28	.11262	.99364	.11335	.8225	.0064	.8790	32
29	.11291	.99360	.11364	.7996	.0064	.8563	31
30	0.11320	0.99357	0.11393	8.7769	1.0065	8.8337	30
31	.11349	.99354	.11423	.7542	.0065	.8112	29
32	.11378	.99350	.11452	.7317	.0065	.7888	28
33	.11407	.99347	.11482	.7093	.0066	.7665	27
34	.11436	.99344	.11511	.6870	.0066	.7444	26
35	0.11465	0.99341	0.11541	8.6648	1.0066	8.7223	25
36	.11494	.99337	.11570	.6427	.0067	.7004	24
37	.11523	.99334	.11600	.6208	.0067	.6786	23
38	.11551	.99330	.11629	.5989	.0067	.6569	22
39	.11580	.99327	.11659	.5772	.0068	.6353	21
40	0.11609	0.99324	0.11688	8.5555	1.0068	8.6138	20
41	.11638	.99320	.11718	.5340	.0068	.5924	19
42	.11667	.99317	.11747	.5126	.0069	.5711	18
43	.11696	.99314	.11777	.4913	.0069	.5499	17
44	.11725	.99310	.11806	.4701	.0069	.5289	16
45	0.11754	0.99307	0.11836	8.4489	1.0070	8.5079	15
46	.11783	.99303	.11865	.4279	.0070	.4871	14
47	.11811	.99300	.11895	.4070	.0070	.4663	13
48	.11840	.99296	.11924	.3862	.0071	.4457	12
49	.11869	.99293	.11954	.3655	.0071	.4251	11
50	0.11898	0.99290	0.11983	8.3449	1.0071	8.4046	10
51	.11927	.99286	.12013	.3244	.0072	.3843	9
52	.11956	.99283	.12042	.3040	.0072	.3640	8
53	.11985	.99279	.12072	.2837	.0073	.3439	7
54	.12014	.99276	.12101	.2635	.0073	.3238	6
55	0.12042	0.99272	0.12131	8.2434	1.0073	8.3039	5
56	.12071	.99269	.12160	.2234	.0074	.2840	4
57	.12100	.99265	.12190	.2035	.0074	.2642	3
58	.12129	.99262	.12219	.1837	.0074	.2446	2
59	.12158	.99258	.12249	.1640	.0075	.2250	1
60	0.12187	0.99255	0.12278	8.1443	1.0075	8.2055	0
M	Cosine	Sine	Cotan.	Tan.	Cosec.	Secant	M

96° 83°

7° Natural Trigonometric Functions 172°

M	Sine	Cosine	Tan.	Cotan.	Secant	Cosec.	M
0	0.12187	0.99255	0.12278	8.1443	1.0075	8.2055	60
1	.12216	.99251	.12308	.1248	.0075	.1861	59
2	.12245	.99247	.12337	.1053	.0076	.1668	58
3	.12273	.99244	.12367	.0860	.0076	.1476	57
4	.12302	.99240	.12396	.0667	.0076	.1285	56
5	0.12331	0.99237	0.12426	8.0476	1.0077	8.1094	55
6	.12360	.99233	.12456	.0285	.0077	.0905	54
7	.12389	.99229	.12485	.0095	.0078	.0717	53
8	.12418	.99226	.12515	7.9906	.0078	.0529	52
9	.12447	.99222	.12544	.9717	.0078	.0342	51
10	0.12476	0.99219	0.12574	7.9530	1.0079	8.0156	50
11	.12504	.99215	.12603	.9344	.0079	7.9971	49
12	.12533	.99211	.12633	.9158	.0079	.9787	48
13	.12562	.99208	.12662	.8973	.0080	.9604	47
14	.12591	.99204	.12692	.8789	.0080	.9421	46
15	0.12620	0.99200	0.12722	7.8606	1.0080	7.9240	45
16	.12649	.99197	.12751	.8424	.0081	.9059	44
17	.12678	.99193	.12781	.8243	.0081	.8879	43
18	.12706	.99189	.12810	.8062	.0082	.8700	42
19	.12735	.99186	.12840	.7882	.0082	.8522	41
20	0.12764	0.99182	0.12869	7.7703	1.0082	7.8344	40
21	.12793	.99178	.12899	.7525	.0083	.8168	39
22	.12822	.99174	.12928	.7348	.0083	.7992	38
23	.12851	.99171	.12958	.7171	.0084	.7817	37
24	.12879	.99167	.12988	.6996	.0084	.7642	36
25	0.12908	0.99163	0.13017	7.6821	1.0084	7.7469	35
26	.12937	.99160	.13047	.6646	.0085	.7296	34
27	.12966	.99156	.13076	.6473	.0085	.7124	33
28	.12995	.99152	.13106	.6300	.0085	.6953	32
29	.13024	.99148	.13136	.6129	.0086	.6783	31
30	0.13053	0.99144	0.13165	7.5957	1.0086	7.6613	30
31	.13081	.99141	.13195	.5787	.0087	.6444	29
32	.13110	.99137	.13224	.5617	.0087	.6276	28
33	.13139	.99133	.13254	.5449	.0087	.6108	27
34	.13168	.99129	.13284	.5280	.0088	.5942	26
35	0.13197	0.99125	0.13313	7.5113	1.0088	7.5776	25
36	.13226	.99121	.13343	.4946	.0089	.5611	24
37	.13254	.99118	.13372	.4780	.0089	.5446	23
38	.13283	.99114	.13402	.4615	.0089	.5282	22
39	.13312	.99110	.13432	.4451	.0090	.5119	21
40	0.13341	0.99106	0.13461	7.4287	1.0090	7.4957	20
41	.13370	.99102	.13491	.4124	.0090	.4795	19
42	.13399	.99098	.13520	.3961	.0091	.4634	18
43	.13427	.99094	.13550	.3800	.0091	.4474	17
44	.13456	.99090	.13580	.3639	.0092	.4315	16
45	0.13485	0.99086	0.13609	7.3479	1.0092	7.4156	15
46	.13514	.99083	.13639	.3319	.0092	.3998	14
47	.13543	.99079	.13669	.3160	.0093	.3840	13
48	.13571	.99075	.13698	.3002	.0093	.3683	12
49	.13600	.99071	.13728	.2844	.0094	.3527	11
50	0.13629	0.99067	0.13757	7.2687	1.0094	7.3372	10
51	.13658	.99063	.13787	.2531	.0094	.3217	9
52	.13687	.99059	.13817	.2375	.0095	.3063	8
53	.13716	.99055	.13846	.2220	.0095	.2909	7
54	.13744	.99051	.13876	.2066	.0096	.2757	6
55	0.13773	0.99047	0.13906	7.1912	1.0096	7.2604	5
56	.13802	.99043	.13935	.1759	.0097	.2453	4
57	.13831	.99039	.13965	.1607	.0097	.2302	3
58	.13860	.99035	.13995	.1455	.0097	.2152	2
59	.13888	.99031	.14024	.1304	.0098	.2002	1
60	0.13917	0.99027	0.14054	7.1154	1.0098	7.1853	0
M	Cosine	Sine	Cotan.	Tan.	Cosec.	Secant	M

97° 82°

8° Natural Trigonometric Functions 171°

M	Sine	Cosine	Tan.	Cotan.	Secant	Cosec.	M
0	0.13917	0.99027	0.14054	7.1154	1.0098	7.1853	60
1	.13946	.99023	.14084	.1004	.0099	.1704	59
2	.13975	.99019	.14113	.0854	.0099	.1557	58
3	.14004	.99015	.14143	.0706	.0099	.1409	57
4	.14032	.99010	.14173	.0558	.0100	.1263	56
5	0.14061	0.99006	0.14202	7.0410	1.0100	7.1117	55
6	.14090	.99002	.14232	.0264	.0101	.0972	54
7	.14119	.98998	.14262	.0117	.0101	.0827	53
8	.14148	.98994	.14291	6.9972	.0102	.0683	52
9	.14176	.98990	.14321	.9827	.0102	.0539	51
10	0.14205	0.98986	0.14351	6.9682	1.0102	7.0396	50
11	.14234	.98982	.14380	.9538	.0103	.0254	49
12	.14263	.98978	.14410	.9395	.0103	.0112	48
13	.14292	.98973	.14440	.9252	.0104	6.9971	47
14	.14320	.98969	.14470	.9110	.0104	.9830	46
15	0.14349	0.98965	0.14499	6.8969	1.0104	6.9690	45
16	.14378	.98961	.14529	.8828	.0105	.9550	44
17	.14407	.98957	.14559	.8687	.0105	.9411	43
18	.14436	.98952	.14588	.8547	.0106	.9273	42
19	.14464	.98948	.14618	.8408	.0106	.9135	41
20	0.14493	0.98944	0.14648	6.8269	1.0107	6.8998	40
21	.14522	.98940	.14677	.8131	.0107	.8861	39
22	.14551	.98936	.14707	.7993	.0107	.8725	38
23	.14579	.98931	.14737	.7856	.0108	.8589	37
24	.14608	.98927	.14767	.7720	.0108	.8454	36
25	0.14637	0.98923	0.14796	6.7584	1.0109	6.8320	35
26	.14666	.98919	.14826	.7448	.0109	.8185	34
27	.14695	.98914	.14856	.7313	.0110	.8052	33
28	.14723	.98910	.14886	.7179	.0110	.7919	32
29	.14752	.98906	.14915	.7045	.0111	.7787	31
30	0.14781	0.98901	0.14945	6.6911	1.0111	6.7655	30
31	.14810	.98897	.14975	.6779	.0111	.7523	29
32	.14838	.98893	.15004	.6646	.0112	.7392	28
33	.14867	.98889	.15034	.6514	.0112	.7262	27
34	.14896	.98884	.15064	.6383	.0113	.7132	26
35	0.14925	0.98880	0.15094	6.6252	1.0113	6.7003	25
36	.14953	.98876	.15123	.6122	.0114	.6874	24
37	.14982	.98871	.15153	.5992	.0114	.6745	23
38	.15011	.98867	.15183	.5863	.0115	.6617	22
39	.15040	.98862	.15213	.5734	.0115	.6490	21
40	0.15068	0.98858	0.15243	6.5605	1.0115	6.6363	20
41	.15097	.98854	.15272	.5478	.0116	.6237	19
42	.15126	.98849	.15302	.5350	.0116	.6111	18
43	.15155	.98845	.15332	.5223	.0117	.5985	17
44	.15183	.98840	.15362	.5097	.0117	.5860	16
45	0.15212	0.98836	0.15391	6.4971	1.0118	6.5736	15
46	.15241	.98832	.15421	.4845	.0118	.5612	14
47	.15270	.98827	.15451	.4720	.0119	.5488	13
48	.15298	.98823	.15481	.4596	.0119	.5365	12
49	.15328	.98818	.15511	.4472	.0119	.5243	11
50	0.15356	0.98814	0.15540	6.4348	1.0120	6.5121	10
51	.15385	.98809	.15570	.4225	.0120	.4999	9
52	.15413	.98805	.15600	.4103	.0121	.4878	8
53	.15442	.98800	.15630	.3980	.0121	.4757	7
54	.15471	.98796	.15659	.3859	.0122	.4637	6
55	0.15500	0.98791	0.15689	6.3737	1.0122	6.4517	5
56	.15528	.98787	.15719	.3616	.0123	.4398	4
57	.15557	.98782	.15749	.3496	.0123	.4279	3
58	.15586	.98778	.15779	.3376	.0124	.4160	2
59	.15615	.98773	.15809	.3257	.0124	.4042	1
60	0.15643	0.98769	0.15838	6.3137	1.0125	6.3924	0
M	Cosine	Sine	Cotan.	Tan.	Cosec.	Secant	M

98° 81°

9° **Natural Trigonometric Functions** 170°

M	Sine	Cosine	Tan.	Cotan.	Secant	Cosec.	M
0	0.15643	0.98769	0.15838	6.3137	1.0125	6.3924	60
1	.15672	.98764	.15868	.3019	.0125	.3807	59
2	.15701	.98760	.15898	.2901	.0125	.3690	58
3	.15730	.98755	.15928	.2783	.0126	.3574	57
4	.15758	.98750	.15958	.2665	.0126	.3458	56
5	0.15787	0.98746	0.15987	6.2548	1.0127	6.3343	55
6	.15816	.98741	.16017	.2432	.0127	.3228	54
7	.15844	.98737	.16047	.2316	.0128	.3113	53
8	.15873	.98732	.16077	.2200	.0128	.2999	52
9	.15902	.98727	.16107	.2085	.0129	.2885	51
10	0.15931	0.98723	0.16137	6.1970	1.0129	6.2772	50
11	.15959	.98718	.16167	.1856	.0130	.2659	49
12	.15988	.98714	.16196	.1742	.0130	.2546	48
13	.16017	.98709	.16226	.1628	.0131	.2434	47
14	.16045	.98704	.16256	.1515	.0131	.2322	46
15	0.16074	0.98700	0.16286	6.1402	1.0132	6.2211	45
16	.16103	.98695	.16316	.1290	.0132	.2100	44
17	.16132	.98690	.16346	.1178	.0133	.1990	43
18	.16160	.98685	.16376	.1066	.0133	.1880	42
19	.16189	.98681	.16405	.0955	.0134	.1770	41
20	0.16218	0.98676	0.16435	6.0844	1.0134	6.1661	40
21	.16246	.98671	.16465	.0734	.0135	.1552	39
22	.16275	.98667	.16495	.0624	.0135	.1443	38
23	.16304	.98662	.16525	.0514	.0136	.1335	37
24	.16333	.98657	.16555	.0405	.0136	.1227	36
25	0.16361	0.98652	0.16585	6.0296	1.0136	6.1120	35
26	.16390	.98648	.16615	.0188	.0137	.1013	34
27	.16419	.98643	.16644	.0080	.0137	.0906	33
28	.16447	.98638	.16674	5.9972	.0138	.0800	32
29	.16476	.98633	.16704	.9865	.0138	.0694	31
30	0.16505	0.98628	0.16734	5.9758	1.0139	6.0588	30
31	.16533	.98624	.16764	.9651	.0139	.0483	29
32	.16562	.98619	.16794	.9545	.0140	.0379	28
33	.16591	.98614	.16824	.9439	.0140	.0274	27
34	.16619	.98609	.16854	.9333	.0141	.0170	26
35	0.16648	0.98604	0.16884	5.9228	1.0141	6.0066	25
36	.16677	.98600	.16914	.9123	.0142	5.9963	24
37	.16705	.98595	.16944	.9019	.0142	.9860	23
38	.16734	.98590	.16973	.8915	.0143	.9758	22
39	.16763	.98585	.17003	.8811	.0143	.9655	21
40	0.16791	0.98580	0.17033	5.8708	1.0144	5.9554	20
41	.16820	.98575	.17063	.8605	.0144	.9452	19
42	.16849	.98570	.17093	.8502	.0145	.9351	18
43	.16878	.98565	.17123	.8400	.0145	.9250	17
44	.16906	.98560	.17153	.8298	.0146	.9150	16
45	0.16935	0.98556	0.17183	5.8196	1.0146	5.9049	15
46	.16964	.98551	.17213	.8095	.0147	.8950	14
47	.16992	.98546	.17243	.7994	.0147	.8850	13
48	.17021	.98541	.17273	.7894	.0148	.8751	12
49	.17050	.98536	.17303	.7794	.0148	.8652	11
50	0.17078	0.98531	0.17333	5.7694	1.0149	5.8554	10
51	.17107	.98526	.17363	.7594	.0150	.8456	9
52	.17136	.98521	.17393	.7495	.0150	.8358	8
53	.17164	.98516	.17423	.7396	.0151	.8261	7
54	.17193	.98511	.17453	.7297	.0151	.8163	6
55	0.17221	0.98506	0.17483	5.7199	1.0152	5.8067	5
56	.17250	.98501	.17513	.7101	.0152	.7970	4
57	.17279	.98496	.17543	.7004	.0153	.7874	3
58	.17307	.98491	.17573	.6906	.0153	.7778	2
59	.17336	.98486	.17603	.6809	.0154	.7683	1
60	0.17365	0.98481	0.17633	5.6713	1.0154	5.7588	0
M	Cosine	Sine	Cotan.	Tan.	Cosec.	Secant	M

99° 80°

10° **Natural Trigonometric Functions** 169°

M	Sine	Cosine	Tan.	Cotan.	Secant	Cosec.	M
0	0.17365	0.98481	0.17633	5.6713	1.0154	5.7588	60
1	.17393	.98476	.17663	.6616	.0155	.7493	59
2	.17422	.98471	.17693	.6520	.0155	.7398	58
3	.17451	.98465	.17723	.6425	.0156	.7304	57
4	.17479	.98460	.17753	.6329	.0156	.7210	56
5	0.17508	0.98455	0.17783	5.6234	1.0157	5.7117	55
6	.17537	.98450	.17813	.6140	.0157	.7023	54
7	.17565	.98445	.17843	.6045	.0158	.6930	53
8	.17594	.98440	.17873	.5951	.0158	.6838	52
9	.17622	.98435	.17903	.5857	.0159	.6745	51
10	0.17651	0.98430	0.17933	5.5764	1.0159	5.6653	50
11	.17680	.98425	.17963	.5670	.0160	.6561	49
12	.17708	.98419	.17993	.5578	.0160	.6470	48
13	.17737	.98414	.18023	.5485	.0161	.6379	47
14	.17766	.98409	.18053	.5393	.0162	.6288	46
15	0.17794	0.98404	0.18083	5.5301	1.0162	5.6197	45
16	.17823	.98399	.18113	.5209	.0163	.6107	44
17	.17852	.98394	.18143	.5117	.0163	.6017	43
18	.17880	.98388	.18173	.5026	.0164	.5928	42
19	.17909	.98383	.18203	.4936	.0164	.5838	41
20	0.17937	0.98378	0.18233	5.4845	1.0165	5.5749	40
21	.17966	.98373	.18263	.4755	.0165	.5660	39
22	.17995	.98368	.18293	.4665	.0166	.5572	38
23	.18023	.98362	.18323	.4575	.0166	.5484	37
24	.18052	.98357	.18353	.4486	.0167	.5396	36
25	0.18080	0.98352	0.18383	5.4396	1.0167	5.5308	35
26	.18109	.98347	.18413	.4308	.0168	.5221	34
27	.18138	.98341	.18444	.4219	.0169	.5134	33
28	.18166	.98336	.18474	.4131	.0169	.5047	32
29	.18195	.98331	.18504	.4043	.0170	.4960	31
30	0.18223	0.98325	0.18534	5.3955	1.0170	5.4874	30
31	.18252	.98320	.18564	.3868	.0171	.4788	29
32	.18281	.98315	.18594	.3780	.0171	.4702	28
33	.18309	.98309	.18624	.3694	.0172	.4617	27
34	.18338	.98304	.18654	.3607	.0172	.4532	26
35	0.18366	0.98299	0.18684	5.3521	1.0173	5.4447	25
36	.18395	.98293	.18714	.3434	.0174	.4362	24
37	.18424	.98288	.18745	.3349	.0174	.4278	23
38	.18452	.98283	.18775	.3263	.0175	.4194	22
39	.18481	.98277	.18805	.3178	.0175	.4110	21
40	0.18509	0.98272	0.18835	5.3093	1.0176	5.4026	20
41	.18538	.98267	.18865	.3008	.0176	.3943	19
42	.18567	.98261	.18895	.2923	.0177	.3860	18
43	.18595	.98256	.18925	.2839	.0177	.3777	17
44	.18624	.98250	.18955	.2755	.0178	.3695	16
45	0.18652	0.98245	0.18985	5.2671	1.0179	5.3612	15
46	.18681	.98240	.19016	.2588	.0179	.3530	14
47	.18709	.98234	.19046	.2505	.0180	.3449	13
48	.18738	.98229	.19076	.2422	.0180	.3367	12
49	.18767	.98223	.19106	.2339	.0181	.3286	11
50	0.18795	0.98218	0.19136	5.2257	1.0181	5.3205	10
51	.18824	.98212	.19166	.2174	.0182	.3124	9
52	.18852	.98207	.19197	.2092	.0182	.3044	8
53	.18881	.98201	.19227	.2011	.0183	.2963	7
54	.18909	.98196	.19257	.1929	.0184	.2883	6
55	0.18938	0.98190	0.19287	5.1848	1.0184	5.2803	5
56	.18967	.98185	.19317	.1767	.0185	.2724	4
57	.18995	.98179	.19347	.1686	.0185	.2645	3
58	.19024	.98174	.19378	.1606	.0186	.2566	2
59	.19052	.98168	.19408	.1525	.0186	.2487	1
60	0.19081	0.98163	0.19438	5.1445	1.0187	5.2408	0
M	Cosine	Sine	Cotan.	Tan.	Cosec.	Secant	M

100° 79°

11° **Natural Trigonometric Functions** 168°

M	Sine	Cosine	Tan.	Cotan.	Secant	Cosec.	M
0	0.19081	0.98163	0.19438	5.1445	1.0187	5.2408	60
1	.19109	.98157	.19468	.1366	.0188	.2330	59
2	.19138	.98152	.19498	.1286	.0188	.2252	58
3	.19166	.98146	.19529	.1207	.0189	.2174	57
4	.19195	.98140	.19559	.1128	.0189	.2097	56
5	0.19224	0.98135	0.19589	5.1049	1.0190	5.2019	55
6	.19252	.98129	.19619	.0970	.0191	.1942	54
7	.19281	.98124	.19649	.0892	.0191	.1865	53
8	.19309	.98118	.19680	.0814	.0192	.1788	52
9	.19338	.98112	.19710	.0736	.0192	.1712	51
10	0.19366	0.93107	0.19740	5.0658	1.0193	5.1636	50
11	.19395	.98101	.19770	.0581	.0193	.1560	49
12	.19423	.98095	.19800	.0504	.0194	.1484	48
13	.19452	.98090	.19831	.0427	.0195	.1409	47
14	.19480	.98084	.19861	.0350	.0195	.1333	46
15	0.19509	0.98078	0.19891	5.0273	1.0196	5.1258	45
16	.19537	.98073	.19921	.0197	.0196	.1183	44
17	.19566	.98067	.19952	.0121	.0197	.1109	43
18	.19595	.98061	.19982	.0045	.0198	.1034	42
19	.19623	.98056	.20012	4.9969	.0198	.0960	41
20	0.19652	0.98050	0.20042	4.9894	1.0199	5.0886	40
21	.19680	.98044	.20073	.9819	.0199	.0812	39
22	.19709	.98039	.20103	.9744	.0200	.0739	38
23	.19737	.98033	.20133	.9669	.0201	.0666	37
24	.19766	.98027	.20163	.9594	.0201	.0593	36
25	0.19794	0.98021	0.20194	4.9520	1.0202	5.0520	35
26	.19823	.98016	.20224	.9446	.0202	.0447	34
27	.19851	.98010	.20254	.9372	.0203	.0375	33
28	.19880	.98004	.20285	.9298	.0204	.0302	32
29	.19908	.97998	.20315	.9225	.0204	.0230	31
30	0.19937	0.97992	0.20345	4.9151	1.0205	5.0158	30
31	.19965	.97987	.20375	.9078	.0205	.0087	29
32	.19994	.97981	.20406	.9006	.0206	.0015	28
33	.20022	.97975	.20436	.8933	.0207	4.9944	27
34	.20051	.97969	.20466	.8860	.0207	.9873	26
35	0.20079	0.97963	0.20497	4.8788	1.0208	4.9802	25
36	.20108	.97957	.20527	.8716	.0208	.9732	24
37	.20136	.97952	.20557	.8644	.0209	.9661	23
38	.20165	.97946	.20588	.8573	.0210	.9591	22
39	.20193	.97940	.20618	.8501	.0210	.9521	21
40	0.20222	0.97934	0.20648	4.8430	1.0211	4.9452	20
41	.20250	.97928	.20679	.8359	.0211	.9382	19
42	.20279	.97922	.20709	.8288	.0212	.9313	18
43	.20307	.97916	.20739	.8217	.0213	.9243	17
44	.20336	.97910	.20770	.8147	.0213	.9175	16
45	0.20364	0.97904	0.20800	4.8077	1.0214	4.9106	15
46	.20393	.97899	.20830	.8007	.0215	.9037	14
47	.20421	.97893	.20861	.7937	.0215	.8969	13
48	.20450	.97887	.20891	.7867	.0216	.8901	12
49	.20478	.97881	.20921	.7798	.0216	.8833	11
50	0.20506	0.97875	0.20952	4.7728	1.0217	4.8765	10
51	.20535	.97869	.20982	.7659	.0218	.8697	9
52	.20563	.97863	.21012	.7591	.0218	.8630	8
53	.20592	.97857	.21043	.7522	.0219	.8563	7
54	.20620	.97851	.21073	.7453	.0220	.8496	6
55	0.20649	0.97845	0.21104	4.7385	1.0220	4.8429	5
56	.20677	.97839	.21134	.7317	.0221	.8362	4
57	.20706	.97833	.21164	.7249	.0221	.8296	3
58	.20734	.97827	.21195	.7181	.0222	.8229	2
59	.20763	.97821	.21225	.7114	.0223	.8163	1
60	0.20791	0.97815	0.21256	4.7046	1.0223	4.8097	0
M	Cosine	Sine	Cotan.	Tan.	Cosec.	Secant	M

101° 78°

12° Natural Trigonometric Functions 167°

M	Sine	Cosine	Tan.	Cotan.	Secant	Cosec.	M
0	0.20791	0.97815	0.21256	4.7046	1.0223	4.8097	60
1	.20820	.97809	.21286	.6979	.0224	.8032	59
2	.20848	.97803	.21316	.6912	.0225	.7966	58
3	.20876	.97797	.21347	.6845	.0225	.7901	57
4	.20905	.97790	.21377	.6778	.0226	.7835	56
5	0.20933	0.97784	0.21408	4.6712	1.0226	4.7770	55
6	.20962	.97778	.21438	.6646	.0227	.7706	54
7	.20990	.97772	.21468	.6580	.0228	.7641	53
8	.21019	.97766	.21499	.6514	.0228	.7576	52
9	.21047	.97760	.21529	.6448	.0229	.7512	51
10	0.21076	0.97754	0.21560	4.6382	1.0230	4.7448	50
11	.21104	.97748	.21590	.6317	.0230	.7384	49
12	.21132	.97741	.21621	.6252	.0231	.7320	48
13	.21161	.97735	.21651	.6187	.0232	.7257	47
14	.21189	.97729	.21682	.6122	.0232	.7193	46
15	0.21218	0.97723	0.21712	4.6057	1.0233	4.7130	45
16	.21246	.97717	.21742	.5993	.0234	.7067	44
17	.21275	.97711	.21773	.5928	.0234	.7004	43
18	.21303	.97704	.21803	.5864	.0235	.6942	42
19	.21331	.97698	.21834	.5800	.0235	.6879	41
20	0.21360	0.97692	0.21864	4.5736	1.0236	4.6817	40
21	.21388	.97686	.21895	.5673	.0237	.6754	39
22	.21417	.97680	.21925	.5609	.0237	.6692	38
23	.21445	.97673	.21956	.5546	.0238	.6631	37
24	.21473	.97667	.21986	.5483	.0239	.6569	36
25	0.21502	0.97661	0.22017	4.5420	1.0239	4.6507	35
26	.21530	.97655	.22047	.5357	.0240	.6446	34
27	.21559	.97648	.22078	.5294	.0241	.6385	33
28	.21587	.97642	.22108	.5232	.0241	.6324	32
29	.21615	.97636	.22139	.5169	.0242	.6263	31
30	0.21644	0.97630	0.22169	4.5107	1.0243	4.6201	30
31	.21672	.97623	.22200	.5045	.0243	.6142	29
32	.21701	.97617	.22230	.4983	.0244	.6081	28
33	.21729	.97611	.22261	.4921	.0245	.6021	27
34	.21757	.97604	.22291	.4860	.0245	.5961	26
35	0.21786	0.97598	0.22322	4.4799	1.0246	4.5901	25
36	.21814	.97592	.22353	.4737	.0247	.5841	24
37	.21843	.97585	.22383	.4676	.0247	.5782	23
38	.21871	.97579	.22414	.4615	.0248	.5722	22
39	.21899	.97573	.22444	.4555	.0249	.5663	21
40	0.21928	0.97566	0.22475	4.4494	1.0249	4.5604	20
41	.21956	.97560	.22505	.4434	.0250	.5545	19
42	.21985	.97553	.22536	.4373	.0251	.5486	18
43	.22013	.97547	.22566	.4313	.0251	.5428	17
44	.22041	.97541	.22597	.4253	.0252	.5369	16
45	0.22070	0.97534	0.22628	4.4194	1.0253	4.5311	15
46	.22098	.97528	.22658	.4134	.0253	.5253	14
47	.22126	.97521	.22689	.4074	.0254	.5195	13
48	.22155	.97515	.22719	.4015	.0255	.5137	12
49	.22183	.97508	.22750	.3956	.0255	.5079	11
50	0.22211	0.97502	0.22781	4.3897	1.0256	4.5021	10
51	.22240	.97495	.22811	.3838	.0257	.4964	9
52	.22268	.97489	.22842	.3779	.0257	.4907	8
53	.22297	.97483	.22872	.3721	.0258	.4850	7
54	.22325	.97476	.22903	.3662	.0259	.4793	6
55	0.22353	0.97470	0.22934	4.3604	1.0260	4.4736	5
56	.22382	.97463	.22964	.3546	.0260	.4679	4
57	.22410	.97457	.22995	.3488	.0261	.4623	3
58	.22438	.97450	.23025	.3430	.0262	.4566	2
59	.22467	.97443	.23056	.3372	.0262	.4510	1
60	0.22495	0.97437	0.23087	4.3315	1.0263	4.4454	0
M	Cosine	Sine	Cotan.	Tan.	Cosec.	Secant	M

102° 77°

13° Natural Trigonometric Functions 166°

M	Sine	Cosine	Tan.	Cotan.	Secant	Cosec.	M
0	0.22495	0.97437	0.23087	4.3315	1.0263	4.4454	60
1	.22523	.97430	.23117	.3257	.0264	.4398	59
2	.22552	.97424	.23148	.3200	.0264	.4342	58
3	.22580	.97417	.23179	.3143	.0265	.4287	57
4	.22608	.97411	.23209	.3086	.0266	.4231	56
5	0.22637	0.97404	0.23240	4.3029	1.0266	4.4176	55
6	.22665	.97398	.23270	.2972	.0267	.4121	54
7	.22693	.97391	.23301	.2916	.0268	.4065	53
8	.22722	.97384	.23332	.2859	.0268	.4011	52
9	.22750	.97378	.23363	.2803	.0269	.3956	51
10	0.22778	0.97371	0.23393	4.2747	1.0270	4.3901	50
11	.22807	.97364	.23424	.2691	.0271	.3847	49
12	.22835	.97358	.23455	.2635	.0271	.3792	48
13	.22863	.97351	.23485	.2579	.0272	.3738	47
14	.22892	.97344	.23516	.2524	.0273	.3684	46
15	0.22920	0.97338	0.23547	4.2468	1.0273	4.3630	45
16	.22948	.97331	.23577	.2413	.0274	.3576	44
17	.22977	.97324	.23608	.2358	.0275	.3522	43
18	.23005	.97318	.23639	.2303	.0276	.3469	42
19	.23033	.97311	.23670	.2248	.0276	.3415	41
20	0.23061	0.97304	0.23700	4.2193	1.0277	4.3362	40
21	.23090	.97298	.23731	.2139	.0278	.3309	39
22	.23118	.97291	.23762	.2084	.0278	.3256	38
23	.23146	.97284	.23793	.2030	.0279	.3203	37
24	.23175	.97277	.23823	.1976	.0280	.3150	36
25	0.23203	0.97271	0.23854	4.1921	1.0280	4.3098	35
26	.23231	.97264	.23885	.1867	.0281	.3045	34
27	.23260	.97257	.23916	.1814	.0282	.2993	33
28	.23288	.97250	.23946	.1760	.0283	.2941	32
29	.23316	.97244	.23977	.1706	.0283	.2888	31
30	0.23344	0.97237	0.24008	4.1653	1.0284	4.2836	30
31	.23373	.97230	.24039	.1600	.0285	.2785	29
32	.23401	.97223	.24069	.1546	.0285	.2733	28
33	.23429	.97216	.24100	.1493	.0286	.2681	27
34	.23458	.97210	.24131	.1440	.0287	.2630	26
35	0.23486	0.97203	0.24162	4.1388	1.0288	4.2579	25
36	.23514	.97196	.24192	.1335	.0288	.2527	24
37	.23542	.97189	.24223	.1282	.0289	.2476	23
38	.23571	.97182	.24254	.1230	.0290	.2425	22
39	.23599	.97175	.24285	.1178	.0291	.2375	21
40	0.23627	0.97169	0.24316	4.1126	1.0291	4.2324	20
41	.23655	.97162	.24346	.1073	.0292	.2273	19
42	.23684	.97155	.24377	.1022	.0293	.2223	18
43	.23712	.97148	.24408	.0970	.0293	.2173	17
44	.23740	.97141	.24439	.0918	.0294	.2122	16
45	0.23768	0.97134	0.24470	4.0867	1.0295	4.2072	15
46	.23797	.97127	.24501	.0815	.0296	.2022	14
47	.23825	.97120	.24531	.0764	.0296	.1972	13
48	.23853	.97113	.24562	.0713	.0297	.1923	12
49	.23881	.97106	.24593	.0662	.0298	.1873	11
50	0.23910	0.97099	0.24624	4.0611	1.0299	4.1824	10
51	.23938	.97092	.24655	.0560	.0299	.1774	9
52	.23966	.97086	.24686	.0509	.0300	.1725	8
53	.23994	.97079	.24717	.0458	.0301	.1676	7
54	.24023	.97072	.24747	.0408	.0302	.1627	6
55	0.24051	0.97065	0.24778	4.0358	1.0302	4.1578	5
56	.24079	.97058	.24809	.0307	.0303	.1529	4
57	.24107	.97051	.24840	.0257	.0304	.1481	3
58	.24136	.97044	.24871	.0207	.0305	.1432	2
59	.24164	.97037	.24902	.0157	.0305	.1384	1
60	0.24192	0.97029	0.24933	4.0108	1.0306	4.1336	0
M	Cosine	Sine	Cotan.	Tan.	Cosec.	Secant	M

103° 76°

14° Natural Trigonometric Functions 165°

M	Sine	Cosine	Tan.	Cotan.	Secant	Cosec.	M
0	0.24192	0.97029	0.24933	4.0108	1.0306	4.1336	60
1	.24220	.97022	.24964	.0058	.0307	.1287	59
2	.24249	.97015	.24995	.0009	.0308	.1239	58
3	.24277	.97008	.25025	3.9959	.0308	.1191	57
4	.24305	.97001	.25056	.9910	.0309	.1144	56
5	0.24333	0.96994	0.25087	3.9861	1.0310	4.1096	55
6	.24361	.96987	.25118	.9812	.0311	.1048	54
7	.24390	.96980	.25149	.9763	.0311	.1001	53
8	.24418	.96973	.25180	.9714	.0312	.0953	52
9	.24446	.96966	.25211	.9665	.0313	.0906	51
10	0.24474	0.96959	0.25242	3.9616	1.0314	4.0859	50
11	.24502	.96952	.25273	.9568	.0314	.0812	49
12	.24531	.96944	.25304	.9520	.0315	.0765	48
13	.24559	.96937	.25335	.9471	.0316	.0718	47
14	.24587	.96930	.25366	.9423	.0317	.0672	46
15	0.24615	0.96923	0.25397	3.9375	1.0317	4.0625	45
16	.24643	.96916	.25428	.9327	.0318	.0579	44
17	.24672	.96909	.25459	.9279	.0319	.0532	43
18	.24700	.96901	.25490	.9231	.0320	.0486	42
19	.24728	.96894	.25521	.9184	.0320	.0440	41
20	0.24756	0.96887	0.25552	3.9136	1.0321	4.0394	40
21	.24784	.96880	.25583	.9089	.0322	.0348	39
22	.24813	.96873	.25614	.9042	.0323	.0302	38
23	.24841	.96865	.25645	.8994	.0323	.0256	37
24	.24869	.96858	.25676	.8947	.0324	.0211	36
25	0.24897	0.96851	0.25707	3.8900	1.0325	4.0165	35
26	.24925	.96844	.25738	.8853	.0326	.0120	34
27	.24953	.96836	.25769	.8807	.0327	.0074	33
28	.24982	.96829	.25800	.8760	.0327	.0029	32
29	.25010	.96822	.25831	.8713	.0328	3.9984	31
30	0.25038	0.96815	0.25862	3.8667	1.0329	3.9939	30
31	.25066	.96807	.25893	.8621	.0330	.9894	29
32	.25094	.96800	.25924	.8574	.0330	.9850	28
33	.25122	.96793	.25955	.8528	.0331	.9805	27
34	.25151	.96785	.25986	.8482	.0332	.9760	26
35	0.25179	0.96778	0.26017	3.8436	1.0333	3.9716	25
36	.25207	.96771	.26048	.8390	.0334	.9672	24
37	.25235	.96763	.26079	.8345	.0334	.9627	23
38	.25263	.96756	.26110	.8299	.0335	.9583	22
39	.25291	.96749	.26141	.8254	.0336	.9539	21
40	0.25319	0.96741	0.26172	3.8208	1.0337	3.9495	20
41	.25348	.96734	.26203	.8163	.0338	.9451	19
42	.25376	.96727	.26234	.8118	.0338	.9408	18
43	.25404	.96719	.26266	.8073	.0339	.9364	17
44	.25432	.96712	.26297	.8027	.0340	.9320	16
45	0.25460	0.96704	0.26328	3.7983	1.0341	3.9277	15
46	.25488	.96697	.26359	.7938	.0341	.9234	14
47	.25516	.96690	.26390	.7893	.0342	.9190	13
48	.25544	.96682	.26421	.7848	.0343	.9147	12
49	.25573	.96675	.26452	.7804	.0344	.9104	11
50	0.25601	0.96667	0.26483	3.7759	1.0345	3.9061	10
51	.25629	.96660	.26514	.7715	.0345	.9018	9
52	.25657	.96652	.26546	.7671	.0346	.8976	8
53	.25685	.96645	.26577	.7627	.0347	.8933	7
54	.25713	.96638	.26608	.7583	.0348	.8890	6
55	0.25741	0.96630	0.26639	3.7539	1.0349	3.8848	5
56	.25769	.96623	.26670	.7495	.0349	.8805	4
57	.25798	.96615	.26701	.7451	.0350	.8763	3
58	.25826	.96608	.26732	.7407	.0351	.8721	2
59	.25854	.96600	.26764	.7364	.0352	.8679	1
60	0.25882	0.96593	0.26795	3.7320	1.0353	3.8637	0
M	Cosine	Sine	Cotan.	Tan.	Cosec.	Secant	M

104° 75°

15° Natural Trigonometric Functions 164°

M	Sine	Cosine	Tan.	Cotan.	Secant	Cosec.	M
0	0.25882	0.96593	0.26795	3.7320	1.0353	3.8637	60
1	.25910	.96585	.26826	.7277	.0353	.8595	59
2	.25938	.96577	.26857	.7234	.0354	.8553	58
3	.25966	.96570	.26888	.7191	.0355	.8512	57
4	.25994	.96562	.26920	.7147	.0356	.8470	56
5	0.26022	0.96555	0.26951	3.7104	1.0357	3.8428	55
6	.26050	.96547	.26982	.7062	.0358	.8387	54
7	.26078	.96540	.27013	.7019	.0358	.8346	53
8	.26107	.96532	.27044	.6976	.0359	.8304	52
9	.26135	.96524	.27076	.6933	.0360	.8263	51
10	0.26163	0.96517	0.27107	3.6891	1.0361	3.8222	50
11	.26191	.96509	.27138	.6848	.0362	.8181	49
12	.26219	.96502	.27169	.6806	.0362	.8140	48
13	.26247	.96494	.27201	.6764	.0363	.8100	47
14	.26275	.96486	.27232	.6722	.0364	.8059	46
15	0.26303	0.96479	0.27263	3.6679	1.0365	3.8018	45
16	.26331	.96471	.27294	.6637	.0366	.7978	44
17	.26359	.96463	.27326	.6596	.0367	.7937	43
18	.26387	.96456	.27357	.6554	.0367	.7897	42
19	.26415	.96448	.27388	.6512	.0368	.7857	41
20	0.26443	0.96440	0.27419	3.6470	1.0369	3.7816	40
21	.26471	.96433	.27451	.6429	.0370	.7776	39
22	.26499	.96425	.27482	.6387	.0371	.7736	38
23	.26527	.96417	.27513	.6346	.0371	.7697	37
24	.26556	.96409	.27544	.6305	.0372	.7657	36
25	0.26584	0.96402	0.27576	3.6263	1.0373	3.7617	35
26	.26612	.96394	.27607	.6222	.0374	.7577	34
27	.26640	.96386	.27638	.6181	.0375	.7538	33
28	.26668	.96378	.27670	.6140	.0376	.7498	32
29	.26696	.96371	.27701	.6100	.0376	.7459	31
30	0.26724	0.96363	0.27732	3.6059	1.0377	3.7420	30
31	.26752	.96355	.27764	.6018	.0378	.7380	29
32	.26780	.96347	.27795	.5977	.0379	.7341	28
33	.26808	.96340	.27826	.5937	.0380	.7302	27
34	.26836	.96332	.27858	.5896	.0381	.7263	26
35	0.26864	0.96324	0.27889	3.5856	1.0382	3.7224	25
36	.26892	.96316	.27920	.5816	.0382	.7186	24
37	.26920	.96308	.27952	.5776	.0383	.7147	23
38	.26948	.96301	.27983	.5736	.0384	.7108	22
39	.26976	.96293	.28014	.5696	.0385	.7070	21
40	0.27004	0.96285	0.28046	3.5656	1.0386	3.7031	20
41	.27032	.96277	.28077	.5616	.0387	.6993	19
42	.27060	.96269	.28109	.5576	.0387	.6955	18
43	.27088	.96261	.28140	.5536	.0388	.6917	17
44	.27116	.96253	.28171	.5497	.0389	.6878	16
45	0.27144	0.96245	0.28203	3.5457	1.0390	3.6840	15
46	.27172	.96238	.28234	.5418	.0391	.6802	14
47	.27200	.96230	.28266	.5378	.0392	.6765	13
48	.27228	.96222	.28297	.5339	.0393	.6727	12
49	.27256	.96214	.28328	.5300	.0393	.6689	11
50	0.27284	0.96206	0.28360	3.5261	1.0394	3.6651	10
51	.27312	.96198	.28391	.5222	.0395	.6614	9
52	.27340	.96190	.28423	.5183	.0396	.6576	8
53	.27368	.96182	.28454	.5144	.0397	.6539	7
54	.27396	.96174	.28486	.5105	.0398	.6502	6
55	0.27424	0.96166	0.28517	3.5066	1.0399	3.6464	5
56	.27452	.96158	.28549	.5028	.0399	.6427	4
57	.27480	.96150	.28580	.4989	.0400	.6390	3
58	.27508	.96142	.28611	.4951	.0401	.6353	2
59	.27536	.96134	.28643	.4912	.0402	.6316	1
60	0.27564	0.96126	0.28674	3.4874	1.0403	3.6279	0
M	Cosine	Sine	Cotan.	Tan.	Cosec.	Secant	M

105° 74°

16° Natural Trigonometric Functions 163°

M	Sine	Cosine	Tan.	Cotan.	Secant	Cosec.	M
0	0.27564	0.96126	0.28674	3.4874	1.0403	3.6279	60
1	.27592	.96118	.28706	.4836	.0404	.6243	59
2	.27620	.96110	.28737	.4798	.0405	.6206	58
3	.27648	.96102	.28769	.4760	.0406	.6169	57
4	.27675	.96094	.28800	.4722	.0406	.6133	56
5	0.27703	0.96086	0.28832	3.4684	1.0407	3.6096	55
6	.27731	.96078	.28863	.4646	.0408	.6060	54
7	.27759	.96070	.28895	.4608	.0409	.6024	53
8	.27787	.96062	.28926	.4570	.0410	.5987	52
9	.27815	.96054	.28958	.4533	.0411	.5951	51
10	0.27843	0.96045	0.28990	3.4495	1.0412	3.5915	50
11	.27871	.96037	.29021	.4458	.0413	.5879	49
12	.27899	.96029	.29053	.4420	.0413	.5843	48
13	.27927	.96021	.29084	.4383	.0414	.5807	47
14	.27955	.96013	.29116	.4346	.0415	.5772	46
15	0.27983	0.96005	0.29147	3.4308	1.0416	3.5736	45
16	.28011	.95997	.29179	.4271	.0417	.5700	44
17	.28039	.95989	.29210	.4234	.0418	.5665	43
18	.28067	.95980	.29242	.4197	.0419	.5629	42
19	.28094	.95972	.29274	.4160	.0420	.5594	41
20	0.28122	0.95964	0.29305	3.4124	1.0420	3.5559	40
21	.28150	.95956	.29337	.4087	.0421	.5523	39
22	.28178	.95948	.29368	.4050	.0422	.5488	38
23	.28206	.95940	.29400	.4014	.0423	.5453	37
24	.28234	.95931	.29432	.3977	.0424	.5418	36
25	0.28262	0.95923	0.29463	3.3941	1.0425	3.5383	35
26	.28290	.95915	.29495	.3904	.0426	.5348	34
27	.28318	.95907	.29526	.3868	.0427	.5313	33
28	.28346	.95898	.29558	.3832	.0428	.5279	32
29	.28374	.95890	.29590	.3795	.0428	.5244	31
30	0.28401	0.95882	0.29621	3.3759	1.0429	3.5209	30
31	.28429	.95874	.29653	.3723	.0430	.5175	29
32	.28457	.95865	.29685	.3687	.0431	.5140	28
33	.28485	.95857	.29716	.3651	.0432	.5106	27
34	.28513	.95849	.29748	.3616	.0433	.5072	26
35	0.28541	0.95840	0.29780	3.3580	1.0434	3.5037	25
36	.28569	.95832	.29811	.3544	.0435	.5003	24
37	.28597	.95824	.29843	.3509	.0436	.4969	23
38	.28624	.95816	.29875	.3473	.0437	.4935	22
39	.28652	.95807	.29906	.3438	.0438	.4901	21
40	0.28680	0.95799	0.29938	3.3402	1.0438	3.4867	20
41	.28708	.95791	.29970	.3367	.0439	.4833	19
42	.28736	.95782	.30001	.3332	.0440	.4799	18
43	.28764	.95774	.30033	.3296	.0441	.4766	17
44	.28792	.95765	.30065	.3261	.0442	.4732	16
45	0.28820	0.95757	0.30096	3.3226	1.0443	3.4698	15
46	.28847	.95749	.30128	.3191	.0444	.4665	14
47	.28875	.95740	.30160	.3156	.0445	.4632	13
48	.28903	.95732	.30192	.3121	.0446	.4598	12
49	.28931	.95724	.30223	.3087	.0447	.4565	11
50	0.28959	0.95715	0.30255	3.3052	1.0448	3.4532	10
51	.28987	.95707	.30287	.3017	.0448	.4498	9
52	.29014	.95698	.30319	.2983	.0449	.4465	8
53	.29042	.95690	.30350	.2948	.0450	.4432	7
54	.29070	.95681	.30382	.2914	.0451	.4399	6
55	0.29098	0.95673	0.30414	3.2879	1.0452	3.4366	5
56	.29126	.95664	.30446	.2845	.0453	.4334	4
57	.29154	.95656	.30478	.2811	.0454	.4301	3
58	.29181	.95647	.30509	.2777	.0455	.4268	2
59	.29209	.95639	.30541	.2742	.0456	.4236	1
60	0.29237	0.95630	0.30573	3.2708	1.0457	3.4203	0
M	Cosine	Sine	Cotan.	Tan.	Cosec.	Secant	M

106° 73°

17° Natural Trigonometric Functions 162°

M	Sine	Cosine	Tan.	Cotan.	Secant	Cosec.	M
0	0.29237	0.95630	0.30573	3.2708	1.0457	3.4203	60
1	.29265	.95622	.30605	.2674	.0458	.4170	59
2	.29293	.95613	.30637	.2640	.0459	.4138	58
3	.29321	.95605	.30668	.2607	.0460	.4106	57
4	.29348	.95596	.30700	.2573	.0461	.4073	56
5	0.29376	0.95588	0.30732	3.2539	1.0461	3.4041	55
6	.29404	.95579	.30764	.2505	.0462	.4009	54
7	.29432	.95571	.30796	.2472	.0463	.3977	53
8	.29460	.95562	.30828	.2438	.0464	.3945	52
9	.29487	.95554	.30859	.2405	.0465	.3913	51
10	0.29515	0.95545	0.30891	3.2371	1.0466	3.3881	50
11	.29543	.95536	.30923	.2338	.0467	.3849	49
12	.29571	.95528	.30955	.2305	.0468	.3817	48
13	.29598	.95519	.30987	.2271	.0469	.3785	47
14	.29626	.95511	.31019	.2238	.0470	.3754	46
15	0.29654	0.95502	0.31051	3.2205	1.0471	3.3722	45
16	.29682	.95493	.31083	.2172	.0472	.3690	44
17	.29710	.95485	.31115	.2139	.0473	.3659	43
18	.29737	.95476	.31146	.2106	.0474	.3627	42
19	.29765	.95467	.31178	.2073	.0475	.3596	41
20	0.29793	0.95459	0.31210	3.2041	1.0476	3.3565	40
21	.29821	.95450	.31242	.2008	.0477	.3534	39
22	.29848	.95441	.31274	.1975	.0478	.3502	38
23	.29876	.95433	.31306	.1942	.0478	.3471	37
24	.29904	.95424	.31338	.1910	.0479	.3440	36
25	0.29932	0.95415	0.31370	3.1877	1.0480	3.3409	35
26	.29959	.95407	.31402	.1845	.0481	.3378	34
27	.29987	.95398	.31434	.1813	.0482	.3347	33
28	.30015	.95389	.31466	.1780	.0483	.3316	32
29	.30043	.95380	.31498	.1748	.0484	.3286	31
30	0.30070	0.95372	0.31530	3.1716	1.0485	3.3255	30
31	.30098	.95363	.31562	.1684	.0486	.3224	29
32	.30126	.95354	.31594	.1652	.0487	.3194	28
33	.30154	.95345	.31626	.1620	.0488	.3163	27
34	.30181	.95337	.31658	.1588	.0489	.3133	26
35	0.30209	0.95328	0.31690	3.1556	1.0490	3.3102	25
36	.30237	.95319	.31722	.1524	.0491	.3072	24
37	.30265	.95310	.31754	.1492	.0492	.3042	23
38	.30292	.95301	.31786	.1460	.0493	.3011	22
39	.30320	.95293	.31818	.1429	.0494	.2981	21
40	0.30348	0.95284	0.31850	3.1397	1.0495	3.2951	20
41	.30375	.95275	.31882	.1366	.0496	.2921	19
42	.30403	.95266	.31914	.1334	.0497	.2891	18
43	.30431	.95257	.31946	.1303	.0498	.2861	17
44	.30459	.95248	.31978	.1271	.0499	.2831	16
45	0.30486	0.95239	0.32010	3.1240	1.0500	3.2801	15
46	.30514	.95231	.32042	.1209	.0501	.2772	14
47	.30542	.95222	.32074	.1177	.0502	.2742	13
48	.30569	.95213	.32106	.1146	.0503	.2712	12
49	.30597	.95204	.32138	.1115	.0504	.2683	11
50	0.30625	0.95195	0.32171	3.1084	1.0505	3.2653	10
51	.30653	.95186	.32203	.1053	.0506	.2624	9
52	.30680	.95177	.32235	.1022	.0507	.2594	8
53	.30708	.95168	.32267	.0991	.0508	.2565	7
54	.30736	.95159	.32299	.0960	.0509	.2535	6
55	0.30763	0.95150	0.32331	3.0930	1.0510	3.2506	5
56	.30791	.95141	.32363	.0899	.0511	.2477	4
57	.30819	.95132	.32395	.0868	.0512	.2448	3
58	.30846	.95124	.32428	.0838	.0513	.2419	2
59	.30874	.95115	.32460	.0807	.0514	.2390	1
60	0.30902	0.95106	0.32492	3.0777	1.0515	3.2361	0
M	Cosine	Sine	Cotan.	Tan.	Cosec.	Secant	M

107° 72°

18° Natural Trigonometric Functions 161°

M	Sine	Cosine	Tan.	Cotan.	Secant	Cosec.	M
0	0.30902	0.95106	0.32492	3.0777	1.0515	3.2361	60
1	.30929	.95097	.32524	.0746	.0516	.2332	59
2	.30957	.95088	.32556	.0716	.0517	.2303	58
3	.30985	.95079	.32588	.0686	.0518	.2274	57
4	.31012	.95070	.32621	.0655	.0519	.2245	56
5	0.31040	0.95061	0.32653	3.0625	1.0520	3.2216	55
6	.31068	.95051	.32685	.0595	.0521	.2188	54
7	.31095	.95042	.32717	.0565	.0522	.2159	53
8	.31123	.95033	.32749	.0535	.0523	.2131	52
9	.31150	.95024	.32782	.0505	.0524	.2102	51
10	0.31178	0.95015	0.32814	3.0475	1.0525	3.2074	50
11	.31206	.95006	.32846	.0445	.0526	.2045	49
12	.31233	.94997	.32878	.0415	.0527	.2017	48
13	.31261	.94988	.32910	.0385	.0528	.1989	47
14	.31289	.94979	.32943	.0356	.0529	.1960	46
15	0.31316	0.94970	0.32975	3.0326	1.0530	3.1932	45
16	.31344	.94961	.33007	.0296	.0531	.1904	44
17	.31372	.94952	.33039	.0267	.0532	.1876	43
18	.31399	.94942	.33072	.0237	.0533	.1848	42
19	.31427	.94933	.33104	.0208	.0534	.1820	41
20	0.31454	0.94924	0.33136	3.0178	1.0535	3.1792	40
21	.31482	.94915	.33169	.0149	.0536	.1764	39
22	.31510	.94906	.33201	.0120	.0537	.1736	38
23	.31537	.94897	.33233	.0090	.0538	.1708	37
24	.31565	.94888	.33265	.0061	.0539	.1681	36
25	0.31592	0.94878	0.33298	3.0032	1.0540	3.1653	35
26	.31620	.94869	.33330	.0003	.0541	.1625	34
27	.31648	.94860	.33362	2.9974	.0542	.1598	33
28	.31675	.94851	.33395	.9945	.0543	.1570	32
29	.31703	.94841	.33427	.9916	.0544	.1543	31
30	0.31730	0.94832	0.33459	2.9887	1.0545	3.1515	30
31	.31758	.94823	.33492	.9858	.0546	.1488	29
32	.31786	.94814	.33524	.9829	.0547	.1461	28
33	.31813	.94805	.33557	.9800	.0548	.1433	27
34	.31841	.94795	.33589	.9772	.0549	.1406	26
35	0.31868	0.94786	0.33621	2.9743	1.0550	3.1379	25
36	.31896	.94777	.33654	.9714	.0551	.1352	24
37	.31923	.94767	.33686	.9686	.0552	.1325	23
38	.31951	.94758	.33718	.9657	.0553	.1298	22
39	.31978	.94749	.33751	.9629	.0554	.1271	21
40	0.32006	0.94740	0.33783	2.9600	1.0555	3.1244	20
41	.32034	.94730	.33816	.9572	.0556	.1217	19
42	.32061	.94721	.33848	.9544	.0557	.1190	18
43	.32089	.94712	.33880	.9515	.0558	.1163	17
44	.32116	.94702	.33913	.9487	.0559	.1137	16
45	0.32144	0.94693	0.33945	2.9459	1.0560	3.1110	15
46	.32171	.94684	.33978	.9431	.0561	.1083	14
47	.32199	.94674	.34010	.9403	.0562	.1057	13
48	.32226	.94665	.34043	.9375	.0563	.1030	12
49	.32254	.94655	.34075	.9347	.0565	.1004	11
50	0.32282	0.94646	0.34108	2.9319	1.0566	3.0977	10
51	.32309	.94637	.34140	.9291	.0567	.0951	9
52	.32337	.94627	.34173	.9263	.0568	.0925	8
53	.32364	.94618	.34205	.9235	.0569	.0898	7
54	.32392	.94608	.34238	.9208	.0570	.0872	6
55	0.32419	0.94599	0.34270	2.9180	1.0571	3.0846	5
56	.32447	.94590	.34303	.9152	.0572	.0820	4
57	.32474	.94580	.34335	.9125	.0573	.0793	3
58	.32502	.94571	.34368	.9097	.0574	.0767	2
59	.32529	.94561	.34400	.9069	.0575	.0741	1
60	0.32557	0.94552	0.34433	2.9042	1.0576	3.0715	0
M	Cosine	Sine	Cotan.	Tan.	Cosec.	Secant	M

108° 71°

19° Natural Trigonometric Functions 160°

M	Sine	Cosine	Tan.	Cotan.	Secant	Cosec.	M
0	0.32557	0.94552	0.34433	2.9042	1.0576	3.0715	60
1	.32584	.94542	.34465	.9015	.0577	.0690	59
2	.32612	.94533	.34498	.8987	.0578	.0664	58
3	.32639	.94523	.34530	.8960	.0579	.0638	57
4	.32667	.94514	.34563	.8933	.0580	.0612	56
5	0.32694	0.94504	0.34595	2.8905	1.0581	3.0586	55
6	.32722	.94495	.34628	.8878	.0582	.0561	54
7	.32749	.94485	.34661	.8851	.0584	.0535	53
8	.32777	.94476	.34693	.8824	.0585	.0509	52
9	.32804	.94466	.34726	.8797	.0586	.0484	51
10	0.32832	0.94457	0.34758	2.8770	1.0587	3.0458	50
11	.32859	.94447	.34791	.8743	.0588	.0433	49
12	.32887	.94438	.34824	.8716	.0589	.0407	48
13	.32914	.94428	.34856	.8689	.0590	.0382	47
14	.32942	.94418	.34889	.8662	.0591	.0357	46
15	0.32969	0.94409	0.34921	2.8636	1.0592	3.0331	45
16	.32996	.94399	.34954	.8609	.0593	.0306	44
17	.33024	.94390	.34987	.8582	.0594	.0281	43
18	.33051	.94380	.35019	.8555	.0595	.0256	42
19	.33079	.94370	.35052	.8529	.0596	.0231	41
20	0.33106	0.94361	0.35085	2.8502	1.0598	3.0206	40
21	.33134	.94351	.35117	.8476	.0599	.0181	39
22	.33161	.94341	.35150	.8449	.0600	.0156	38
23	.33189	.94332	.35183	.8423	.0601	.0131	37
24	.33216	.94322	.35215	.8396	.0602	.0106	36
25	0.33243	0.94313	0.35248	2.8370	1.0603	3.0081	35
26	.33271	.94303	.35281	.8344	.0604	.0056	34
27	.33298	.94293	.35314	.8318	.0605	.0031	33
28	.33326	.94283	.35346	.8291	.0606	.0007	32
29	.33353	.94274	.35379	.8265	.0607	2.9982	31
30	0.33381	0.94264	0.35412	2.8239	1.0608	2.9957	30
31	.33408	.94254	.35445	.8213	.0609	.9933	29
32	.33435	.94245	.35477	.8187	.0611	.9908	28
33	.33463	.94235	.35510	.8161	.0612	.9884	27
34	.33490	.94225	.35543	.8135	.0613	.9859	26
35	0.33518	0.94215	0.35576	2.8109	1.0614	2.9835	25
36	.33545	.94206	.35608	.8083	.0615	.9810	24
37	.33572	.94196	.35641	.8057	.0616	.9786	23
38	.33600	.94186	.35674	.8032	.0617	.9762	22
39	.33627	.94176	.35707	.8006	.0618	.9738	21
40	0.33655	0.94167	0.35739	2.7980	1.0619	2.9713	20
41	.33682	.94157	.35772	.7954	.0620	.9689	19
42	.33709	.94147	.35805	.7929	.0622	.9665	18
43	.33737	.94137	.35838	.7903	.0623	.9641	17
44	.33764	.94127	.35871	.7878	.0624	.9617	16
45	0.33792	0.94118	0.35904	2.7852	1.0625	2.9593	15
46	.33819	.94108	.35936	.7827	.0626	.9569	14
47	.33846	.94098	.35969	.7801	.0627	.9545	13
48	.33874	.94088	.36002	.7776	.0628	.9521	12
49	.33901	.94078	.36035	.7751	.0629	.9497	11
50	0.33928	0.94068	0.36068	2.7725	1.0630	2.9474	10
51	.33956	.94058	.36101	.7700	.0632	.9450	9
52	.33983	.94049	.36134	.7675	.0633	.9426	8
53	.34011	.94039	.36167	.7650	.0634	.9402	7
54	.34038	.94029	.36199	.7625	.0635	.9379	6
55	0.34065	0.94019	0.36232	2.7600	1.0636	2.9355	5
56	.34093	.94009	.36265	.7575	.0637	.9332	4
57	.34120	.93999	.36298	.7550	.0638	.9308	3
58	.34147	.93989	.36331	.7525	.0639	.9285	2
59	.34175	.93979	.36364	.7500	.0641	.9261	1
60	0.34202	0.93969	0.36397	2.7475	1.0642	2.9238	0
M	Cosine	Sine	Cotan.	Tan.	Cosec.	Secant	M

109° 70°

20° Natural Trigonometric Functions 159°

M	Sine	Cosine	Tan.	Cotan.	Secant	Cosec.	M
0	0.34202	0.93969	0.36397	2.7475	1.0642	2.9238	60
1	.34229	.93959	.36430	.7450	.0643	.9215	59
2	.34257	.93949	.36463	.7425	.0644	.9191	58
3	.34284	.93939	.36496	.7400	.0645	.9168	57
4	.34311	.93929	.36529	.7376	.0646	.9145	56
5	0.34339	0.93919	0.36562	2.7351	1.0647	2.9122	55
6	.34366	.93909	.36595	.7326	.0648	.9098	54
7	.34393	.93899	.36628	.7302	.0650	.9075	53
8	.34421	.93889	.36661	.7277	.0651	.9052	52
9	.34448	.93879	.36694	.7252	.0652	.9029	51
10	0.34475	0.93869	0.36727	2.7228	1.0653	2.9006	50
11	.34502	.93859	.36760	.7204	.0654	.8983	49
12	.34530	.93849	.36793	.7179	.0655	.8960	48
13	.34557	.93839	.36826	.7155	.0656	.8937	47
14	.34584	.93829	.36859	.7130	.0658	.8915	46
15	0.34612	0.93819	0.36892	2.7106	1.0659	2.8892	45
16	.34639	.93809	.36925	.7082	.0660	.8869	44
17	.34666	.93799	.36958	.7058	.0661	.8846	43
18	.34693	.93789	.36991	.7033	.0662	.8824	42
19	.34721	.93779	.37024	.7009	.0663	.8801	41
20	0.34748	0.93769	0.37057	2.6985	1.0664	2.8778	40
21	.34775	.93758	.37090	.6961	.0666	.8756	39
22	.34803	.93748	.37123	.6937	.0667	.8733	38
23	.34830	.93738	.37156	.6913	.0668	.8711	37
24	.34857	.93728	.37190	.6889	.0669	.8688	36
25	0.34884	0.93718	0.37223	2.6865	1.0670	2.8666	35
26	.34912	.93708	.37256	.6841	.0671	.8644	34
27	.34939	.93698	.37289	.6817	.0673	.8621	33
28	.34966	.93687	.37322	.6794	.0674	.8599	32
29	.34993	.93677	.37355	.6770	.0675	.8577	31
30	0.35021	0.93667	0.37388	2.6746	1.0676	2.8554	30
31	.35048	.93657	.37422	.6722	.0677	.8532	29
32	.35075	.93647	.37455	.6699	.0678	.8510	28
33	.35102	.93637	.37488	.6675	.0679	.8488	27
34	.35130	.93626	.37521	.6652	.0681	.8466	26
35	0.35157	0.93616	0.37554	2.6628	1.0682	2.8444	25
36	.35184	.93606	.37587	.6604	.0683	.8422	24
37	.35211	.93596	.37621	.6581	.0684	.8400	23
38	.35239	.93585	.37654	.6558	.0685	.8378	22
39	.35266	.93575	.37687	.6534	.0686	.8356	21
40	0.35293	0.93565	0.37720	2.6511	1.0688	2.8334	20
41	.35320	.93555	.37754	.6487	.0689	.8312	19
42	.35347	.93544	.37787	.6464	.0690	.8290	18
43	.35375	.93534	.37820	.6441	.0691	.8269	17
44	.35402	.93524	.37853	.6418	.0692	.8247	16
45	0.35429	0.93513	0.37887	2.6394	1.0694	2.8225	15
46	.35456	.93503	.37920	.6371	.0695	.8204	14
47	.35483	.93493	.37953	.6348	.0696	.8182	13
48	.35511	.93482	.37986	.6325	.0697	.8160	12
49	.35538	.93472	.38020	.6302	.0698	.8139	11
50	0.35565	0.93462	0.38053	2.6279	1.0699	2.8117	10
51	.35592	.93451	.38086	.6256	.0701	.8096	9
52	.35619	.93441	.38120	.6233	.0702	.8074	8
53	.35647	.93431	.38153	.6210	.0703	.8053	7
54	.35674	.93420	.38186	.6187	.0704	.8032	6
55	0.35701	0.93410	0.38220	2.6164	1.0705	2.8010	5
56	.35728	.93400	.38253	.6142	.0707	.7989	4
57	.35755	.93389	.38286	.6119	.0708	.7968	3
58	.35782	.93379	.38320	.6096	.0709	.7947	2
59	.35810	.93368	.38353	.6073	.0710	.7925	1
60	0 35837	0.93358	0.38386	2.6051	1.0711	2.7904	0
M	Cosine	Sine	Cotan.	Tan.	Cosec.	Secant	M

110° 69°

21° **Natural Trigonometric Functions** 158°

M	Sine	Cosine	Tan.	Cotan.	Secant	Cosec.	M
0	0.35837	0.93358	0.38386	2.6051	1.0711	2.7904	60
1	.35864	.93348	.38420	.6028	.0713	.7883	59
2	.35891	.93337	.38453	.6006	.0714	.7862	58
3	.35918	.93327	.38486	.5983	.0715	.7841	57
4	.35945	.93316	.38520	.5960	.0716	.7820	56
5	0.35972	0.93306	0.38553	2.5938	1.0717	2.7799	55
6	.36000	.93295	.38587	.5916	.0719	.7778	54
7	.36027	.93285	.38620	.5893	.0720	.7757	53
8	.36054	.93274	.38654	.5871	.0721	.7736	52
9	.36081	.93264	.38687	.5848	.0722	.7715	51
10	0.36108	0.93253	0.38720	2.5826	1.0723	2.7694	50
11	.36135	.93243	.38754	.5804	.0725	.7674	49
12	.36162	.93232	.38787	.5781	.0726	.7653	48
13	.36189	.93222	.38821	.5759	.0727	.7632	47
14	.36217	.93211	.38854	.5737	.0728	.7611	46
15	0.36244	0.93201	0.38888	2.5715	1.0729	2.7591	45
16	.36271	.93190	.38921	.5693	.0731	.7570	44
17	.36298	.93180	.38955	.5671	.0732	.7550	43
18	.36325	.93169	.38988	.5649	.0733	.7529	42
19	.36352	.93158	.39022	.5627	.0734	.7509	41
20	0.36379	0.93148	0.39055	2.5605	1.0736	2.7488	40
21	.36406	.93137	.39089	.5583	.0737	.7468	39
22	.36433	.93127	.39122	.5561	.0738	.7447	38
23	.36460	.93116	.39156	.5539	.0739	.7427	37
24	.36488	.93105	.39189	.5517	.0740	.7406	36
25	0.36515	0.93095	0.39223	2.5495	1.0742	2.7386	35
26	.36542	.93084	.39257	.5473	.0743	.7366	34
27	.36569	.93074	.39290	.5451	.0744	.7346	33
28	.36596	.93063	.39324	.5430	.0745	.7325	32
29	.36623	.93052	.39357	.5408	.0747	.7305	31
30	0.36650	0.93042	0.39391	2.5386	1.0748	2.7285	30
31	.36677	.93031	.39425	.5365	.0749	.7265	29
32	.36704	.93020	.39458	.5343	.0750	.7245	28
33	.36731	.93010	.39492	.5322	.0751	.7225	27
34	.36758	.92999	.39525	.5300	.0753	.7205	26
35	0.36785	0.92988	0.39559	2.5278	1.0754	2.7185	25
36	.36812	.92978	.39593	.5257	.0755	.7165	24
37	.36839	.92967	.39626	.5236	.0756	.7145	23
38	.36866	.92956	.39660	.5214	.0758	.7125	22
39	.36893	.92945	.39694	.5193	.0759	.7105	21
40	0.36921	0.92935	0.39727	2.5171	1.0760	2.7085	20
41	.36948	.92924	.39761	.5150	.0761	.7065	19
42	.36975	.92913	.39795	.5129	.0763	.7045	18
43	.37002	.92902	.39828	.5108	.0764	.7026	17
44	.37029	.92892	.39862	.5086	.0765	.7006	16
45	0.37056	0.92881	0.39896	2.5065	1.0766	2.6986	15
46	.37083	.92870	.39930	.5044	.0768	.6967	14
47	.37110	.92859	.39963	.5023	.0769	.6947	13
48	.37137	.92848	.39997	.5002	.0770	.6927	12
49	.37164	.92838	.40031	.4981	.0771	.6908	11
50	0.37191	0.92827	0.40065	2.4960	1.0773	2.6888	10
51	.37218	.92816	.40098	.4939	.0774	.6869	9
52	.37245	.92805	.40132	.4918	.0775	.6849	8
53	.37272	.92794	.40166	.4897	.0776	.6830	7
54	.37299	.92784	.40200	.4876	.0778	.6810	6
55	0.37326	0.92773	0.40233	2.4855	1.0779	2.6791	5
56	.37353	.92762	.40267	.4834	.0780	.6772	4
57	.37380	.92751	.40301	.4813	.0781	.6752	3
58	.37407	.92740	.40335	.4792	.0783	.6733	2
59	.37434	.92729	.40369	.4772	.0784	.6714	1
60	0.37461	0.92718	0.40403	2.4751	1.0785	2.6695	0
M	Cosine	Sine	Cotan.	Tan.	Cosec.	Secant	M

111° 68°

22° **Natural Trigonometric Functions** 157°

M	Sine	Cosine	Tan.	Cotan.	Secant	Cosec.	M
0	0.37461	0.92718	0.40403	2.4751	1.0785	2.6695	60
1	.37488	.92707	.40436	.4730	.0787	.6675	59
2	.37514	.92696	.40470	.4709	.0788	.6656	58
3	.37541	.92686	.40504	.4689	.0789	.6637	57
4	.37568	.92675	.40538	.4668	.0790	.6618	56
5	0.37595	0.92664	0.40572	2.4647	1.0792	2.6599	55
6	.37622	.92653	.40606	.4627	.0793	.6580	54
7	.37649	.92642	.40640	.4606	.0794	.6561	53
8	.37676	.92631	.40673	.4586	.0795	.6542	52
9	.37703	.92620	.40707	.4565	.0797	.6523	51
10	0.37730	0.92609	0.40741	2.4545	1.0798	2.6504	50
11	.37757	.92598	.40775	.4525	.0799	.6485	49
12	.37784	.92587	.40809	.4504	.0801	.6466	48
13	.37811	.92576	.40843	.4484	.0802	.6447	47
14	.37838	.92565	.40877	.4463	.0803	.6428	46
15	0.37865	0.92554	0.40911	2.4443	1.0804	2.6410	45
16	.37892	.92543	.40945	.4423	.0806	.6391	44
17	.37919	.92532	.40979	.4403	.0807	.6372	43
18	.37946	.92521	.41013	.4382	.0808	.6353	42
19	.37972	.92510	.41047	.4362	.0810	.6335	41
20	0.37999	0.92499	0.41081	2.4342	1.0811	2.6316	40
21	.38026	.92488	.41115	.4322	.0812	.6297	39
22	.38053	.92477	.41149	.4302	.0813	.6279	38
23	.38080	.92466	.41183	.4282	.0815	.6260	37
24	.38107	.92455	.41217	.4262	.0816	.6242	36
25	0.38134	0.92443	0.41251	2.4242	1.0817	2.6223	35
26	.38161	.92432	.41285	.4222	.0819	.6205	34
27	.38188	.92421	.41319	.4202	.0820	.6186	33
28	.38214	.92410	.41353	.4182	.0821	.6168	32
29	.38241	.92399	.41387	.4162	.0823	.6150	31
30	0.38268	0.92388	0.41421	2.4142	1.0824	2.6131	30
31	.38295	.92377	.41455	.4122	.0825	.6113	29
32	.38322	.92366	.41489	.4102	.0826	.6095	28
33	.38349	.92354	.41524	.4083	.0828	.6076	27
34	.38376	.92343	.41558	.4063	.0829	.6058	26
35	0.38403	0.92332	0.41592	2.4043	1.0830	2.6040	25
36	.38429	.92321	.41626	.4023	.0832	.6022	24
37	.38456	.92310	.41660	.4004	.0833	.6003	23
38	.38483	.92299	.41694	.3984	.0834	.5985	22
39	.38510	.92287	.41728	.3964	.0836	.5967	21
40	0.38537	0.92276	0.41762	2.3945	1.0837	2.5949	20
41	.38564	.92265	.41797	.3925	.0838	.5931	19
42	.38591	.92254	.41831	.3906	.0840	.5913	18
43	.38617	.92242	.41865	.3886	.0841	.5895	17
44	.38644	.92231	.41899	.3867	.0842	.5877	16
45	0.38671	0.92220	0.41933	2.3847	1.0844	2.5859	15
46	.38698	.92209	.41968	.3828	.0845	.5841	14
47	.38725	.92197	.42002	.3808	.0846	.5823	13
48	.38751	.92186	.42036	.3789	.0847	.5805	12
49	.38778	.92175	.42070	.3770	.0849	.5787	11
50	0.38805	0.92164	0.42105	2.3750	1.0850	2.5770	10
51	.38832	.92152	.42139	.3731	.0851	.5752	9
52	.38859	.92141	.42173	.3712	.0853	.5734	8
53	.38886	.92130	.42207	.3692	.0854	.5716	7
54	.38912	.92118	.42242	.3673	.0855	.5699	6
55	0.38939	0.92107	0.42276	2.3654	1.0857	2.5681	5
56	.38966	.92096	.42310	.3635	.0858	.5663	4
57	.38993	.92084	.42344	.3616	.0859	.5646	3
58	.39019	.92073	.42379	.3597	.0861	.5628	2
59	.39046	.92062	.42413	.3577	.0862	.5610	1
60	0.39073	0.92050	0.42447	2.3558	1.0864	2.5593	0
M	Cosine	Sine	Cotan.	Tan.	Cosec.	Secant	M

112° 67°

23° **Natural Trigonometric Functions** 156°

M	Sine	Cosine	Tan.	Cotan.	Secant	Cosec.	M
0	0.39073	0.92050	0.42447	2.3558	1.0864	2.5593	60
1	.39100	.92039	.42482	.3539	.0865	.5575	59
2	.39126	.92028	.42516	.3520	.0866	.5558	58
3	.39153	.92016	.42550	.3501	.0868	.5540	57
4	.39180	.92005	.42585	.3482	.0869	.5523	56
5	0.39207	0.91993	0.42619	2.3463	1.0870	2.5506	55
6	.39234	.91982	.42654	.3445	.0872	.5488	54
7	.39260	.91971	.42688	.3426	.0873	.5471	53
8	.39287	.91959	.42722	.3407	.0874	.5453	52
9	.39314	.91948	.42757	.3388	.0876	.5436	51
10	0.39341	0.91936	0.42791	2.3369	1.0877	2.5419	50
11	.39367	.91925	.42826	.3350	.0878	.5402	49
12	.39394	.91913	.42860	.3332	.0880	.5384	48
13	.39421	.91902	.42894	.3313	.0881	.5367	47
14	.39448	.91891	.42929	.3294	.0882	.5350	46
15	0.39474	0.91879	0.42963	2.3276	1.0884	2.5333	45
16	.39501	.91868	.42998	.3257	.0885	.5316	44
17	.39528	.91856	.43032	.3238	.0886	.5299	43
18	.39554	.91845	.43067	.3220	.0888	.5281	42
19	.39581	.91833	.43101	.3201	.0889	.5264	41
20	0.39608	0.91822	0.43136	2.3183	1.0891	2.5247	40
21	.39635	.91810	.43170	.3164	.0892	.5230	39
22	.39661	.91798	.43205	.3145	.0893	.5213	38
23	.39688	.91787	.43239	.3127	.0895	.5196	37
24	.39715	.91775	.43274	.3109	.0896	.5179	36
25	0.39741	0.91764	0.43308	2.3090	1.0897	2.5163	35
26	.39768	.91752	.43343	.3072	.0899	.5146	34
27	.39795	.91741	.43377	.3053	.0900	.5129	33
28	.39821	.91729	.43412	.3035	.0902	.5112	32
29	.39848	.91718	.43447	.3017	.0903	.5095	31
30	0.39875	0.91706	0.43481	2.2998	1.0904	2.5078	30
31	.39901	.91694	.43516	.2980	.0906	.5062	29
32	.39928	.91683	.43550	.2962	.0907	.5045	28
33	.39955	.91671	.43585	.2944	.0908	.5028	27
34	.39981	.91659	.43620	.2925	.0910	.5011	26
35	0.40008	0.91648	0.43654	2.2907	1.0911	2.4995	25
36	.40035	.91636	.43689	.2889	.0913	.4978	24
37	.40061	.91625	.43723	.2871	.0914	.4961	23
38	.40088	.91613	.43758	.2853	.0915	.4945	22
39	.40115	.91601	.43793	.2835	.0917	.4928	21
40	0.40141	0.91590	0.43827	2.2817	1.0918	2.4912	20
41	.40168	.91578	.43862	.2799	.0920	.4895	19
42	.40195	.91566	.43897	.2781	.0921	.4879	18
43	.40221	.91554	.43932	.2763	.0922	.4862	17
44	.40248	.91543	.43966	.2745	.0924	.4846	16
45	0.40275	0.91531	0.44001	2.2727	1.0925	2.4829	15
46	.40301	.91519	.44036	.2709	.0927	.4813	14
47	.40328	.91508	.44070	.2691	.0928	.4797	13
48	.40354	.91496	.44105	.2673	.0929	.4780	12
49	.40381	.91484	.44140	.2655	.0931	.4764	11
50	0.40408	0.91472	0.44175	2.2637	1.0932	2.4748	10
51	.40434	.91461	.44209	.2619	.0934	.4731	9
52	.40461	.91449	.44244	.2602	.0935	.4715	8
53	.40487	.91437	.44279	.2584	.0936	.4699	7
54	.40514	.91425	.44314	.2566	.0938	.4683	6
55	0.40541	0.91414	0.44349	2.2548	1.0939	2.4666	5
56	.40567	.91402	.44383	.2531	.0941	.4650	4
57	.40594	.91390	.44418	.2513	.0942	.4634	3
58	.40620	.91378	.44453	.2495	.0943	.4618	2
59	.40647	.91366	.44488	.2478	.0945	.4602	1
60	0.40674	0.91354	0.44523	2.2460	1.0946	2.4586	0
M	Cosine	Sine	Cotan.	Tan.	Cosec.	Secant	M

113° 66°

24° Natural Trigonometric Functions 155°

M	Sine	Cosine	Tan.	Cotan.	Secant	Cosec.	M
0	0.40674	0.91354	0.44523	2.2460	1.0946	2.4586	60
1	.40700	.91343	.44558	.2443	.0948	.4570	59
2	.40727	.91331	.44593	.2425	.0949	.4554	58
3	.40753	.91319	.44627	.2408	.0951	.4538	57
4	.40780	.91307	.44662	.2390	.0952	.4522	56
5	0.40806	0.91295	0.44697	2.2373	1.0953	2.4506	55
6	.40833	.91283	.44732	.2355	.0955	.4490	54
7	.40860	.91271	.44767	.2338	.0956	.4474	53
8	.40886	.91260	.44802	.2320	.0958	.4458	52
9	.40913	.91248	.44837	.2303	.0959	.4442	51
10	0.40939	0.91236	0.44872	2.2286	1.0961	2.4426	50
11	.40966	.91224	.44907	.2268	.0962	.4411	49
12	.40992	.91212	.44942	.2251	.0963	.4395	48
13	.41019	.91200	.44977	.2234	.0965	.4379	47
14	.41045	.91188	.45012	.2216	.0966	.4363	46
15	0.41072	0.91176	0.45047	2.2199	1.0968	2.4347	45
16	.41098	.91164	.45082	.2182	.0969	.4332	44
17	.41125	.91152	.45117	.2165	.0971	.4316	43
18	.41151	.91140	.45152	.2147	.0972	.4300	42
19	.41178	.91128	.45187	.2130	.0973	.4285	41
20	0.41204	0.91116	0.45222	2.2113	1.0975	2.4269	40
21	.41231	.91104	.45257	.2096	.0976	.4254	39
22	.41257	.91092	.45292	.2079	.0978	.4238	38
23	.41284	.91080	.45327	.2062	.0979	.4222	37
24	.41310	.91068	.45362	.2045	.0981	.4207	36
25	0.41337	0.91056	0.45397	2.2028	1.0982	2.4191	35
26	.41363	.91044	.45432	.2011	.0984	.4176	34
27	.41390	.91032	.45467	.1994	.0985	.4160	33
28	.41416	.91020	.45502	.1977	.0986	.4145	32
29	.41443	.91008	.45537	.1960	.0988	.4130	31
30	0.41469	0.90996	0.45573	2.1943	1.0989	2.4114	30
31	.41496	.90984	.45608	.1926	.0991	.4099	29
32	.41522	.90972	.45643	.1909	.0992	.4083	28
33	.41549	.90960	.45678	.1892	.0994	.4068	27
34	.41575	.90948	.45713	.1875	.0995	.4053	26
35	0.41602	0.90936	0.45748	2.1859	1.0997	2.4037	25
36	.41628	.90924	.45783	.1842	.0998	.4022	24
37	.41654	.90911	.45819	.1825	.1000	.4007	23
38	.41681	.90899	.45854	.1808	.1001	.3992	22
39	.41707	.90887	.45889	.1792	.1003	.3976	21
40	0.41734	0.90875	0.45924	2.1775	1.1004	2.3961	20
41	.41760	.90863	.45960	.1758	.1005	.3946	19
42	.41787	.90851	.45995	.1741	.1007	.3931	18
43	.41813	.90839	.46030	.1725	.1008	.3916	17
44	.41839	.90826	.46065	.1708	.1010	.3901	16
45	0.41866	0.90814	0.46101	2.1692	1.1011	2.3886	15
46	.41892	.90802	.46136	.1675	.1013	.3871	14
47	.41919	.90790	.46171	.1658	.1014	.3856	13
48	.41945	.90778	.46206	.1642	.1016	.3841	12
49	.41972	.90765	.46242	.1625	.1017	.3826	11
50	0.41998	0.90753	0.46277	2.1609	1.1019	2.3811	10
51	.42024	.90741	.46312	.1592	.1020	.3796	9
52	.42051	.90729	.46348	.1576	.1022	.3781	8
53	.42077	.90717	.46383	.1559	.1023	.3766	7
54	.42103	.90704	.46418	.1543	.1025	.3751	6
55	0.42130	0.90692	0.46454	2.1527	1.1026	2.3736	5
56	.42156	.90680	.46489	.1510	.1028	.3721	4
57	.42183	.90668	.46524	.1494	.1029	.3706	3
58	.42209	.90655	.46560	.1478	.1031	.3691	2
59	.42235	.90643	.46595	.1461	.1032	.3677	1
60	0.42262	0.90631	0.46631	2.1445	1.1034	2.3662	0
M	Cosine	Sine	Cotan.	Tan.	Cosec.	Secant	M

114° 65°

25° Natural Trigonometric Functions 154°

M	Sine	Cosine	Tan.	Cotan.	Secant	Cosec.	M
0	0.42262	0.90631	0.46631	2.1445	1.1034	2.3662	60
1	.42288	.90618	.46666	.1429	.1035	.3647	59
2	.42314	.90606	.46702	.1412	.1037	.3632	58
3	.42341	.90594	.46737	.1396	.1038	.3618	57
4	.42367	.90581	.46772	.1380	.1040	.3603	56
5	0.42394	0.90569	0.46808	2.1364	1.1041	2.3588	55
6	.42420	.90557	.46843	.1348	.1043	.3574	54
7	.42446	.90544	.46879	.1331	.1044	.3559	53
8	.42473	.90532	.46914	.1315	.1046	.3544	52
9	.42499	.90520	.46950	.1299	.1047	.3530	51
10	0.42525	0.90507	0.46985	2.1283	1.1049	2.3515	50
11	.42552	.90495	.47021	.1267	.1050	.3501	49
12	.42578	.90483	.47056	.1251	.1052	.3486	48
13	.42604	.90470	.47092	.1235	.1053	.3472	47
14	.42630	.90458	.47127	.1219	.1055	.3457	46
15	0.42657	0.90445	0.47163	2.1203	1.1056	2.3443	45
16	.42683	.90433	.47199	.1187	.1058	.3428	44
17	.42709	.90421	.47234	.1171	.1059	.3414	43
18	.42736	.90408	.47270	.1155	.1061	.3399	42
19	.42762	.90396	.47305	.1139	.1062	.3385	41
20	0.42788	0.90383	0.47341	2.1123	1.1064	2.3371	40
21	.42815	.90371	.47376	.1107	.1065	.3356	39
22	.42841	.90358	.47412	.1092	.1067	.3342	38
23	.42867	.90346	.47448	.1076	.1068	.3328	37
24	.42893	.90333	.47483	.1060	.1070	.3313	36
25	0.42920	0.90321	0.47519	2.1044	1.1072	2.3299	35
26	.42946	.90308	.47555	.1028	.1073	.3285	34
27	.42972	.90296	.47590	.1013	.1075	.3271	33
28	.42998	.90283	.47626	.0997	.1076	.3256	32
29	.43025	.90271	.47662	.0981	.1078	.3242	31
30	0.43051	0.90258	0.47697	2.0965	1.1079	2.3228	30
31	.43077	.90246	.47733	.0950	.1081	.3214	29
32	.43104	.90233	.47769	.0934	.1082	.3200	28
33	.43130	.90221	.47805	.0918	.1084	.3186	27
34	.43156	.90208	.47840	.0903	.1085	.3172	26
35	0.43182	0.90196	0.47876	2.0887	1.1087	2.3158	25
36	.43208	.90183	.47912	.0872	.1088	.3143	24
37	.43235	.90171	.47948	.0856	.1090	.3129	23
38	.43261	.90158	.47983	.0840	.1092	.3115	22
39	.43287	.90145	.48019	.0825	.1093	.3101	21
40	0.43313	0.90133	0.48055	2.0809	1.1095	2.3087	20
41	.43340	.90120	.48091	.0794	.1096	.3073	19
42	.43366	.90108	.48127	.0778	.1098	.3059	18
43	.43392	.90095	.48162	.0763	.1099	.3046	17
44	.43418	.90082	.48198	.0747	.1101	.3032	16
45	0.43444	0.90070	0.48234	2.0732	1.1102	2.3018	15
46	.43471	.90057	.48270	.0717	.1104	.3004	14
47	.43497	.90044	.48306	.0701	.1106	.2990	13
48	.43523	.90032	.48342	.0686	.1107	.2976	12
49	.43549	.90019	.48378	.0671	.1109	.2962	11
50	0.43575	0.90006	0.48414	2.0655	1.1110	2.2949	10
51	.43602	.89994	.48449	.0640	.1112	.2935	9
52	.43628	.89981	.48485	.0625	.1113	.2921	8
53	.43654	.89968	.48521	.0609	.1115	.2907	7
54	.43680	.89956	.48557	.0594	.1116	.2894	6
55	0.43706	0.89943	0.48593	2.0579	1.1118	2.2880	5
56	.43732	.89930	.48629	.0564	.1120	.2866	4
57	.43759	.89918	.48665	.0548	.1121	.2853	3
58	.43785	.89905	.48701	.0533	.1123	.2839	2
59	.43811	.89892	.48737	.0518	.1124	.2825	1
60	0.43837	0.89879	0.48773	2.0503	1.1126	2.2812	0
M	Cosine	Sine	Cotan.	Tan.	Cosec.	Secant	M

115° 64°

26° Natural Trigonometric Functions 153°

M	Sine	Cosine	Tan.	Cotan.	Secant	Cosec.	M
0	0.43837	0.89879	0.48773	2.0503	1.1126	2.2812	60
1	.43863	.89867	.48809	.0488	.1127	.2798	59
2	.43889	.89854	.48845	.0473	.1129	.2784	58
3	.43915	.89841	.48881	.0458	.1131	.2771	57
4	.43942	.89828	.48917	.0443	.1132	.2757	56
5	0.43968	0.89815	0.48953	2.0427	1.1134	2.2744	55
6	.43994	.89803	.48989	.0412	.1135	.2730	54
7	.44020	.89790	.49025	.0397	.1137	.2717	53
8	.44046	.89777	.49062	.0382	.1139	.2703	52
9	.44072	.89764	.49098	.0367	.1140	.2690	51
10	0.44098	0.89751	0.49134	2.0352	1.1142	2.2676	50
11	.44124	.89739	.49170	.0338	.1143	.2663	49
12	.44150	.89726	.49206	.0323	.1145	.2650	48
13	.44177	.89713	.49242	.0308	.1147	.2636	47
14	.44203	.89700	.49278	.0293	.1148	.2623	46
15	0.44229	0.89687	0.49314	2.0278	1.1150	2.2610	45
16	.44255	.89674	.49351	.0263	.1151	.2596	44
17	.44281	.89661	.49387	.0248	.1153	.2583	43
18	.44307	.89649	.49423	.0233	.1155	.2570	42
19	.44333	.89636	.49459	.0219	.1156	.2556	41
20	0.44359	0.89623	0.49495	2.0204	1.1158	2.2543	40
21	.44385	.89610	.49532	.0189	.1159	.2530	39
22	.44411	.89597	.49568	.0174	.1161	.2517	38
23	.44437	.89584	.49604	.0159	.1163	.2503	37
24	.44463	.89571	.49640	.0145	.1164	.2490	36
25	0.44489	0.89558	0.49677	2.0130	1.1166	2.2477	35
26	.44516	.89545	.49713	.0115	.1167	.2464	34
27	.44542	.89532	.49749	.0101	.1169	.2451	33
28	.44568	.89519	.49785	.0086	.1171	.2438	32
29	.44594	.89506	.49822	.0071	.1172	.2425	31
30	0.44620	0.89493	0.49858	2.0057	1.1174	2.2411	30
31	.44646	.89480	.49894	.0042	.1176	.2398	29
32	.44672	.89467	.49931	.0028	.1177	.2385	28
33	.44698	.89454	.49967	.0013	.1179	.2372	27
34	.44724	.89441	.50003	1.9998	.1180	.2359	26
35	0.44750	0.89428	0.50040	1.9984	1.1182	2.2346	25
36	.44776	.89415	.50076	.9969	.1184	.2333	24
37	.44802	.89402	.50113	.9955	.1185	.2320	23
38	.44828	.89389	.50149	.9940	.1187	.2307	22
39	.44854	.89376	.50185	.9926	.1189	.2294	21
40	0.44880	0.89363	0.50222	1.9912	1.1190	2.2282	20
41	.44906	.89350	.50258	.9897	.1192	.2269	19
42	.44932	.89337	.50295	.9883	.1193	.2256	18
43	.44958	.89324	.50331	.9868	.1195	.2243	17
44	.44984	.89311	.50368	.9854	.1197	.2230	16
45	0.45010	0.89298	0.50404	1.9840	1.1198	2.2217	15
46	.45036	.89285	.50441	.9825	.1200	.2204	14
47	.45062	.89272	.50477	.9811	.1202	.2192	13
48	.45088	.89258	.50514	.9797	.1203	.2179	12
49	.45114	.89245	.50550	.9782	.1205	.2166	11
50	0.45140	0.89232	0.50587	1.9768	1.1207	2.2153	10
51	.45166	.89219	.50623	.9754	.1208	.2141	9
52	.45191	.89206	.50660	.9739	.1210	.2128	8
53	.45217	.89193	.50696	.9725	.1212	.2115	7
54	.45243	.89180	.50733	.9711	.1213	.2103	6
55	0.45269	0.89166	0.50769	1.9697	1.1215	2.2090	5
56	.45295	.89153	.50806	.9683	.1217	.2077	4
57	.45321	.89140	.50843	.9668	.1218	.2065	3
58	.45347	.89127	.50879	.9654	.1220	.2052	2
59	.45373	.89114	.50916	.9640	.1222	.2039	1
60	0.45399	0.89101	0.50952	1.9626	1.1223	2.2027	0
M	Cosine	Sine	Cotan.	Tan.	Cosec.	Secant	M

116° 63°

27° Natural Trigonometric Functions 152°

M	Sine	Cosine	Tan.	Cotan.	Secant	Cosec.	M
0	0.45399	0.89101	0.50952	1.9626	1.1223	2.2027	60
1	.45425	.89087	.50989	.9612	.1225	.2014	59
2	.45451	.89074	.51026	.9598	.1226	.2002	58
3	.45477	.89061	.51062	.9584	.1228	.1989	57
4	.45503	.89048	.51099	.9570	.1230	.1977	56
5	0.45528	0.89034	0.51136	1.9556	1.1231	2.1964	55
6	.45554	.89021	.51172	.9542	.1233	.1952	54
7	.45580	.89008	.51209	.9528	.1235	.1939	53
8	.45606	.88995	.51246	.9514	.1237	.1927	52
9	.45632	.88981	.51283	.9500	.1238	.1914	51
10	0.45658	0.88968	0.51319	1.9486	1.1240	2.1902	50
11	.45684	.88955	.51356	.9472	.1242	.1889	49
12	.45710	.88942	.51393	.9458	.1243	.1877	48
13	.45736	.88928	.51430	.9444	.1245	.1865	47
14	.45761	.88915	.51466	.9430	.1247	.1852	46
15	0.45787	0.88902	0.51503	1.9416	1.1248	2.1840	45
16	.45813	.88888	.51540	.9402	.1250	.1828	44
17	.45839	.88875	.51577	.9388	.1252	.1815	43
18	.45865	.88862	.51614	.9375	.1253	.1803	42
19	.45891	.88848	.51651	.9361	.1255	.1791	41
20	0.45917	0.88835	0.51687	1.9347	1.1257	2.1778	40
21	.45942	.88822	.51724	.9333	.1258	.1766	39
22	.45968	.88808	.51761	.9319	.1260	.1754	38
23	.45994	.88795	.51798	.9306	.1262	.1742	37
24	.46020	.88781	.51835	.9292	.1264	.1730	36
25	0.46046	0.88768	0.51872	1.9278	1.1265	2.1717	35
26	.46072	.88755	.51909	.9264	.1267	.1705	34
27	.46097	.88741	.51946	.9251	.1269	.1693	33
28	.46123	.88728	.51983	.9237	.1270	.1681	32
29	.46149	.88714	.52020	.9223	.1272	.1669	31
30	0.46175	0.88701	0.52057	1.9210	1.1274	2.1657	30
31	.46201	.88688	.52094	.9196	.1275	.1645	29
32	.46226	.88674	.52131	.9182	.1277	.1633	28
33	.46252	.88661	.52168	.9169	.1279	.1620	27
34	.46278	.88647	.52205	.9155	.1281	.1608	26
35	0.46304	0.88634	0.52242	1.9142	1.1282	2.1596	25
36	.46330	.88620	.52279	.9128	.1284	.1584	24
37	.46355	.88607	.52316	.9115	.1286	.1572	23
38	.46381	.88593	.52353	.9101	.1287	.1560	22
39	.46407	.88580	.52390	.9088	.1289	.1548	21
40	0.46433	0.88566	0.52427	1.9074	1.1291	2.1536	20
41	.46458	.88553	.52464	.9061	.1293	.1525	19
42	.46484	.88539	.52501	.9047	.1294	.1513	18
43	.46510	.88526	.52538	.9034	.1296	.1501	17
44	.46536	.88512	.52575	.9020	.1298	.1489	16
45	0.46561	0.88499	0.52612	1.9007	1.1299	2.1477	15
46	.46587	.88485	.52650	.8993	.1301	.1465	14
47	.46613	.88472	.52687	.8980	.1303	.1453	13
48	.46639	.88458	.52724	.8967	.1305	.1441	12
49	.46664	.88444	.52761	.8953	.1306	.1430	11
50	0.46690	0.88431	0.52798	1.8940	1.1308	2.1418	10
51	.46716	.88417	.52836	.8927	.1310	.1406	9
52	.46741	.88404	.52873	.8913	.1312	.1394	8
53	.46767	.88390	.52910	.8900	.1313	.1382	7
54	.46793	.88376	.52947	.8887	.1315	.1371	6
55	0.46819	0.88363	0.52984	1.8873	1.1317	2.1359	5
56	.46844	.88349	.53022	.8860	.1319	.1347	4
57	.46870	.88336	.53059	.8847	.1320	.1335	3
58	.46896	.88322	.53096	.8834	.1322	.1324	2
59	.46921	.88308	.53134	.8820	.1324	.1312	1
60	0.46947	0.88295	0.53171	1.8807	1.1326	2.1300	0
M	Cosine	Sine	Cotan.	Tan.	Cosec.	Secant	M

117° 62°

28° Natural Trigonometric Functions 151°

M	Sine	Cosine	Tan.	Cotan.	Secant	Cosec.	M
0	0.46947	0.88295	0.53171	1.8807	1.1326	2.1300	60
1	.46973	.88281	.53208	.8794	.1327	.1289	59
2	.46998	.88267	.53245	.8781	.1329	.1277	58
3	.47024	.88254	.53283	.8768	.1331	.1266	57
4	.47050	.88240	.53320	.8754	.1333	.1254	56
5	0.47075	0.88226	0.53358	1.8741	1.1334	2.1242	55
6	.47101	.88213	.53395	.8728	.1336	.1231	54
7	.47127	.88199	.53432	.8715	.1338	.1219	53
8	.47152	.88185	.53470	.8702	.1340	.1208	52
9	.47178	.88171	.53507	.8689	.1341	.1196	51
10	0.47204	0.88158	0.53545	1.8676	1.1343	2.1185	50
11	.47229	.88144	.53582	.8663	.1345	.1173	49
12	.47255	.88130	.53619	.8650	.1347	.1162	48
13	.47281	.88117	.53657	.8637	.1349	.1150	47
14	.47306	.88103	.53694	.8624	.1350	.1139	46
15	0.47332	0.88089	0.53732	1.8611	1.1352	2.1127	45
16	.47357	.88075	.53769	.8598	.1354	.1116	44
17	.47383	.88061	.53807	.8585	.1356	.1104	43
18	.47409	.88048	.53844	.8572	.1357	.1093	42
19	.47434	.88034	.53882	.8559	.1359	.1082	41
20	0.47460	0.88020	0.53919	1.8546	1.1361	2.1070	40
21	.47486	.88006	.53957	.8533	.1363	.1059	39
22	.47511	.87992	.53995	.8520	.1365	.1048	38
23	.47537	.87979	.54032	.8507	.1366	.1036	37
24	.47562	.87965	.54070	.8495	.1368	.1025	36
25	0.47588	0.87951	0.54107	1.8482	1.1370	2.1014	35
26	.47613	.87937	.54145	.8469	.1372	.1002	34
27	.47639	.87923	.54183	.8456	.1373	.0991	33
28	.47665	.87909	.54220	.8443	.1375	.0980	32
29	.47690	.87895	.54258	.8430	.1377	.0969	31
30	0.47716	0.87882	0.54295	1.8418	1.1379	2.0957	30
31	.47741	.87868	.54333	.8405	.1381	.0946	29
32	.47767	.87854	.54371	.8392	.1382	.0935	28
33	.47792	.87840	.54409	.8379	.1384	.0924	27
34	.47818	.87826	.54446	.8367	.1386	.0912	26
35	0.47844	0.87812	0.54484	1.8354	1.1388	2.0901	25
36	.47869	.87798	.54522	.8341	.1390	.0890	24
37	.47895	.87784	.54559	.8329	.1391	.0879	23
38	.47920	.87770	.54597	.8316	.1393	.0868	22
39	.47946	.87756	.54635	.8303	.1395	.0857	21
40	0.47971	0.87742	0.54673	1.8291	1.1397	2.0846	20
41	.47997	.87728	.54711	.8278	.1399	.0835	19
42	.48022	.87715	.54748	.8265	.1401	.0824	18
43	.48048	.87701	.54786	.8253	.1402	.0812	17
44	.48073	.87687	.54824	.8240	.1404	.0801	16
45	0.48099	0.87673	0.54862	1.8227	1.1406	2.0790	15
46	.48124	.87659	.54900	.8215	.1408	.0779	14
47	.48150	.87645	.54937	.8202	.1410	.0768	13
48	.48175	.87631	.54975	.8190	.1411	.0757	12
49	.48201	.87617	.55013	.8177	.1413	.0746	11
50	0.48226	0.87603	0.55051	1.8165	1.1415	2.0735	10
51	.48252	.87588	.55089	.8152	.1417	.0725	9
52	.48277	.87574	.55127	.8140	.1419	.0714	8
53	.48303	.87560	.55165	.8127	.1421	.0703	7
54	.48328	.87546	.55203	.8115	.1422	.0692	6
55	0.48354	0.87532	0.55241	1.8102	1.1424	2.0681	5
56	.48379	.87518	.55279	.8090	.1426	.0670	4
57	.48405	.87504	.55317	.8078	.1428	.0659	3
58	.48430	.87490	.55355	.8065	.1430	.0648	2
59	.48455	.87476	.55393	.8053	.1432	.0637	1
60	0.48481	0.87462	0.55431	1.8040	1.1433	2.0627	0
M	Cosine	Sine	Cotan.	Tan.	Cosec.	Secant	M

118° 61°

29° Natural Trigonometric Functions 150°

M	Sine	Cosine	Tan.	Cotan.	Secant	Cosec.	M
0	0.48481	0.87462	0.55431	1.8040	1.1433	2.0627	60
1	.48506	.87448	.55469	.8028	.1435	.0616	59
2	.48532	.87434	.55507	.8016	.1437	.0605	58
3	.48557	.87420	.55545	.8003	.1439	.0594	57
4	.48583	.87405	.55583	.7991	.1441	.0583	56
5	0.48608	0.87391	0.55621	1.7979	1.1443	2.0573	55
6	.48633	.87377	.55659	.7966	.1445	.0562	54
7	.48659	.87363	.55697	.7954	.1446	.0551	53
8	.48684	.87349	.55735	.7942	.1448	.0540	52
9	.48710	.87335	.55774	.7930	.1450	.0530	51
10	0.48735	0.87320	0.55812	1.7917	1.1452	2.0519	50
11	.48760	.87306	.55850	.7905	.1454	.0508	49
12	.48786	.87292	.55888	.7893	.1456	.0498	48
13	.48811	.87278	.55926	.7881	.1458	.0487	47
14	.48837	.87264	.55964	.7868	.1459	.0476	46
15	0.48862	0.87250	0.56003	1.7856	1.1461	2.0466	45
16	.48887	.87235	.56041	.7844	.1463	.0455	44
17	.48913	.87221	.56079	.7832	.1465	.0444	43
18	.48938	.87207	.56117	.7820	.1467	.0434	42
19	.48964	.87193	.56156	.7808	.1469	.0423	41
20	0.48989	0.87178	0.56194	1.7795	1.1471	2.0413	40
21	.49014	.87164	.56232	.7783	.1473	.0402	39
22	.49040	.87150	.56270	.7771	.1474	.0392	38
23	.49065	.87136	.56309	.7759	.1476	.0381	37
24	.49090	.87121	.56347	.7747	.1478	.0370	36
25	0.49116	0.87107	0.56385	1.7735	1.1480	2.0360	35
26	.49141	.87093	.56424	.7723	.1482	.0349	34
27	.49166	.87078	.56462	.7711	.1484	.0339	33
28	.49192	.87064	.56500	.7699	.1486	.0329	32
29	.49217	.87050	.56539	.7687	.1488	.0318	31
30	0.49242	0.87035	0.56577	1.7675	1.1489	2.0308	30
31	.49268	.87021	.56616	.7663	.1491	.0297	29
32	.49293	.87007	.56654	.7651	.1493	.0287	28
33	.49318	.86992	.56692	.7639	.1495	.0276	27
34	.49343	.86978	.56731	.7627	.1497	.0266	26
35	0.49369	0.86964	0.56769	1.7615	1.1499	2.0256	25
36	.49394	.86949	.56808	.7603	.1501	.0245	24
37	.49419	.86935	.56846	.7591	.1503	.0235	23
38	.49445	.86921	.56885	.7579	.1505	.0224	22
39	.49470	.86906	.56923	.7567	.1507	.0214	21
40	0.49495	0.86892	0.56962	1.7555	1.1508	2.0204	20
41	.49521	.86877	.57000	.7544	.1510	.0194	19
42	.49546	.86863	.57039	.7532	.1512	.0183	18
43	.49571	.86849	.57077	.7520	.1514	.0173	17
44	.49596	.86834	.57116	.7508	.1516	.0163	16
45	0.49622	0.86820	0.57155	1.7496	1.1518	2.0152	15
46	.49647	.86805	.57193	.7484	.1520	.0142	14
47	.49672	.86791	.57232	.7473	.1522	.0132	13
48	.49697	.86776	.57270	.7461	.1524	.0122	12
49	.49723	.86762	.57309	.7449	.1526	.0111	11
50	0.49748	0.86748	0.57348	1.7437	1.1528	2.0101	10
51	.49773	.86733	.57386	.7426	.1530	.0091	9
52	.49798	.86719	.57425	.7414	.1531	.0081	8
53	.49823	.86704	.57464	.7402	.1533	.0071	7
54	.49849	.86690	.57503	.7390	.1535	.0061	6
55	0.49874	0.86675	0.57541	1.7379	1.1537	2.0050	5
56	.49899	.86661	.57580	.7367	.1539	.0040	4
57	.49924	.86646	.57619	.7355	.1541	.0030	3
58	.49950	.86632	.57657	.7344	.1543	.0020	2
59	.49975	.86617	.57696	.7332	.1545	.0010	1
60	0.50000	0.86603	0.57735	1.7320	1.1547	2.0000	0
M	Cosine	Sine	Cotan.	Tan.	Cosec.	Secant	M

119° 60°

30° Natural Trigonometric Functions 149°

M	Sine	Cosine	Tan.	Cotan.	Secant	Cosec.	M
0	0.50000	0.86603	0.57735	1.7320	1.1547	2.0000	60
1	.50025	.86588	.57774	.7309	.1549	1.9990	59
2	.50050	.86573	.57813	.7297	.1551	.9980	58
3	.50075	.86559	.57851	.7286	.1553	.9970	57
4	.50101	.86544	.57890	.7274	.1555	.9960	56
5	0.50126	0.86530	0.57929	1.7262	1.1557	1.9950	55
6	.50151	.86515	.57968	.7251	.1559	.9940	54
7	.50176	.86500	.58007	.7239	.1561	.9930	53
8	.50201	.86486	.58046	.7228	.1562	.9920	52
9	.50226	.86471	.58085	.7216	.1564	.9910	51
10	0.50252	0.86457	0.58123	1.7205	1.1566	1.9900	50
11	.50277	.86442	.58162	.7193	.1568	.9890	49
12	.50302	.86427	.58201	.7182	.1570	.9880	48
13	.50327	.86413	.58240	.7170	.1572	.9870	47
14	.50352	.86398	.58279	.7159	.1574	.9860	46
15	0.50377	0.86383	0.58318	1.7147	1.1576	1.9850	45
16	.50402	.86369	.58357	.7136	.1578	.9840	44
17	.50428	.86354	.58396	.7124	.1580	.9830	43
18	.50453	.86339	.58435	.7113	.1582	.9820	42
19	.50478	.86325	.58474	.7101	.1584	.9811	41
20	0.50503	0.86310	0.58513	1.7090	1.1586	1.9801	40
21	.50528	.86295	.58552	.7079	.1588	.9791	39
22	.50553	.86281	.58591	.7067	.1590	.9781	38
23	.50578	.86266	.58630	.7056	.1592	.9771	37
24	.50603	.86251	.58670	.7044	.1594	.9761	36
25	0.50628	0.86237	0.58709	1.7033	1.1596	1.9752	35
26	.50653	.86222	.58748	.7022	.1598	.9742	34
27	.50679	.86207	.58787	.7010	.1600	.9732	33
28	.50704	.86192	.58826	.6999	.1602	.9722	32
29	.50729	.86178	.58865	.6988	.1604	.9713	31
30	0.50754	0.86163	0.58904	1.6977	1.1606	1.9703	30
31	.50779	.86148	.58944	.6965	.1608	.9693	29
32	.50804	.86133	.58983	.6954	.1610	.9683	28
33	.50829	.86118	.59022	.6943	.1612	.9674	27
34	.50854	.86104	.59061	.6931	.1614	.9664	26
35	0.50879	0.86089	0.59100	1.6920	1.1616	1.9654	25
36	.50904	.86074	.59140	.6909	.1618	.9645	24
37	.50929	.86059	.59179	.6898	.1620	.9635	23
38	.50954	.86044	.59218	.6887	.1622	.9625	22
39	.50979	.86030	.59258	.6875	.1624	.9616	21
40	0.51004	0.86015	0.59297	1.6864	1.1626	1.9606	20
41	.51029	.86000	.59336	.6853	.1628	.9596	19
42	.51054	.85985	.59376	.6842	.1630	.9587	18
43	.51079	.85970	.59415	.6831	.1632	.9577	17
44	.51104	.85955	.59454	.6820	.1634	.9568	16
45	0.51129	0.85941	0.59494	1.6808	1.1636	1.9558	15
46	.51154	.85926	.59533	.6797	.1638	.9549	14
47	.51179	.85911	.59572	.6786	.1640	.9539	13
48	.51204	.85896	.59612	.6775	.1642	.9530	12
49	.51229	.85881	.59651	.6764	.1644	.9520	11
50	0.51254	0.85866	0.59691	1.6753	1.1646	1.9510	10
51	.51279	.85851	.59730	.6742	.1648	.9501	9
52	.51304	.85836	.59770	.6731	.1650	.9491	8
53	.51329	.85821	.59809	.6720	.1652	.9482	7
54	.51354	.85806	.59849	.6709	.1654	.9473	6
55	0.51379	0.85791	0.59888	1.6698	1.1656	1.9463	5
56	.51404	.85777	.59928	.6687	.1658	.9454	4
57	.51429	.85762	.59967	.6676	.1660	.9444	3
58	.51454	.85747	.60007	.6665	.1662	.9435	2
59	.51479	.85732	.60046	.6654	.1664	.9425	1
60	0.51504	0.85717	0.60086	1.6643	1.1666	1.9416	0
M	Cosine	Sine	Cotan.	Tan.	Cosec.	Secant	M

120° 59°

31° Natural Trigonometric Functions 148°

M	Sine	Cosine	Tan.	Cotan.	Secant	Cosec.	M
0	0.51504	0.85717	0.60086	1.6643	1.1666	1.9416	60
1	.51529	.85702	.60126	.6632	.1668	.9407	59
2	.51554	.85687	.60165	.6621	.1670	.9397	58
3	.51578	.85672	.60205	.6610	.1672	.9388	57
4	.51603	.85657	.60244	.6599	.1674	.9378	56
5	0.51628	0.85642	0.60284	1.6588	1.1676	1.9369	55
6	.51653	.85627	.60324	.6577	.1678	.9360	54
7	.51678	.85612	.60363	.6566	.1681	.9350	53
8	.51703	.85597	.60403	.6555	.1683	.9341	52
9	.51728	.85582	.60443	.6544	.1685	.9332	51
10	0.51753	0.85566	0.60483	1.6534	1.1687	1.9322	50
11	.51778	.85551	.60522	.6523	.1689	.9313	49
12	.51803	.85536	.60562	.6512	.1691	.9304	48
13	.51827	.85521	.60602	.6501	.1693	.9295	47
14	.51852	.85506	.60642	.6490	.1695	.9285	46
15	0.51877	0.85491	0.60681	1.6479	1.1697	1.9276	45
16	.51902	.85476	.60721	.6469	.1699	.9267	44
17	.51927	.85461	.60761	.6458	.1701	.9258	43
18	.51952	.85446	.60801	.6447	.1703	.9248	42
19	.51977	.85431	.60841	.6436	.1705	.9239	41
20	0.52002	0.85416	0.60881	1.6425	1.1707	1.9230	40
21	.52026	.85400	.60920	.6415	.1709	.9221	39
22	.52051	.85385	.60960	.6404	.1712	.9212	38
23	.52076	.85370	.61000	.6393	.1714	.9203	37
24	.52101	.85355	.61040	.6383	.1716	.9193	36
25	0.52126	0.85340	0.61080	1.6372	1.1718	1.9184	35
26	.52151	.85325	.61120	.6361	.1720	.9175	34
27	.52175	.85309	.61160	.6350	.1722	.9166	33
28	.52200	.85294	.61200	.6340	.1724	.9157	32
29	.52225	.85279	.61240	.6329	.1726	.9148	31
30	0.52250	0.85264	0.61280	1.6318	1.1728	1.9139	30
31	.52275	.85249	.61320	.6308	.1730	.9130	29
32	.52299	.85234	.61360	.6297	.1732	.9121	28
33	.52324	.85218	.61400	.6286	.1734	.9112	27
34	.52349	.85203	.61440	.6276	.1737	.9102	26
35	0.52374	0.85188	0.61480	1.6265	1.1739	1.9093	25
36	.52398	.85173	.61520	.6255	.1741	.9084	24
37	.52423	.85157	.61560	.6244	.1743	.9075	23
38	.52448	.85142	.61601	.6233	.1745	.9066	22
39	.52473	.85127	.61641	.6223	.1747	.9057	21
40	0.52498	0.85112	0.61681	1.6212	1.1749	1.9048	20
41	.52522	.85096	.61721	.6202	.1751	.9039	19
42	.52547	.85081	.61761	.6191	.1753	.9030	18
43	.52572	.85066	.61801	.6181	.1756	.9021	17
44	.52597	.85050	.61842	.6170	.1758	.9013	16
45	0.52621	0.85035	0.61882	1.6160	1.1760	1.9004	15
46	.52646	.85020	.61922	.6149	.1762	.8995	14
47	.52671	.85004	.61962	.6139	.1764	.8986	13
48	.52695	.84989	.62003	.6128	.1766	.8977	12
49	.52720	.84974	.62043	.6118	.1768	.8968	11
50	0.52745	0.84959	0.62083	1.6107	1.1770	1.8959	10
51	.52770	.84943	.62123	.6097	.1772	.8950	9
52	.52794	.84928	.62164	.6086	.1775	.8941	8
53	.52819	.84912	.62204	.6076	.1777	.8932	7
54	.52844	.84897	.62244	.6066	.1779	.8924	6
55	0.52868	0.84882	0.62285	1.6055	1.1781	1.8915	5
56	.52893	.84866	.62325	.6045	.1783	.8906	4
57	.52918	.84851	.62366	.6034	.1785	.8897	3
58	.52942	.84836	.62406	.6024	.1787	.8888	2
59	.52967	.84820	.62446	.6014	.1790	.8879	1
60	0.52992	0.84805	0.62487	1.6003	1.1792	1.8871	0
M	Cosine	Sine	Cotan.	Tan.	Cosec.	Secant	M

121° 58°

32° Natural Trigonometric Functions 147°

M	Sine	Cosine	Tan.	Cotan.	Secant	Cosec.	M
0	0.52992	0.84805	0.62487	1.6003	1.1792	1.8871	60
1	.53016	.84789	.62527	.5993	.1794	.8862	59
2	.53041	.84774	.62568	.5983	.1796	.8853	58
3	.53066	.84758	.62608	.5972	.1798	.8844	57
4	.53090	.84743	.62649	.5962	.1800	.8836	56
5	0.53115	0.84728	0.62689	1.5952	1.1802	1.8827	55
6	.53140	.84712	.62730	.5941	.1805	.8818	54
7	.53164	.84697	.62770	.5931	.1807	.8809	53
8	.53189	.84681	.62811	.5921	.1809	.8801	52
9	.53214	.84666	.62851	.5910	.1811	.8792	51
10	0.53238	0.84650	0.62892	1.5900	1.1813	1.8783	50
11	.53263	.84635	.62933	.5890	.1815	.8775	49
12	.53288	.84619	.62973	.5880	.1818	.8766	48
13	.53312	.84604	.63014	.5869	.1820	.8757	47
14	.53337	.84588	.63055	.5859	.1822	.8749	46
15	0.53361	0.84573	0.63095	1.5849	1.1824	1.8740	45
16	.53386	.84557	.63136	.5839	.1826	.8731	44
17	.53411	.84542	.63177	.5829	.1828	.8723	43
18	.53435	.84526	.63217	.5818	.1831	.8714	42
19	.53460	.84511	.63258	.5808	.1833	.8706	41
20	0.53484	0.84495	0.63299	1.5798	1.1835	1.8697	40
21	.53509	.84479	.63339	.5788	.1837	.8688	39
22	.53533	.84464	.63380	.5778	.1839	.8680	38
23	.53558	.84448	.63421	.5768	.1841	.8671	37
24	.53583	.84433	.63462	.5757	.1844	.8663	36
25	0.53607	0.84417	0.63503	1.5747	1.1846	1.8654	35
26	.53632	.84402	.63543	.5737	.1848	.8646	34
27	.53656	.84386	.63584	.5727	.1850	.8637	33
28	.53681	.84370	.63625	.5717	.1852	.8629	32
29	.53705	.84355	.63666	.5707	.1855	.8620	31
30	0.53730	0.84339	0.63707	1.5697	1.1857	1.8611	30
31	.53754	.84323	.63748	.5687	.1859	.8603	29
32	.53779	.84308	.63789	.5677	.1861	.8595	28
33	.53803	.84292	.63830	.5667	.1863	.8586	27
34	.53828	.84276	.63871	.5657	.1866	.8578	26
35	0.53852	0.84261	0.63912	1.5646	1.1868	1.8569	25
36	.53877	.84245	.63953	.5636	.1870	.8561	24
37	.53901	.84229	.63994	.5626	.1872	.8552	23
38	.53926	.84214	.64035	.5616	.1874	.8544	22
39	.53950	.84198	.64076	.5606	.1877	.8535	21
40	0.53975	0.84182	0.64117	1.5596	1.1879	1.8527	20
41	.53999	.84167	.64158	.5586	.1881	.8519	19
42	.54024	.84151	.64199	.5577	.1883	.8510	18
43	.54048	.84135	.64240	.5567	.1886	.8502	17
44	.54073	.84120	.64281	.5557	.1888	.8493	16
45	0.54097	0.84104	0.64322	1.5547	1.1890	1.8485	15
46	.54122	.84088	.64363	.5537	.1892	.8477	14
47	.54146	.84072	.64404	.5527	.1894	.8468	13
48	.54171	.84057	.64446	.5517	.1897	.8460	12
49	.54195	.84041	.64487	.5507	.1899	.8452	11
50	0.54220	0.84025	0.64528	1.5497	1.1901	1.8443	10
51	.54244	.84009	.64569	.5487	.1903	.8435	9
52	.54268	.83993	.64610	.5477	.1906	.8427	8
53	.54293	.83978	.64652	.5467	.1908	.8418	7
54	.54317	.83962	.64693	.5458	.1910	.8410	6
55	0.54342	0.83946	0.64734	1.5448	1.1912	1.8402	5
56	.54366	.83930	.64775	.5438	.1915	.8394	4
57	.54391	.83914	.64817	.5428	.1917	.8385	3
58	.54415	.83899	.64858	.5418	.1919	.8377	2
59	.54439	.83883	.64899	.5408	.1921	.8369	1
60	0.54464	0.83867	0.64941	1.5399	1.1924	1.8361	0
M	Cosine	Sine	Cotan.	Tan.	Cosec.	Secant	M

122° 57°

33° **Natural Trigonometric Functions** 146°

M	Sine	Cosine	Tan.	Cotan.	Secant	Cosec.	M
0	0.54464	0.83867	0.64941	1.5399	1.1924	1.8361	60
1	.54488	.83851	.64982	.5389	.1926	.8352	59
2	.54513	.83835	.65023	.5379	.1928	.8344	58
3	.54537	.83819	.65065	.5369	.1930	.8336	57
4	.54561	.83804	.65106	.5359	.1933	.8328	56
5	0.54586	0.83788	0.65148	1.5350	1.1935	1.8320	55
6	.54610	.83772	.65189	.5340	.1937	.8311	54
7	.54634	.83756	.65231	.5330	.1939	.8303	53
8	.54659	.83740	.65272	.5320	.1942	.8295	52
9	.54683	.83724	.65314	.5311	.1944	.8287	51
10	0.54708	0.83708	0.65355	1.5301	1.1946	1.8279	50
11	.54732	.83692	.65397	.5291	.1948	.8271	49
12	.54756	.83676	.65438	.5282	.1951	.8263	48
13	.54781	.83660	.65480	.5272	.1953	.8255	47
14	.54805	.83644	.65521	.5262	.1955	.8246	46
15	0.54829	0.83629	0.65563	1.5252	1.1958	1.8238	45
16	.54854	.83613	.65604	.5243	.1960	.8230	44
17	.54878	.83597	.65646	.5233	.1962	.8222	43
18	.54902	.83581	.65688	.5223	.1964	.8214	42
19	.54926	.83565	.65729	.5214	.1967	.8206	41
20	0.54951	0.83549	0.65771	1.5204	1.1969	1.8198	40
21	.54975	.83533	.65813	.5195	.1971	.8190	39
22	.54999	.83517	.65854	.5185	.1974	.8182	38
23	.55024	.83501	.65896	.5175	.1976	.8174	37
24	.55048	.83485	.65938	.5166	.1978	.8166	36
25	0.55072	0.83469	0.65980	1.5156	1.1980	1.8158	35
26	.55097	.83453	.66021	.5147	.1983	.8150	34
27	.55121	.83437	.66063	.5137	.1985	.8142	33
28	.55145	.83421	.66105	.5127	.1987	.8134	32
29	.55169	.83405	.66147	.5118	.1990	.8126	31
30	0.55194	0.83388	0.66188	1.5108	1.1992	1.8118	30
31	.55218	.83372	.66230	.5099	.1994	.8110	29
32	.55242	.83356	.66272	.5089	.1997	.8102	28
33	.55266	.83340	.66314	.5080	.1999	.8094	27
34	.55291	.83324	.66356	.5070	.2001	.8086	26
35	0.55315	0.83308	0.66398	1.5061	1.2004	1.8078	25
36	.55339	.83292	.66440	.5051	.2006	.8070	24
37	.55363	.83276	.66482	.5042	.2008	.8062	23
38	.55388	.83260	.66524	.5032	.2010	.8054	22
39	.55412	.83244	.66566	.5023	.2013	.8047	21
40	0.55436	0.83228	0.66608	1.5013	1.2015	1.8039	20
41	.55460	.83211	.66650	.5004	.2017	.8031	19
42	.55484	.83195	.66692	.4994	.2020	.8023	18
43	.55509	.83179	.66734	.4985	.2022	.8015	17
44	.55533	.83163	.66776	.4975	.2024	.8007	16
45	0.55557	0.83147	0.66818	1.4966	1.2027	1.7999	15
46	.55581	.83131	.66860	.4957	.2029	.7992	14
47	.55605	.83115	.66902	.4947	.2031	.7984	13
48	.55629	.83098	.66944	.4938	.2034	.7976	12
49	.55654	.83082	.66986	.4928	.2036	.7968	11
50	0.55678	0.83066	0.67028	1.4919	1.2039	1.7960	10
51	.55702	.83050	.67071	.4910	.2041	.7953	9
52	.55726	.83034	.67113	.4900	.2043	.7945	8
53	.55750	.83017	.67155	.4891	.2046	.7937	7
54	.55774	.83001	.67197	.4881	.2048	.7929	6
55	0.55799	0.82985	0.67239	1.4872	1.2050	1.7921	5
56	.55823	.82969	.67282	.4863	.2053	.7914	4
57	.55847	.82952	.67324	.4853	.2055	.7906	3
58	.55871	.82936	.67366	.4844	.2057	.7898	2
59	.55895	.82920	.67408	.4835	.2060	.7891	1
60	0.55919	0.82904	0.67451	1.4826	1.2062	1.7883	0
M	Cosine	Sine	Cotan.	Tan.	Cosec.	Secant	M

123° 56°

34° **Natural Trigonometric Functions** 145°

M	Sine	Cosine	Tan.	Cotan.	Secant	Cosec.	M
0	0.55919	0.82904	0.67451	1.4826	1.2062	1.7883	60
1	.55943	.82887	.67493	.4816	.2064	.7875	59
2	.55967	.82871	.67535	.4807	.2067	.7867	58
3	.55992	.82855	.67578	.4798	.2069	.7860	57
4	.56016	.82839	.67620	.4788	.2072	.7852	56
5	0.56040	0.82822	0.67663	1.4779	1.2074	1.7844	55
6	.56064	.82806	.67705	.4770	.2076	.7837	54
7	.56088	.82790	.67747	.4761	.2079	.7829	53
8	.56112	.82773	.67790	.4751	.2081	.7821	52
9	.56136	.82757	.67832	.4742	.2083	.7814	51
10	0.56160	0.82741	0.67875	1.4733	1.2086	1.7806	50
11	.56184	.82724	.67917	.4724	.2088	.7798	49
12	.56208	.82708	.67960	.4714	.2091	.7791	48
13	.56232	.82692	.68002	.4705	.2093	.7783	47
14	.56256	.82675	.68045	.4696	.2095	.7776	46
15	0.56280	0.82659	0.68087	1.4687	1.2098	1.7768	45
16	.56304	.82643	.68130	.4678	.2100	.7760	44
17	.56328	.82626	.68173	.4669	.2103	.7753	43
18	.56353	.82610	.68215	.4659	.2105	.7745	42
19	.56377	.82593	.68258	.4650	.2107	.7738	41
20	0.56401	0.82577	0.68301	1.4641	1.2110	1.7730	40
21	.56425	.82561	.68343	.4632	.2112	.7723	39
22	.56449	.82544	.68386	.4623	.2115	.7715	38
23	.56473	.82528	.68429	.4614	.2117	.7708	37
24	.56497	.82511	.68471	.4605	.2119	.7700	36
25	0.56521	0.82495	0.68514	1.4595	1.2122	1.7693	35
26	.56545	.82478	.68557	.4586	.2124	.7685	34
27	.56569	.82462	.68600	.4577	.2127	.7678	33
28	.56593	.82445	.68642	.4568	.2129	.7670	32
29	.56617	.82429	.68685	.4559	.2132	.7663	31
30	0.56641	0.82413	0.68728	1.4550	1.2134	1.7655	30
31	.56664	.82396	.68771	.4541	.2136	.7648	29
32	.56688	.82380	.68814	.4532	.2139	.7640	28
33	.56712	.82363	.68857	.4523	.2141	.7633	27
34	.56736	.82347	.68899	.4514	.2144	.7625	26
35	0.56760	0.82330	0.68942	1.4505	1.2146	1.7618	25
36	.56784	.82314	.68985	.4496	.2149	.7610	24
37	.56808	.82297	.69028	.4487	.2151	.7603	23
38	.56832	.82280	.69071	.4478	.2153	.7596	22
39	.56856	.82264	.69114	.4469	.2156	.7588	21
40	0.56880	0.82247	0.69157	1.4460	1.2158	1.7581	20
41	.56904	.82231	.69200	.4451	.2161	.7573	19
42	.56928	.82214	.69243	.4442	.2163	.7566	18
43	.56952	.82198	.69286	.4433	.2166	.7559	17
44	.56976	.82181	.69329	.4424	.2168	.7551	16
45	0.57000	0.82165	0.69372	1.4415	1.2171	1.7544	15
46	.57023	.82148	.69415	.4406	.2173	.7537	14
47	.57047	.82131	.69459	.4397	.2175	.7529	13
48	.57071	.82115	.69502	.4388	.2178	.7522	12
49	.57095	.82098	.69545	.4379	.2180	.7514	11
50	0.57119	0.82082	0.69588	1.4370	1.2183	1.7507	10
51	.57143	.82065	.69631	.4361	.2185	.7500	9
52	.57167	.82048	.69674	.4352	.2188	.7493	8
53	.57191	.82032	.69718	.4343	.2190	.7485	7
54	.57214	.82015	.69761	.4335	.2193	.7478	6
55	0.57238	0.81998	0.69804	1.4326	1.2195	1.7471	5
56	.57262	.81982	.69847	.4317	.2198	.7463	4
57	.57286	.81965	.69891	.4308	.2200	.7456	3
58	.57310	.81948	.69934	.4299	.2203	.7449	2
59	.57334	.81932	.69977	.4290	.2205	.7442	1
60	0.57358	0.81915	0.70021	1.4281	1.2208	1.7434	0
M	Cosine	Sine	Cotan.	Tan.	Cosec.	Secant	M

124° 55°

35° **Natural Trigonometric Functions** 144°

M	Sine	Cosine	Tan.	Cotan.	Secant	Cosec.	M
0	0.57358	0.81915	0.70021	1.4281	1.2208	1.7434	60
1	.57381	.81898	.70064	.4273	.2210	.7427	59
2	.57405	.81882	.70107	.4264	.2213	.7420	58
3	.57429	.81865	.70151	.4255	.2215	.7413	57
4	.57453	.81848	.70194	.4246	.2218	.7405	56
5	0.57477	0.81832	0.70238	1.4237	1.2220	1.7398	55
6	.57500	.81815	.70281	.4228	.2223	.7391	54
7	.57524	.81798	.70325	.4220	.2225	.7384	53
8	.57548	.81781	.70368	.4211	.2228	.7377	52
9	.57572	.81765	.70412	.4202	.2230	.7369	51
10	0.57596	0.81748	0.70455	1.4193	1.2233	1.7362	50
11	.57619	.81731	.70499	.4185	.2235	.7355	49
12	.57643	.81714	.70542	.4176	.2238	.7348	48
13	.57667	.81698	.70586	.4167	.2240	.7341	47
14	.57691	.81681	.70629	.4158	.2243	.7334	46
15	0.57714	0.81664	0.70673	1.4150	1.2245	1.7327	45
16	.57738	.81647	.70717	.4141	.2248	.7319	44
17	.57762	.81630	.70760	.4132	.2250	.7312	43
18	.57786	.81614	.70804	.4123	.2253	.7305	42
19	.57809	.81597	.70848	.4115	.2255	.7298	41
20	0.57833	0.81580	0.70891	1.4106	1.2258	1.7291	40
21	.57857	.81563	.70935	.4097	.2260	.7284	39
22	.57881	.81546	.70979	.4089	.2263	.7277	38
23	.57904	.81530	.71022	.4080	.2265	.7270	37
24	.57928	.81513	.71066	.4071	.2268	.7263	36
25	0.57952	0.81496	0.71110	1.4063	1.2270	1.7256	35
26	.57975	.81479	.71154	.4054	.2273	.7249	34
27	.57999	.81462	.71198	.4045	.2276	.7242	33
28	.58023	.81445	.71241	.4037	.2278	.7234	32
29	.58047	.81428	.71285	.4028	.2281	.7227	31
30	0.58070	0.81411	0.71329	1.4019	1.2283	1.7220	30
31	.58094	.81395	.71373	.4011	.2286	.7213	29
32	.58118	.81378	.71417	.4002	.2288	.7206	28
33	.58141	.81361	.71461	.3994	.2291	.7199	27
34	.58165	.81344	.71505	.3985	.2293	.7192	26
35	0.58189	0.81327	0.71549	1.3976	1.2296	1.7185	25
36	.58212	.81310	.71593	.3968	.2298	.7178	24
37	.58236	.81293	.71637	.3959	.2301	.7171	23
38	.58259	.81276	.71681	.3951	.2304	.7164	22
39	.58283	.81259	.71725	.3942	.2306	.7157	21
40	0.58307	0.81242	0.71769	1.3933	1.2309	1.7151	20
41	.58330	.81225	.71813	.3925	.2311	.7144	19
42	.58354	.81208	.71857	.3916	.2314	.7137	18
43	.58378	.81191	.71901	.3908	.2316	.7130	17
44	.58401	.81174	.71945	.3899	.2319	.7123	16
45	0.58425	0.81157	0.71990	1.3891	1.2322	1.7116	15
46	.58448	.81140	.72034	.3882	.2324	.7109	14
47	.58472	.81123	.72078	.3874	.2327	.7102	13
48	.58496	.81106	.72122	.3865	.2329	.7095	12
49	.58519	.81089	.72166	.3857	.2332	.7088	11
50	0.58543	0.81072	0.72211	1.3848	1.2335	1.7081	10
51	.58566	.81055	.72255	.3840	.2337	.7075	9
52	.58590	.81038	.72299	.3831	.2340	.7068	8
53	.58614	.81021	.72344	.3823	.2342	.7061	7
54	.58637	.81004	.72388	.3814	.2345	.7054	6
55	0.58661	0.80987	0.72432	1.3806	1.2348	1.7047	5
56	.58684	.80970	.72477	.3797	.2350	.7040	4
57	.58708	.80953	.72521	.3789	.2353	.7033	3
58	.58731	.80936	.72565	.3781	.2355	.7027	2
59	.58755	.80919	.72610	.3772	.2358	.7020	1
60	0.58778	0.80902	0.72654	1.3764	1.2361	1.7013	0
M	Cosine	Sine	Cotan.	Tan.	Cosec.	Secant	M

125° 54°

36° Natural Trigonometric Functions 143°

M	Sine	Cosine	Tan.	Cotan.	Secant	Cosec.	M
0	0.58778	0.80902	0.72654	1.3764	1.2361	1.7013	60
1	58802	80885	72699	3755	2363	7006	59
2	58825	80867	72743	3747	2366	6999	58
3	58849	80850	72788	3738	2368	6993	57
4	58873	80833	72832	3730	2371	6986	56
5	58896	80816	72877	3722	2374	6979	55
6	58920	80799	72921	3713	2376	6973	54
7	58943	80782	72966	3705	2379	6966	53
8	58967	80765	73010	3697	2382	6959	52
9	58990	80748	73055	3688	2384	6952	51
10	59014	80730	73100	3680	2387	6945	50
11	59037	80713	73144	3672	2389	6938	49
12	59060	80696	73189	3663	2392	6932	48
13	59084	80679	73234	3655	2395	6925	47
14	59107	80662	73278	3647	2397	6918	46
15	59131	80644	73323	3639	2400	6912	45
16	59154	80627	73368	3630	2403	6905	44
17	59178	80610	73413	3622	2405	6898	43
18	59201	80593	73457	3613	2408	6892	42
19	59225	80576	73502	3605	2411	6885	41
20	59248	80558	73547	3597	2413	6878	40
21	59272	80541	73592	3588	2416	6871	39
22	59295	80524	73637	3580	2419	6865	38
23	59318	80507	73681	3572	2421	6858	37
24	59342	80489	73726	3564	2424	6851	36
25	59365	80472	73771	3555	2427	6845	35
26	59389	80455	73816	3547	2429	6838	34
27	59412	80438	73861	3539	2432	6831	33
28	59436	80420	73906	3531	2435	6825	32
29	59459	80403	73951	3522	2437	6818	31
30	59482	80386	73996	3514	2440	6812	30
31	59506	80368	74041	3506	2443	6805	29
32	59529	80351	74086	3498	2445	6798	28
33	59552	80334	74131	3489	2448	6792	27
34	59576	80316	74176	3481	2451	6785	26
35	59599	80299	74221	3473	2453	6779	25
36	59622	80282	74266	3465	2456	6772	24
37	59646	80264	74312	3457	2459	6766	23
38	59669	80247	74357	3449	2461	6759	22
39	59692	80230	74402	3440	2464	6752	21
40	59716	80212	74447	3432	2467	6746	20
41	59739	80195	74492	3424	2470	6739	19
42	59763	80178	74538	3416	2472	6733	18
43	59786	80160	74583	3408	2475	6726	17
44	59809	80143	74628	3400	2478	6720	16
45	59832	80125	74673	3392	2480	6713	15
46	59856	80108	74719	3383	2483	6707	14
47	59879	80091	74764	3375	2486	6700	13
48	59902	80073	74809	3367	2488	6694	12
49	59926	80056	74855	3359	2491	6687	11
50	59949	80038	74900	3351	2494	6681	10
51	59972	80021	74946	3343	2497	6674	9
52	59995	80003	74991	3335	2499	6668	8
53	60019	79986	75037	3327	2502	6661	7
54	60042	79968	75082	3319	2505	6655	6
55	60065	79951	75128	3311	2508	6648	5
56	60088	79934	75173	3303	2510	6642	4
57	60112	79916	75219	3294	2513	6635	3
58	60135	79898	75264	3286	2516	6629	2
59	60158	79881	75310	3278	2519	6623	1
60	0.60181	0.79863	0.75355	1.3270	1.2521	1.6616	0
M	Cosine	Sine	Cotan.	Tan.	Cosec.	Secant	M

126°   53°

37° Natural Trigonometric Functions 142°

M	Sine	Cosine	Tan.	Cotan.	Secant	Cosec.	M
0	0.60181	0.79863	0.75355	1.3270	1.2521	1.6616	60
1	60205	79846	75401	3262	2524	6610	59
2	60228	79829	75447	3254	2527	6603	58
3	60251	79811	75492	3246	2530	6597	57
4	60274	79793	75538	3238	2532	6591	56
5	60298	79776	75584	3230	2535	6584	55
6	60321	79758	75629	3222	2538	6578	54
7	60344	79741	75675	3214	2541	6572	53
8	60367	79723	75721	3206	2543	6565	52
9	60390	79706	75767	3198	2546	6559	51
10	60413	79688	75813	3190	2549	6553	50
11	60437	79671	75858	3182	2552	6546	49
12	60460	79653	75904	3174	2554	6540	48
13	60483	79635	75950	3166	2557	6533	47
14	60506	79618	75996	3159	2560	6527	46
15	60529	79600	76042	3151	2563	6521	45
16	60552	79582	76088	3143	2565	6514	44
17	60576	79565	76134	3135	2568	6508	43
18	60599	79547	76179	3127	2571	6502	42
19	60622	79530	76225	3119	2574	6496	41
20	60645	79512	76271	3111	2577	6490	40
21	60668	79494	76317	3103	2580	6483	39
22	60691	79477	76364	3095	2582	6477	38
23	60714	79459	76410	3087	2585	6470	37
24	60737	79441	76456	3079	2588	6464	36
25	60761	79424	76502	3071	2591	6458	35
26	60784	79406	76548	3064	2593	6452	34
27	60807	79388	76594	3056	2596	6445	33
28	60830	79371	76640	3048	2599	6439	32
29	60853	79353	76686	3040	2602	6433	31
30	60876	79335	76733	3032	2605	6427	30
31	60899	79318	76779	3024	2607	6420	29
32	60922	79300	76825	3017	2610	6414	28
33	60945	79282	76871	3009	2613	6408	27
34	60968	79264	76918	3001	2616	6402	26
35	60991	79247	76964	2993	2619	6396	25
36	61014	79229	77010	2985	2622	6389	24
37	61037	79211	77057	2977	2624	6383	23
38	61061	79193	77103	2970	2627	6377	22
39	61084	79176	77149	2962	2630	6371	21
40	61107	79158	77196	2954	2633	6365	20
41	61130	79140	77242	2946	2636	6359	19
42	61153	79122	77289	2938	2639	6352	18
43	61176	79104	77335	2931	2641	6346	17
44	61199	79087	77382	2923	2644	6340	16
45	61222	79069	77428	2915	2647	6334	15
46	61245	79051	77475	2907	2650	6328	14
47	61268	79033	77521	2900	2653	6322	13
48	61291	79015	77568	2892	2656	6316	12
49	61314	78998	77614	2884	2659	6309	11
50	61337	78980	77661	2876	2661	6303	10
51	61360	78962	77708	2869	2664	6297	9
52	61383	78944	77754	2861	2667	6291	8
53	61405	78926	77801	2853	2670	6285	7
54	61428	78908	77848	2845	2673	6279	6
55	61451	78891	77895	2838	2676	6273	5
56	61474	78873	77941	2830	2679	6267	4
57	61497	78855	77988	2822	2681	6261	3
58	61520	78837	78035	2815	2684	6255	2
59	61543	78819	78082	2807	2687	6249	1
60	0.61566	0.78801	0.78128	1.2799	1.2690	1.6243	0
M	Cosine	Sine	Cotan.	Tan.	Cosec.	Secant	M

127°   52°

38° Natural Trigonometric Functions 141°

M	Sine	Cosine	Tan.	Cotan.	Secant	Cosec.	M
0	0.61566	0.78801	0.78128	1.2799	1.2690	1.6243	60
1	61589	78783	78175	2792	2693	6237	59
2	61612	78765	78222	2784	2696	6231	58
3	61635	78747	78269	2776	2699	6224	57
4	61658	78729	78316	2769	2702	6218	56
5	61681	78711	78363	2761	2705	6212	55
6	61703	78693	78410	2753	2707	6206	54
7	61726	78675	78457	2746	2710	6200	53
8	61749	78657	78504	2738	2713	6194	52
9	61772	78640	78551	2730	2716	6188	51
10	61795	78622	78598	2723	2719	6182	50
11	61818	78604	78645	2715	2722	6176	49
12	61841	78586	78692	2708	2725	6170	48
13	61864	78568	78739	2700	2728	6164	47
14	61886	78550	78786	2692	2731	6159	46
15	61909	78532	78831	2685	2734	6153	45
16	61932	78514	78881	2677	2737	6147	44
17	61955	78496	78928	2670	2739	6141	43
18	61978	78478	78975	2662	2742	6135	42
19	62001	78460	79022	2654	2745	6129	41
20	62024	78441	79070	2647	2748	6123	40
21	62046	78423	79117	2639	2751	6117	39
22	62069	78405	79164	2632	2754	6111	38
23	62092	78387	79212	2624	2757	6105	37
24	62115	78369	79259	2617	2760	6099	36
25	62137	78351	79306	2609	2763	6093	35
26	62160	78333	79354	2602	2766	6087	34
27	62183	78315	79401	2594	2769	6081	33
28	62206	78297	79449	2587	2772	6077	32
29	62229	78279	79496	2579	2775	6070	31
30	62251	78261	79543	2572	2778	6058	30
31	62274	78243	79591	2564	2781	6058	29
32	62297	78225	79638	2557	2784	6052	28
33	62320	78206	79686	2549	2787	6046	27
34	62342	78188	79734	2542	2790	6040	26
35	62365	78170	79781	2534	2793	6034	25
36	62388	78152	79829	2527	2795	6029	24
37	62411	78134	79877	2519	2798	6023	23
38	62433	78116	79924	2512	2801	6017	22
39	62456	78097	79972	2504	2804	6011	21
40	62479	78079	80020	2497	2807	6005	20
41	62501	78061	80067	2489	2810	6000	19
42	62524	78043	80115	2482	2813	5994	18
43	62547	78025	80163	2475	2816	5988	17
44	62570	78007	80211	2467	2819	5982	16
45	62592	77988	80258	2460	2822	5976	15
46	62615	77970	80306	2452	2825	5971	14
47	62638	77952	80354	2445	2828	5965	13
48	62660	77934	80402	2437	2831	5959	12
49	62683	77915	80450	2430	2834	5953	11
50	62706	77897	80498	2423	2837	5947	10
51	62728	77879	80546	2415	2840	5942	9
52	62751	77861	80594	2408	2843	5936	8
53	62774	77842	80642	2400	2846	5930	7
54	62796	77824	80690	2393	2849	5924	6
55	62819	77806	80738	2386	2852	5919	5
56	62842	77788	80786	2378	2855	5913	4
57	62864	77769	80834	2371	2858	5907	3
58	62887	77751	80882	2364	2861	5901	2
59	62909	77733	80930	2356	2864	5896	1
60	0.62932	0.77715	0.80978	1.2349	1.2867	1.5890	0
M	Cosine	Sine	Cotan.	Tan.	Cosec.	Secant	M

128°   51°

39° Natural Trigonometric Functions 140°

M	Sine	Cosine	Tan.	Cotan.	Secant	Cosec.	M
0	0.62932	0.77715	0.80978	1.2349	1.2867	1.5890	60
1	.62955	.77696	.81026	.2342	.2871	.5884	59
2	.62977	.77678	.81075	.2334	.2874	.5879	58
3	.63000	.77660	.81123	.2327	.2877	.5873	57
4	.63022	.77641	.81171	.2320	.2880	.5867	56
5	0.63045	0.77623	0.81219	1.2312	1.2883	1.5862	55
6	.63067	.77605	.81268	.2305	.2886	.5856	54
7	.63090	.77586	.81316	.2297	.2889	.5850	53
8	.63113	.77568	.81364	.2290	.2892	.5845	52
9	.63135	.77549	.81413	.2283	.2895	.5839	51
10	0.63158	0.77531	0.81461	1.2276	1.2898	1.5833	50
11	.63180	.77513	.81509	.2268	.2901	.5828	49
12	.63203	.77494	.81558	.2261	.2904	.5822	48
13	.63225	.77476	.81606	.2254	.2907	.5816	47
14	.63248	.77458	.81655	.2247	.2910	.5811	46
15	0.63270	0.77439	0.81703	1.2239	1.2913	1.5805	45
16	.63293	.77421	.81752	.2232	.2916	.5799	44
17	.63315	.77402	.81800	.2225	.2919	.5794	43
18	.63338	.77384	.81849	.2218	.2922	.5788	42
19	.63360	.77365	.81898	.2210	.2926	.5783	41
20	0.63383	0.77347	0.81946	1.2203	1.2929	1.5777	40
21	.63405	.77329	.81995	.2196	.2932	.5771	39
22	.63428	.77310	.82043	.2189	.2935	.5766	38
23	.63450	.77292	.82092	.2181	.2938	.5760	37
24	.63473	.77273	.82141	.2174	.2941	.5755	36
25	0.63495	0.77255	0.82190	1.2167	1.2944	1.5749	35
26	.63518	.77236	.82238	.2160	.2947	.5743	34
27	.63540	.77218	.82287	.2152	.2950	.5738	33
28	.63563	.77199	.82336	.2145	.2953	.5732	32
29	.63585	.77181	.82385	.2138	.2956	.5727	31
30	0.63608	0.77162	0.82434	1.2131	1.2960	1.5721	30
31	.63630	.77144	.82482	.2124	.2963	.5716	29
32	.63653	.77125	.82531	.2117	.2966	.5710	28
33	.63675	.77107	.82580	.2109	.2969	.5705	27
34	.63697	.77088	.82629	.2102	.2972	.5699	26
35	0.63720	0.77070	0.82678	1.2095	1.2975	1.5694	25
36	.63742	.77051	.82727	.2088	.2978	.5688	24
37	.63765	.77033	.82776	.2081	.2981	.5683	23
38	.63787	.77014	.82825	.2074	.2985	.5677	22
39	.63810	.76996	.82874	.2066	.2988	.5672	21
40	0.63832	0.76977	0.82923	1.2059	1.2991	1.5666	20
41	.63854	.76958	.82972	.2052	.2994	.5661	19
42	.63877	.76940	.83022	.2045	.2997	.5655	18
43	.63899	.76921	.83071	.2038	.3000	.5650	17
44	.63921	.76903	.83120	.2031	.3003	.5644	16
45	0.63944	0.76884	0.83169	1.2024	1.3006	1.5639	15
46	.63966	.76865	.83218	.2016	.3010	.5633	14
47	.63989	.76847	.83267	.2009	.3013	.5628	13
48	.64011	.76828	.83317	.2002	.3016	.5622	12
49	.64033	.76810	.83366	.1995	.3019	.5617	11
50	0.64056	0.76791	0.83415	1.1988	1.3022	1.5611	10
51	.64078	.76772	.83465	.1981	.3025	.5606	9
52	.64100	.76754	.83514	.1974	.3029	.5600	8
53	.64123	.76735	.83563	.1967	.3032	.5595	7
54	.64145	.76716	.83613	.1960	.3035	.5590	6
55	0.64167	0.76698	0.83662	1.1953	1.3038	1.5584	5
56	.64189	.76679	.83712	.1946	.3041	.5579	4
57	.64212	.76660	.83761	.1939	.3044	.5573	3
58	.64234	.76642	.83811	.1932	.3048	.5568	2
59	.64256	.76623	.83860	.1924	.3051	.5563	1
60	0.64279	0.76604	0.83910	1.1917	1.3054	1.5557	0
M	Cosine	Sine	Cotan.	Tan.	Cosec.	Secant	M

129° 50°

40° Natural Trigonometric Functions 139°

M	Sine	Cosine	Tan.	Cotan.	Secant	Cosec.	M
0	0.64279	0.76604	0.83910	1.1917	1.3054	1.5557	60
1	.64301	.76586	.83959	.1910	.3057	.5552	59
2	.64323	.76567	.84009	.1903	.3060	.5546	58
3	.64345	.76548	.84059	.1896	.3064	.5541	57
4	.64368	.76530	.84108	.1889	.3067	.5536	56
5	0.64390	0.76511	0.84158	1.1882	1.3070	1.5530	55
6	.64412	.76492	.84208	.1875	.3073	.5525	54
7	.64435	.76473	.84257	.1868	.3076	.5520	53
8	.64457	.76455	.84307	.1861	.3080	.5514	52
9	.64479	.76436	.84357	.1854	.3083	.5509	51
10	0.64501	0.76417	0.84407	1.1847	1.3086	1.5503	50
11	.64523	.76398	.84457	.1840	.3089	.5498	49
12	.64546	.76380	.84506	.1833	.3092	.5493	48
13	.64568	.76361	.84556	.1826	.3096	.5487	47
14	.64590	.76342	.84606	.1819	.3099	.5482	46
15	0.64612	0.76323	0.84656	1.1812	1.3102	1.5477	45
16	.64635	.76304	.84706	.1805	.3105	.5471	44
17	.64657	.76286	.84756	.1798	.3109	.5466	43
18	.64679	.76267	.84806	.1791	.3112	.5461	42
19	.64701	.76248	.84856	.1785	.3115	.5456	41
20	0.64723	0.76229	0.84906	1.1778	1.3118	1.5450	40
21	.64745	.76210	.84956	.1771	.3121	.5445	39
22	.64768	.76191	.85006	.1764	.3125	.5440	38
23	.64790	.76173	.85056	.1757	.3128	.5434	37
24	.64812	.76154	.85107	.1750	.3131	.5429	36
25	0.64834	0.76135	0.85157	1.1743	1.3134	1.5424	35
26	.64856	.76116	.85207	.1736	.3138	.5419	34
27	.64878	.76097	.85257	.1729	.3141	.5413	33
28	.64900	.76078	.85307	.1722	.3144	.5408	32
29	.64923	.76059	.85358	.1715	.3148	.5403	31
30	0.64945	0.76041	0.85408	1.1708	1.3151	1.5398	30
31	.64967	.76022	.85458	.1702	.3154	.5392	29
32	.64989	.76003	.85509	.1695	.3157	.5387	28
33	.65011	.75984	.85559	.1688	.3161	.5382	27
34	.65033	.75965	.85609	.1681	.3164	.5377	26
35	0.65055	0.75946	0.85660	1.1674	1.3167	1.5371	25
36	.65077	.75927	.85710	.1667	.3170	.5366	24
37	.65100	.75908	.85761	.1660	.3174	.5361	23
38	.65121	.75889	.85811	.1653	.3177	.5356	22
39	.65144	.75870	.85862	.1647	.3180	.5351	21
40	0.65166	0.75851	0.85912	1.1640	1.3184	1.5345	20
41	.65188	.75832	.85963	.1633	.3187	.5340	19
42	.65210	.75813	.86013	.1626	.3190	.5335	18
43	.65232	.75794	.86064	.1619	.3193	.5330	17
44	.65254	.75775	.86115	.1612	.3197	.5325	16
45	0.65276	0.75756	0.86165	1.1605	1.3200	1.5319	15
46	.65298	.75737	.86216	.1599	.3203	.5314	14
47	.65320	.75718	.86267	.1592	.3207	.5309	13
48	.65342	.75700	.86318	.1585	.3210	.5304	12
49	.65364	.75680	.86368	.1578	.3213	.5299	11
50	0.65386	0.75661	0.86419	1.1571	1.3217	1.5294	10
51	.65408	.75642	.86470	.1565	.3220	.5289	9
52	.65430	.75623	.86521	.1558	.3223	.5283	8
53	.65452	.75604	.86572	.1551	.3227	.5278	7
54	.65474	.75585	.86623	.1544	.3230	.5273	6
55	0.65496	0.75566	0.86674	1.1537	1.3233	1.5268	5
56	.65518	.75547	.86725	.1531	.3237	.5263	4
57	.65540	.75528	.86775	.1524	.3240	.5258	3
58	.65562	.75509	.86826	.1517	.3243	.5253	2
59	.65584	.75490	.86878	.1510	.3247	.5248	1
60	0.65606	0.75471	0.86929	1.1504	1.3250	1.5242	0
M	Cosine	Sine	Cotan.	Tan.	Cosec.	Secant	M

130° 49°

41° Natural Trigonometric Functions 138°

M	Sine	Cosine	Tan.	Cotan.	Secant	Cosec.	M
0	0.65606	0.75471	0.86929	1.1504	1.3250	1.5242	60
1	.65628	.75452	.86980	.1497	.3253	.5237	59
2	.65650	.75433	.87031	.1490	.3257	.5232	58
3	.65672	.75414	.87082	.1483	.3260	.5227	57
4	.65694	.75394	.87133	.1477	.3263	.5222	56
5	0.65716	0.75375	0.87184	1.1470	1.3267	1.5217	55
6	.65737	.75356	.87235	.1463	.3270	.5212	54
7	.65759	.75337	.87287	.1456	.3274	.5207	53
8	.65781	.75318	.87338	.1450	.3277	.5202	52
9	.65803	.75299	.87389	.1443	.3280	.5197	51
10	0.65825	0.75280	0.87441	1.1436	1.3284	1.5192	50
11	.65847	.75261	.87492	.1430	.3287	.5187	49
12	.65869	.75241	.87543	.1423	.3290	.5182	48
13	.65891	.75222	.87595	.1416	.3294	.5177	47
14	.65913	.75203	.87646	.1409	.3297	.5171	46
15	0.65934	0.75184	0.87698	1.1403	1.3301	1.5166	45
16	.65956	.75165	.87749	.1396	.3304	.5161	44
17	.65978	.75146	.87801	.1389	.3307	.5156	43
18	.66000	.75126	.87852	.1383	.3311	.5151	42
19	.66022	.75107	.87904	.1376	.3314	.5146	41
20	0.66044	0.75088	0.87955	1.1369	1.3318	1.5141	40
21	.66066	.75069	.88007	.1363	.3321	.5136	39
22	.66087	.75049	.88058	.1356	.3324	.5131	38
23	.66109	.75030	.88110	.1349	.3328	.5126	37
24	.66131	.75011	.88162	.1343	.3331	.5121	36
25	0.66153	0.74992	0.88213	1.1336	1.3335	1.5116	35
26	.66175	.74973	.88265	.1329	.3338	.5111	34
27	.66197	.74953	.88317	.1323	.3342	.5106	33
28	.66218	.74934	.88369	.1316	.3345	.5101	32
29	.66240	.74915	.88421	.1309	.3348	.5096	31
30	0.66262	0.74895	0.88472	1.1303	1.3352	1.5092	30
31	.66284	.74876	.88524	.1296	.3355	.5087	29
32	.66305	.74857	.88576	.1290	.3359	.5082	28
33	.66327	.74838	.88628	.1283	.3362	.5077	27
34	.66349	.74818	.88680	.1276	.3366	.5072	26
35	0.66371	0.74799	0.88732	1.1270	1.3369	1.5067	25
36	.66393	.74780	.88784	.1263	.3372	.5062	24
37	.66414	.74760	.88836	.1257	.3376	.5057	23
38	.66436	.74741	.88888	.1250	.3379	.5052	22
39	.66458	.74722	.88940	.1243	.3383	.5047	21
40	0.66479	0.74702	0.88992	1.1237	1.3386	1.5042	20
41	.66501	.74683	.89044	.1230	.3390	.5037	19
42	.66523	.74664	.89097	.1224	.3393	.5032	18
43	.66545	.74644	.89149	.1217	.3397	.5027	17
44	.66566	.74625	.89201	.1211	.3400	.5022	16
45	0.66588	0.74606	0.89253	1.1204	1.3404	1.5018	15
46	.66610	.74586	.89306	.1197	.3407	.5013	14
47	.66631	.74567	.89358	.1191	.3411	.5008	13
48	.66653	.74548	.89410	.1184	.3414	.5003	12
49	.66675	.74528	.89463	.1178	.3418	.4998	11
50	0.66697	0.74509	0.89515	1.1171	1.3421	1.4993	10
51	.66718	.74489	.89567	.1165	.3425	.4988	9
52	.66740	.74470	.89620	.1158	.3428	.4983	8
53	.66762	.74450	.89672	.1152	.3432	.4979	7
54	.66783	.74431	.89725	.1145	.3435	.4974	6
55	0.66805	0.74412	0.89777	1.1139	1.3439	1.4969	5
56	.66826	.74392	.89830	.1132	.3442	.4964	4
57	.66848	.74373	.89882	.1126	.3446	.4959	3
58	.66870	.74353	.89935	.1119	.3449	.4954	2
59	.66891	.74334	.89988	.1113	.3453	.4949	1
60	0.66913	0.74314	0.90040	1.1106	1.3456	1.4945	0
M	Cosine	Sine	Cotan.	Tan.	Cosec.	Secant	M

131° 48°

135° / 45° — Natural Trigonometric Functions

M	Cosec.	Secant	Cosec.	Cotan.	Tan.	Cosine	Sine	M
0	1.4395	1.3902	1.3902	1.0355	.96166	0.71934	0.69466	60
1	.4391	.3905		.0349	.96925	.71914	.69487	59
2	.4387	.3909		.0343	.96981	.71893	.69508	58
3	.4383	.3913		.0337	.96138	.71873	.69528	57
4	.4378	.3917		.0331		.71853	.69549	56
5	.4374	.3921		.0325	.96634	.71833	.69570	55
6	.4370	.3925		.0319	.96607	.71813	.69591	54
7	.4365	.3929		.0313	.96663	.71792	.69612	53
8	.4361	.3933		.0307	.96720	.71772	.69633	52
9	.4357	.3937		.0301	.96776	.71752	.69654	51
10	.4352	1.3941		1.0295	.97133	.71732	.69675	50
11	.4348	.3945		.0289	.97189	.71711	.69696	49
12	.4344	.3949		.0283	.97246	.71691	.69716	48
13	.4339	.3953		.0277	.97302	.71671	.69737	47
14	.4335	.3956		.0271	.97359	.71650	.69758	46
15	.4331	.3960		.0265	.97416	.71630	.69779	45
16	.4327	.3964		.0259	.97472	.71610	.69800	44
17	.4322	.3968		.0253	.97529	.71590	.69821	43
18	.4318	.3972		.0247	.97586	.71559	.69842	42
19	.4314	.3976		.0241	.97643	.71549	.69862	41
20	.4310	1.3980		1.0235	.97700	.71529	.69883	40
21	.4305	.3984		.0229	.97756	.71508	.69904	39
22	.4301	.3988		.0223	.97813	.71488	.69925	38
23	.4297	.3992		.0218	.97870	.71468	.69946	37
24	.4292	.3996		.0212	.97927	.71447	.69966	36
25	.4288	1.4000		1.0206	.97984	.71427	.69987	35
26	.4284	.4004		.0200	.98041	.71407	.70008	34
27	.4280	.4008		.0194	.98098	.71386	.70029	33
28	.4276	.4012		.0188	.98155	.71366	.70049	32
29	.4271	.4016		.0182	.98212	.71345	.70070	31
30	.4267	1.4020		1.0176	.98270	.71325	.70091	30
31	.4263	.4024		.0170	.98327	.71305	.70112	29
32	.4259	.4028		.0164	.98384	.71284	.70132	28
33	.4254	.4032		.0158	.98441	.71264	.70153	27
34	.4250	.4036		.0152	.98499	.71243	.70174	26
35	.4246	1.4040		1.0146	.98556	.71223	.70194	25
36	.4242	.4044		.0141	.98613	.71203	.70215	24
37	.4238	.4048		.0135	.98671	.71182	.70236	23
38	.4233	.4052		.0129	.98728	.71162	.70257	22
39	.4229	.4056		.0123	.98786	.71141	.70277	21
40	.4225	1.4060		1.0117	.98843	.71121	.70298	20
41	.4221	.4065		.0111	.98901	.71100	.70319	19
42	.4217	.4069		.0105	.98958	.71080	.70339	18
43	.4212	.4073		.0099	.99016	.71059	.70360	17
44	.4208	.4077		.0093	.99073	.71039	.70381	16
45	.4204	.4081		.0088	.99131	.71019	.70401	15
46	.4200	.4085		.0082	.99189	.70998	.70422	14
47	.4196	.4089		.0076	.99246	.70978	.70443	13
48	.4192	.4093		.0070	.99304	.70957	.70463	12
49	.4188	.4097		.0064	.99362	.70936	.70484	11
50	.4183	1.4101		1.0058	.99420	.70916	.70505	10
51	.4179	.4105		.0052	.99478	.70895	.70525	9
52	.4175	.4109		.0047	.99536	.70875	.70546	8
53	.4171	.4113		.0041	.99594	.70854	.70567	7
54	.4167	.4117		.0035	.99652	.70834	.70587	6
55	.4163	.4122		.0029	.99710	.70813	.70608	5
56	.4159	.4126		.0023	.99767	.70793	.70628	4
57	.4154	.4130		.0017	.99826	.70772	.70649	3
58	.4150	.4134		.0011	.99884	.70752	.70669	2
59	.4146	.4138		.0006	.99942	.70731	.70690	1
60	1.4142	1.4142		1.0000	1.00000	0.70711	0.70711	0

(Column headers reversed at foot: M | Cosec. | Secant | Cosec. | Tan. | Cotan. | Sine | Cosine | M — 44° / 134°)

136° / 46° — Natural Trigonometric Functions

M	Cosec.	Secant	Cosec.	Cotan.	Tan.	Cosine	Sine	M
0	1.4663	1.3673	1.0724		0.93251	0.73135	0.68200	60
...								...
60	1.4395	1.3902	1.0355		0.96569	0.71934	0.69466	0

(43° / 133°)

137° / 47° — Natural Trigonometric Functions

M	Cosec.	Secant	Cotan.	Tan.	Cotan.	Sine	Cosine	M
0	1.4945	1.3456	1.1106	0.90040			0.66913	60
...								...
60	1.4663	1.3673	1.0724	0.93251			0.68200	0

(42° / 132°)

Index

501